COMPUTATIONAL METHODS
IN WATER RESOURCES X

Water Science and Technology Library

VOLUME 12

The titles published in this series are listed at the end of this volume.

COMPUTATIONAL METHODS IN WATER RESOURCES X

Volume 1

edited by

ALEXANDER PETERS
IBM Heidelberg, Germany

GABRIEL WITTUM
Universität Stuttgart, Germany

BRUNO HERRLING
Universität Karlsruhe, Germany

UDO MEISSNER
Technische Hochschule Darmstadt, Germany

CARLOS A. BREBBIA
Wessex Institute of Technology, UK

WILLIAM G. GRAY
University of Notre Dame, USA

and

GEORGE F. PINDER
University of Vermont, USA

SPRINGER-SCIENCE+BUSINESS MEDIA, B.V.

Library of Congress Cataloging-in-Publication Data

Computational methods in water resources X / editors, Alexander Peters
... [et al.].
     p.   cm. -- (Water science and technology library : v. 12)
   Edited proceedings of the Tenth International Conference on
Computational methods in Water Resources, held at Universität
Heidelberg, Germany, Jul. 1994.

     1. Hydrology--Data processing--Congresses. 2. Hydrology-
-Mathematical models--Congresses.   I. Peter, A. (Alexander), 1956-
. II. International Conference on Computational Methods in Water
Resources (10th : 1994 : Universität Heidelberg)   III. Series.
GB656.2.E42C653   1994
551.49'0285--dc20   94-17902

ISBN 978-94-010-9206-7   ISBN 978-94-010-9204-3 (eBook)
DOI 10.1007/978-94-010-9204-3

Printed on acid-free paper

EDITORS:

A. Peters
IBM Heidelberg STSS
Vangerowstr. 18
69020 Heidelberg
Germany

G. Wittum
ICA / Numerik
Universität Stuttgart
Pfaffenwaldring 27
70550 Stuttgart
Germany

B. Herrling
Institut für Hydromechanik
Universität Karlsruhe
Kaiserstr. 12
76131 Karlsruhe
Germany

U. Meissner
Technische Hochschule Darmstadt
Institut für Numerische Methoden und Informatik im Bauwesen
Petersenstr. 13
64287 Darmstadt
Germany

C.A. Brebbia
Wessex Institute of Technology
University of Portsmouth
Ashurst Lodge
Southhampton SO4 2AA
UK

W.G. Gray
University of Notre Dame
Department of Civil Engineering and Geological Sciences
Notre Dame, IN 46566-0767
USA

G.F. Pinder
College of Engineering and Mathematics
University of Vermont
101 Votey Building
Burlington, VT 05405
USA

PREFACE

These volumes constitute the edited proceedings of the Tenth International Conference on Computational Methods in Water Resources (formerly Finite Elements in Water Resources), held at Universität Heidelberg, Germany in July 1994. The biennial series began in 1976 at Princeton University, U.S.A., as a forum for researchers in the expanding field of applications of finite element methods to problems in water resources. Alternating between the U.S.A. and Europe, meetings have been held at Imperial College, U.K. (1978); the University of Mississippi (1980); Hannover University, Germany (1982); the University of Vermont (1984); the Laboratorio Nacional de Engenharia Civil, Portugal (1986); the Massachusetts Institute of Technology (1988); the Giorgio Cini Foundation, Italy (1990); and the University of Colorado at Denver (1992). The Heidelberg conference is organized jointly by Interdisziplinäres Zentrum für Wissenschaftliches Rechnen (Interdisciplinary Center for Scientific Computing) and Sonderforschungsbereich 359 of Universität Heidelberg and Institute of Supercomputing and Applied Mathematics of IBM Heidelberg.

The 1994 proceedings present the work of authors from 23 countries. Numerical methods, mathematical modeling and applications to subsurface and surface hydrology are covered by a wide variety of papers. Issues of formation description and modeling, including parameter estimation, heterogeneity, and scaling up continue to attract the attention of a large number of researchers. It is significant to mention that several papers edited in this book concern the solution of the Navier-Stokes equations.

The organizers of the Heidelberg meeting greatly appreciate the efforts of featured lecturers A.J. Baker, M.A. Celia, W. Jäger, and W. Kinzelbach. We wish to thank the invited speakers P. Ackerer, H.G. Bock, J. Carrera, G. Dagan, H. Daniels, R.E. Ewing, E.O. Frind, P. Gresho, W. Hackbusch, I. Herrera, G. Gambolati, P. Knabner, H. Kobus, M. Kawahara, S.P. Neuman, K. Pruess, R. Rannacher, J. Troesch, T. van Genuchten, P. Wesseling, J.J. Westerink, and W.G. Yeh. We are also indebted to Anja McKellar, who did most of the secretarial work.

The papers appearing in this volume have been reproduced directly from material submitted by the authors, who are wholly responsible for their content.

The Editors

CONTENTS

x

2. SUBSURFACE TRANSPORT

3. SCALING AND HETEROGENEITY

4. GEOSTATISTICS

5. REACTIVE FLOW

6. FRACTURED POROUS MEDIA

7. PARAMETER ESTIMATION

VOLUME 2

8. REMEDIATION OPTIMIZATION

9. SUBSURFACE MULTIPHASE FLOW

10. SALTWATER INTRUSION

11. SHALLOW WATER EQUATIONS

12. FLOW AND TRANSPORT IN RIVERS

13. NAVIER-STOKES EQUATIONS

14. COASTAL FLOW

15. SEDIMENT TRANSPORT

16. ALGEBRAIC METHODS

17. SOFTWARE DEVELOPMENT

18. PARALLEL METHODS

xxvi

1. GROUNDWATER AND FLOW IN POROUS MEDIA

EFFICIENT MIXED METHODS FOR GROUNDWATER FLOW ON TRIANGULAR OR TETRAHEDRAL MESHES

TODD ARBOGAST, CLINT N. DAWSON, and PHILIP T. KEENAN
Department of Computational and Applied Mathematics, Rice University,
P.O. Box 1892, Houston, TX 77251–1892, U.S.A.

Simulating flow in porous media requires the solution of elliptic or parabolic partial differential equations. When the computational domain is irregularly shaped, applying finite element methods with triangular elements offers great flexibility.

The mixed finite element method has proven useful for solving flow equations. The difficulty with mixed methods in general is in solving the linear algebraic systems that arise. On rectangular elements, the mixed method with lowest-order approximating spaces can be reduced to a simple finite difference method in the primary variable, thus reducing the linear system to a sparse, symmetric, positive (semi-)definite matrix, for which many solution techniques are known. This type of reduction is not straightforward for triangular elements. In this paper we outline new mixed type methods which generalize finite differences in a manner suitable for use with triangular elements. Numerical examples illustrate the accuracy and efficiency of these new methods.

INTRODUCTION

Simulating flow in porous media requires the solution of elliptic or parabolic partial differential equations. In this paper, we discuss several variants of the mixed finite element method for solving elliptic equations of the form

$$-\nabla \cdot (K(\mathbf{x})\nabla p(\mathbf{x})) \equiv \nabla \cdot \mathbf{u}(\mathbf{x}) = f(\mathbf{x}), \quad \mathbf{x} \in \Omega, \tag{1}$$

where the aquifer region is represented by a polygonal domain Ω in $\mathbb{R}^2$ partitioned into triangular elements, and K is a general tensor. We remark briefly on tetrahedral meshes defined on $\Omega \subset \mathbb{R}^3$ at the end of the paper. For simplicity, we assume the boundary condition

$$p(\mathbf{x}) = 0, \quad \mathbf{x} \in \partial\Omega, \tag{2}$$

but the methods apply to general boundary conditions. We solve (1) and (2) using the lowest-order Raviart-Thomas approximating spaces (RT_0) on triangular elements [8]. A C++ package which implements all of the methods presented here has been developed by the third author [6].

A. Peters et al. (eds.), Computational Methods in Water Resources X, 3–10.
© 1994 Kluwer Academic Publishers. Printed in the Netherlands.

The mixed finite element method (MFEM) for solving elliptic equations was first described in [8]. The MFEM is especially useful in groundwater flow problems because mixed methods approximate the velocity u and the pressure p to the same order of accuracy. Furthermore, the approximate velocity calculated by the mixed method is locally mass conservative, satisfying $\nabla \cdot \mathbf{u} = f$ almost everywhere. In contrast, Galerkin finite element methods only conserve mass globally.

On rectangular elements, if K in (1) is a scalar or a diagonal matrix, one finds that numerical quadrature reduces the mixed method to a five-point finite difference method for pressure [10]. The new AWYM (Arbogast-Wheeler-Yotov method) extends these results to general coefficient tensors [2, 3, 4]. On triangular elements, however, the linear system arising from the MFEM is sparse but indefinite, making it expensive to solve. We have developed a cell-centered stencil method (CCSM) which reduces the MFEM on triangular elements to a ten-point finite difference type stencil. This provides a locally conservative finite difference method on triangles that is efficient and simple to implement. The CCSM is based on the AWYM with special quadrature rules and reduces to the usual finite difference method when applied to rectangular elements. The method is highly accurate on smooth triangulations. In more general cases the method loses accuracy, which can however be avoided by adding Lagrange multipliers on element faces where the discontinuities appear.

THE NUMERICAL MODELS

This section briefly defines some variants of the mixed method. See [1] for more details.

Let $(\cdot, \cdot)$ denote the $L^2(\Omega)$ inner product: for functions ϕ and ψ,

$$(\phi, \psi) = \int_\Omega \phi(\mathbf{x}) \psi(\mathbf{x}) \, dx \quad \text{and} \quad L^2(\Omega) = \{\phi : \|\phi\| \equiv (\phi, \phi)^{1/2} < \infty\}.$$

Let $H(\text{div}; \Omega) = \{\mathbf{v} \in (L^2(\Omega))^2 : \text{div} \, \mathbf{v} \in L^2(\Omega)\}$. Let $W_h \subset L^2(\Omega)$ and $\mathbf{V}_h \subset H(\text{div}; \Omega)$ be the lowest-order Raviart-Thomas (RT_0) [8] approximating spaces defined on a triangulation of Ω into elements with maximum diameter $h > 0$. We note that W_h consists of functions that are constant on each element in the triangulation. Any function $\mathbf{v} \in \mathbf{V}_h$ is completely determined by the values of $\mathbf{v} \cdot \eta$ on the edges of the triangulation, where η is the unit normal to an edge. In the standard MFEM, we seek $\mathbf{U} \in V_h$ and $P \in W_h$ that satisfy

$$(K^{-1}\mathbf{U}, \mathbf{v}) - (P, \nabla \cdot \mathbf{v}) = 0, \quad \mathbf{v} \in V_h, \tag{3}$$

$$(\nabla \cdot \mathbf{U}, w) = (f, w), \quad w \in W_h. \tag{4}$$

It is well-known that for the RT_0 spaces on rectangular elements, numerical quadrature can be used in (3) to reduce the linear system of equations to a sparse, symmetric system in P only. Below we describe similar approaches for triangular elements.

The variants of the mixed method we investigate are based on a new formulation developed by Arbogast, Wheeler, and Yotov [2, 4]. In this approach, a mesh-dependent matrix function $\mathbf{s}_g$ and an additional variable $\mathbf{y}$ are introduced, with

$$\mathbf{s}_g \mathbf{y} = -\nabla p, \tag{5}$$

$$\mathbf{s}_g \mathbf{u} = \mathbf{s}_g K \mathbf{s}_g \mathbf{y}, \tag{6}$$

$$\nabla \cdot \mathbf{u} = q. \tag{7}$$

If $\mathbf{s}_g(x)$ is invertible, (5)–(7) is equivalent to (1). Let T denote any triangle in T_h, and let T_{ref} denote a reference element, which we assume is the equilateral triangle with vertices at $(-1, 0)$, $(1, 0)$, and $(0, \sqrt{3})$. Let D_T denote the Jacobian matrix of the affine mapping between element T and T_{ref}, and let $J_T = |\det D_T|$. On T, define $\mathbf{s}_g$ by

$$\mathbf{s}_g|_T = J_T (D_T)^T D_T.$$

Note that $\mathbf{s}_g$ on each element is symmetric and positive definite. We approximate $\mathbf{y}$ by $\mathbf{Y} \in \mathbf{V}_h$, $\mathbf{u}$ by $\mathbf{U} \in \mathbf{V}_h$, and p by $P \in W_h$, where $\mathbf{Y}$, $\mathbf{U}$, and P satisfy

$$(\mathbf{s}_g \mathbf{Y}, \mathbf{v}) = (P, \nabla \cdot \mathbf{v}), \quad \mathbf{v} \in \mathbf{V}_h, \tag{8}$$

$$(\mathbf{s}_g \mathbf{U}, \mathbf{z}) = (\mathbf{s}_g K \mathbf{s}_g \mathbf{Y}, \mathbf{z}), \quad \mathbf{z} \in \mathbf{V}_h, \tag{9}$$

$$(\nabla \cdot \mathbf{U}, w) = (f, w), \quad w \in W_h. \tag{10}$$

In matrix form, (8)–(10) can be written as

$$\begin{bmatrix} S & 0 & -B \\ C & -S & 0 \\ 0 & B^T & 0 \end{bmatrix} \begin{bmatrix} \bar{\mathbf{Y}} \\ \bar{\mathbf{U}} \\ \bar{P} \end{bmatrix} = \begin{bmatrix} 0 \\ 0 \\ F \end{bmatrix}. \tag{11}$$

Reducing to an equation in $\bar{P}$ only, we find

$$A_{\text{awym}} \bar{P} \equiv (B^T S^{-1} C S^{-1} B) \bar{P} = R.$$

One advantage of the AWYM over the MFEM is that it does not require calculating K^{-1}, and thus is definable in cases where $K(x) = 0$. For time-dependent problems where $K = K(\cdot, t)$, the AWYM has the advantage over the MFEM that only S^{-1} is needed in the computation, not M^{-1}. Unlike M, S is not time-dependent and hence can be factored once at the beginning of the computation, whereas M must be factored at each time-step. In general, however, the computational expense of the AWYM and the MFEM are roughly equivalent.

For the two methods described above, assuming the RT_0 approximating spaces are used, the pressure and velocity are globally first order accurate. Superconvergence of pressure to second order is observed at the center of mass of each cell, and for rectangular meshes, superconvergence of velocities to second order is observed at certain Gauss points [2, 3].

We note that if S were a diagonal matrix, then A_{awym} would be sparse. Recently we developed the cell-centered stencil method (CCSM) for triangular meshes [1] which reduces S to a diagonal matrix by numerical integration of the left side of (8). In two space dimensions, the resulting matrix has at most ten nonzero entries on any row.

On any triangle T, let $\mathbf{v}_k$ denote the basis function of $\mathbf{V}_h$ associated with edge k, denoted by e_k, $k = 1, 2, 3$. Assume that e_k is mapped into edge k of T_{ref}, denoted by $\hat{e}_k$. Define the Piola transformation [9] on vectors by

$$\hat{\mathbf{v}}_k = J_T D_T \mathbf{v}_k, \quad k = 1, 2, 3. \tag{12}$$

It can be shown that

$$\int_T (\mathbf{s}_g \mathbf{v}_k) \cdot \mathbf{v}_l \, dx = \int_{T_{\text{ref}}} \hat{\mathbf{v}}_k \cdot \hat{\mathbf{v}}_l \, dx, \tag{13}$$

for $k, l = 1, 2, 3$. Moreover, from (12), one can show that $\hat{\mathbf{v}}_k$ is a scalar multiple of the standard basis function for $\mathbf{V}_h$ corresponding to edge k of T_{ref}.

Using these facts, we define a quadrature rule $Q_T(g)$ on T_{ref} by

$$Q_T(g) = \frac{\sqrt{3}}{6} \left[g(-1, 0) + g(1, 0) + g(0, \sqrt{3}) + 3g\left(0, \frac{\sqrt{3}}{3}\right) \right]. \tag{14}$$

It has the two properties that $Q_T(g)$ is exact for polynomials of degree one, and $Q_T(\hat{\mathbf{v}}_k \cdot \hat{\mathbf{v}}_l) = 0$ for $k \neq l$. Thus, by (13),

$$\int_T (\mathbf{s}_g \mathbf{v}_k) \cdot \mathbf{v}_l \, dx \approx Q_T(\hat{\mathbf{v}}_k \cdot \hat{\mathbf{v}}_l) = \begin{cases} 0, & k \neq l, \\ \frac{\sqrt{3}}{6}(\text{length}(e_k))^2, & k = l. \end{cases} \tag{15}$$

Let S_D denote the diagonal approximation to S determined by the quadrature rule given above. Then the CCSM reduces to finding $\bar{P}$ which satisfies

$$A_{\text{ccsm}} \bar{P} = (B^T S_D^{-1} C S_D^{-1} B) \bar{P} = R. \tag{16}$$

Note that A_{ccsm} is a sparse approximation to A_{awym}.

The CCSM has been implemented for elliptic problems in two space dimensions [1] and compared to the results given by the AWYM and MFEM. Recall that the AWYM and MFEM give $\mathcal{O}(h^2)$ errors for pressures at element centers and $\mathcal{O}(h)$ global L^2 errors for velocities. For smooth triangulations, the CCSM gives the same orders of convergence, but for domains where a smooth triangulation is not feasible, these errors can be reduced by a full order of h. The reason is that the function $\mathbf{s}_g$ can be nonsmooth across element edges, depending on the triangulation. In these regions, the accuracy of the quadrature rule is reduced by one order (from h^2 to h), leading to a corresponding loss of accuracy in the solution. Thus it appears the CCSM may not be useful for totally unstructured, nonsmooth meshes, however, for meshes with some structure, we can modify the CCSM slightly to regain accuracy.

One way to regain accuracy in the solution is to smooth the mesh as it is refined. If this is for some reason not possible, one can use the Enhanced Cell-Centered Stencil Method (ECCSM) [1], which is a combination of the CCSM with the standard mixed-hybrid finite element method (MHFEM) [5]; in particular, pressure unknowns are added at element edges where s_g is not smooth. The result is a system of equations in velocity unknowns $\bar{Y}$ and $\bar{U}$ and pressure unknowns $\bar{P}$ and $\bar{\lambda}$, where $\bar{\lambda}$ represents pressure unknowns on element boundaries. This system can be reduced to an equation for $\bar{P}$ and $\bar{\lambda}$ only. Depending on the structure of the mesh, the ECCSM lies somewhere between the MHFEM and the CCSM in total number of unknowns. For domains which can be divided into a relatively small number of regular domains, where smooth triangulations can be used, the ECCSM is roughly equivalent in number of unknowns to the CCSM.

As mentioned above, the purpose of introducing the boundary pressures is to regain the accuracy in the numerical solution lost by the CCSM on nonsmooth meshes. We have implemented the ECCSM for problems where the CCSM lost accuracy and observed $\mathcal{O}(h^2)$ errors for pressure at cell centers and $\mathcal{O}(h)$ global L^2 errors for velocity [1], which are the same rates obtained by the MFEM, MHFEM, and AWYM. Moreover, numerical experiments indicate that a new postprocessing scheme developed by Keenan [7] can improve the convergence rate for the velocities almost to $\mathcal{O}(h^2)$.

NUMERICAL RESULTS

We created a large suite of test problems to examine the behavior of the numerical methods described above. We varied the shape of the domain, the coefficient tensor K, and the analytic solution. In each case, the boundary conditions and the forcing term f were constructed to match the prescribed solution. We report in detail on one typical case and then summarize the results from the full test suite.

Among the domains considered was the one shown in Figure 1. This figure illustrates the initial decomposition of the domain into elements. The domain is neither simply connected nor convex; moreover we chose to use both rectangles and triangles in subdividing it, to illustrate the flexibility of the C++ program.

In the convergence study, the domain was refined uniformly to generate progressively finer meshes. Each application of uniform refinement replaced each triangle or rectangle with 4 smaller but geometrically similar ones. The finest mesh had 2432 elements. Uniform refinement generates hierarchical meshes: each new mesh contains all the edges of the previous one. However, discontinuities in the geometry mapping across edges of the original coarse triangulation are not smoothed out by refinement.

In Tables 1 and 2, we give detailed results for a test problem using Dirichlet boundary conditions, a non-diagonal tensor, and a known analytic solution. We

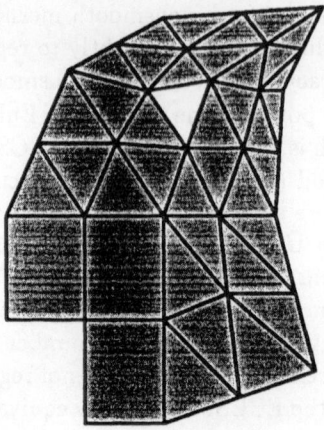

Figure 1: Example Domain

h	ECCSM	CCSM	MHFEM	MFEM
0.16	0.59	0.48	0.39	0.39
0.08	0.11	0.12	0.11	0.11
0.04	0.026	0.043	0.029	0.029
0.02	0.0062	0.019	0.0076	0.0076
Rate	h^2	$h^{1.4}$	h^2	h^2

Table 1: l^2 error in p for domain in Figure 1

report the l^2 norm of the error in the pressure p and the flux $K\nabla p$. (The l^2 norm is the discrete two-norm taken at the centers of elements.)

In the full test suite, the condition number of the linear system for most methods was $O(h^{-1})$, as estimated by the number of conjugate gradient iterations used. However, the ECCSM combined with uniform refinement produced better conditioned systems, with condition numbers around $O(h^{-0.9})$. Using a conjugate gradient solver with no preconditioning, the MFEM took much longer than the other three methods (approximately 50 times longer than the MHFEM on 2000 elements). On a typical smooth mesh problem the CCSM took approximately half as much CPU time as the MHFEM. The ECCSM was somewhat slower than the MHFEM on coarse meshes, since it solves for both pressures and Lagrange multipliers. By around four levels of mesh refinement it had caught up to the MHFEM, since it did not need Lagrange multipliers on every edge, and it should outperform it when additional refinement is used.

h	ECCSM	CCSM	MHFEM	MFEM	Post-Proc
0.16	6.4	9.3	6.0	6.0	6.4
0.08	3.5	5.9	3.1	3.1	1.8
0.04	1.6	3.7	1.5	1.5	0.60
0.02	0.80	2.5	0.77	0.77	0.20
Rate	h	$h^{0.6}$	h	h	$h^{1.6}$

Table 2: l^2 error in $K\nabla p$ for domain in Figure 1

The error in the pressure converged approximately like $O(h^2)$ for all methods except the CCSM in situations like the domain of Figure 1, where the geometry matrix changes discontinuously because of uniform refinement. Similarly the error in the flux converged like $O(h)$, except for CCSM with geometry discontinuities. Using smooth mesh refinement the CCSM achieved the same convergence orders as the other methods.

On rectangles one finds that the velocities are superconvergent at special points and can be post-processed to yield second order accurate vector approximations everywhere. Empirical evidence indicates that a new post processing scheme developed by the third author can recover close to second order accuracy for the velocities on triangular meshes as well [7]. The postprocessing method can be applied to any of the mixed method variants. The "Post-Proc" column in Table 2 shows the errors obtained with the ECCSM when post processing was used. The convergence rate for the post processed flux is generally between $h^{1.5}$ and $h^{1.9}$, depending in part on the smoothness of the mesh refinement process.

The C++ program can handle three dimensional elements such as bricks and tetrahedra. We observed numerically that the stencil approach breaks down on tetrahedral meshes. This appears to be due to the fact that regular tetrahedra do not fill space, whereas equilateral triangles do tile the plane. This means that the geometry matrix s_g is unavoidably discontinuous everywhere, no matter how much one attempts to smooth the tetrahedral mesh. Therefore, the MHFEM seems to be the best choice for tetrahedral meshes. The CCSM could however be used with prismatic elements on meshes constructed from the tensor product of a triangular mesh in two dimensions and a one dimensional collection of intervals.

CONCLUSIONS

The CCSM is an accurate and efficient method for smooth meshes of triangular elements, which appears to be about twice as fast as competing methods. On meshes of tetrahedral elements, however, the CCSM loses accuracy, so the MHFEM should be used instead.

ACKNOWLEDGMENTS

This research was supported in part by the Department of Energy, the State of Texas Governor's Energy Office, and project grants from the National Science Foundation. The third author was supported in part by an NSF Postdoctoral Fellowship.

REFERENCES

[1] Arbogast, T., Dawson, C., and Keenan, P. T. (1993) "Mixed Finite Element Methods as Finite Difference Methods for Solving Elliptic Equations on Triangular Elements", Technical Report #93–53, Department of Computational and Applied Mathematics, Rice University.

[2] Arbogast, T., Wheeler, M.F., and Yotov, I. (1994) "Logically rectangular mixed methods for groundwater flow and transport on general geometry", this volume.

[3] Arbogast, T., Wheeler, M.F., and Yotov, I. (in preparation) "Mixed finite elements for elliptic problems with tensor coefficients as finite differences".

[4] Arbogast, T., Wheeler, M.F., and Yotov, I. (in preparation) "Mixed finite element methods on general geometry".

[5] Brezzi, F. and Fortin, M. (1991) "Mixed and hybrid finite elements", Springer Series in Computational Mathematics, Vol. 15, Springer-Verlag, Berlin.

[6] Keenan, P. T. (1993) "Instructions for my C++ elliptic PDE solver programs using mixed methods on general geometry", Technical Report #93-56, Department of Computational and Applied Mathematics, Rice University.

[7] Keenan, P. T. (in preparation) "An efficient postprocessor for velocities from mixed methods on triangular and tetrahedral elements".

[8] Raviart, P. A., and Thomas, J. M. (1977) "A mixed finite element method for 2nd order elliptic problems", in Mathematical Aspects of the Finite Element Method, Lecture Notes in Math., Springer-Verlag, Berlin.

[9] Thomas, J. M. (1977) "Sur l'analyse numérique des méthodes d'éléments finis hybrides et mixtes", Thèse d'Etat, Université Pierre et Marie Curie.

[10] Russell, T.F., and Wheeler, M.F. (1983) "Finite element and finite difference methods for continuous flows in porous media", in Ewing, R.E. (ed.), The Mathematics of Reservoir Simulation, Frontiers in Applied Mathematics 1, Society for Industrial and Applied Mathematics, Philadelphia, Chapter II, pp. 35–106.

ASYMPTOTIC CONVERGENCE OF A NEW SPECTRAL METHOD FOR THE SOLUTION IN TIME OF FINITE ELEMENT FLOW EQUATIONS

LUCA BERGAMASCHI[†], GIUSEPPE GAMBOLATI[†], and GIORGIO PINI[†]

[†]Dipartimento di Metodi e Modelli Matematici per le Scienze Applicate
University of Padua, ITALY

An efficient method (DACG, Deflated Accelerated Conjugate Gradient) has recently been developed, based on the conjugate gradient (CG) minimization of the Rayleigh quotient over successive deflated subspaces of decreasing size. A new version of this algorithm is presented and discussed to evaluate the 40 leftmost eigenpairs of FE problems of increasing size up to 10500. Its asymptotic convergence rate is shown to be inversely proportional to the square root of the spectral condition number of the Hessian of the Rayleigh quotient in the current restricted subspace. Numerical results show that preconditioning the CG method with the inverse of the incomplete factors greatly improves convergence which may prove to be slow only toward very close eigenvalues.

INTRODUCTION

Spectral (or reduction) techniques are often used in combination with the finite differences (FD) or the finite elements (FE) method in engineering problems. A typical application is the integration in time of large sets of FE subsurface flow equations [Gambolati, 1993a; Gambolati, 1993b]. The spectral approach requires the solution of a generalized symmetric eigenproblem involving the matrix pencil A (stiffness) and B (capacity). The transient FE response is primarily described by the leftmost eigenpairs whose evaluation represents the basic computational burden of the spectral method.

Recently an efficient optimization method, DACG, has been proposed to compute the m leftmost eigenpairs of the generalized eigenproblem $Ax = \lambda Bx$. This method performs a conjugate gradient minimization of the Rayleigh quotient over convenient subspaces of decreasing dimension. In a previous paper [Bergamaschi et al., 1993a] the theoretical asymptotic convergence rate of DACG was analyzed, and shown to be strictly related to the spectral condition number of the Hessian of the Rayleigh quotient. In a more recent paper [Bergamaschi et al., 1993b] the optimal choice of coefficient β_k in the recurrence equation defining the new search direction is discussed.

In the present paper, the DACG performance with different preconditioners: A^{-1}, D^{-1} (where $D =\text{diag}(A)$) and $(LU)^{-1}$, L and U being the incomplete factors of A, is studied for the computation of the 40 leftmost eigenpairs of six large eigenproblems.

A. Peters et al. (eds.), Computational Methods in Water Resources X, 11–18.

More precisely the incomplete LU factorization without fill-in (ILU(0)) [*Meijerink and van der Vorst, 1977*], and an incomplete factorization technique based on a two parameters strategy for dropping elements (ILUT(p,τ)) [*Saad, 1991*] have been employed. For different preconditioners the convergence profiles and total CPU times are derived and discussed.

DESCRIPTION OF DACG METHOD

Consider the generalized eigenproblem

$$Ax = \lambda Bx \tag{1}$$

where A and B are sparse symmetric positive definite $N \times N$ matrices.

Let $\lambda_1 \geq \lambda_2 \ldots \geq \lambda_N$ be the real positive eigenvalues, and $v_1, v_2 \ldots, v_N$ the corresponding eigenvectors.

Assume that the j leftmost eigenpairs of (1) have been evaluated. Then the $N - j$ smallest eigenpair is obtained by the following DACG procedure:

1. Start with an initial eigenvector estimate x_0 such that $V_j^T B x_0 = 0$,

$$V_j = [v_N, \ldots, v_{N-j+1}].$$

   Set $k = 0$ (iteration index) and p_{-1} as an arbitrary vector.

2. If $k = 0$ set $\beta_k = 0$, otherwise calculate β_k by [*Polak, 1971*], [*Gambolati et al., 1992*], [*Bergamaschi et al., 1993b*]:

$$\beta_k = \frac{g_k^T K^{-1}(g_k - g_{k-1})}{g_{k-1}^T K^{-1} g_{k-1}} \tag{2}$$

   where K^{-1} is the symmetric, positive definite preconditioning matrix and

$$g_k = \frac{2}{x_k^T B x_k}[Ax_k - R(x_k)Bx_k] \tag{3}$$

   is the gradient of the Rayleigh quotient $R(x) = x^T A x / x^T B x$ assessed at the current iterate x_k.

3. Calculate:

$$\tilde{p}_k = K^{-1} g_k + \beta_k p_{k-1} \tag{4}$$

4. Evaluate p_k by B-orthogonalization of $\tilde{p}_k$ against to V_j, using a Gram-Schmidt B-orthogonalization process.

5. Compute the coefficient α_k by minimizing the Rayleigh quotient $R(x_k + \alpha_k p_k)$ [*Perdon and Gambolati, 1986*].

6. Set:

$$\tilde{x}_{k+1} = x_k + \alpha_k p_k$$

   The new approximate eigenvector x_{k+1} is evaluated by B-normalization of $\tilde{x}_{k+1}$.

7. Increment the iteration counter and go back to step 2). The iteration is completed whenever k is larger than the allowed maximum number of iterations IMAX or when:

$$er_{j,k+1} = \frac{|q(x_{k+1}) - q(x_k)|}{q(x_k)} < \text{TOL1}. \tag{5}$$

DEFINITION OF SOME PRECONDITIONERS

The choice of an efficient preconditioner is crucial towards the convergence of the DACG algorithm. In [*Bergamaschi et al., 1993*b] it has been found that using A^{-1} as the preconditioner the asymptotic convergence rate ρ_j is directly proportional to the square root of the relative separation between the eigenvalue being sought and the next higher one, i. e. :

$$\rho_j = 4\sqrt{\frac{\lambda_{N-j-1} - \lambda_{N-j}}{\lambda_{N-j-1}}}.$$

Without preconditioning this result cannot be obtained and the convergence rate decreases by a factor of $\sqrt{\lambda_1/\lambda_N}$ which is normally very high in finite element problems of practical interest. Unfortunately the A^{-1} preconditioner is expensive since it needs either the inversion of matrix A, which is often an unfeasible operation, or the equivalent solution of the linear system $Ay = g_k$, to provide the y vector required in the computation of eq. (4).

The need for a cheap preconditioner may suggest the choice: $K^{-1} = D^{-1}$, D being the main diagonal of A. D may certainly be inverted since A is positive definite. In the solution of linear systems, however, the diagonal preconditioner proves too 'poor' to ensure a good asymptotic convergence, the condition number of the corresponding iteration matrix being $\gg 1$. Therefore we are looking for a better and at the same time inexpensive preconditioner K^{-1} yielding the approximation cond($K^{-1}A$) ≈ 1.

THE ILUT(p,τ) PRECONDITIONER

The incomplete LU factorization without fill-in [*Meijerink and van der Vorst, 1977*] is one of the most popular techniques of preconditioning. When dealing with symmetric matrices, the idea is to find a matrix L (in this case $U^T = L$), lower triangular, such that $L + L^T$ has the same pattern as A, and $(LL^T)_{ij} = a_{ij}$, for every i,j so that $a_{ij} \neq 0$. The usual way of defining L (which is not uniquely defined), is to perform the standard Gauss elimination and replace by zero any fill-in coefficient during the process. *Saad [1991]* introduces an incomplete factorization technique, based on a two-parameter strategy for eliminating the new elements generated by the factorization. The parameter p represents the largest allowed value of fill-in on every row; moreover every coefficient which is smaller than the tolerance τ (relative to the original Euclidean norm of the row of A) is automatically dropped in the decomposition algorithm. Once this factorization is completed, we take $(LL^T)^{-1}$ as the preconditioner. Note that the product $K^{-1}g_k$ in eq. (4) is now computed by sequentially solving two triangular systems.

NUMERICAL RESULTS

N	p	τ	NCOEFU	# it	time (sec.) ILUT	iter	tot	speed-up
441	ILU(0)		1681	1019	0.03	3.85	3.88	–
	1	1E-6	1657	795	0.04	3.16	3.20	1.20
	3	1E-6	2485	835	0.06	3.52	3.58	1.08
	5	1E-6	3305	832	0.09	3.66	3.75	1.03
812	ILU(0)		3135	1130	0.05	8.05	8.10	–
	1	1E-6	3186	781	0.13	5.64	5.77	1.40
	5	1E-6	6287	624	0.22	5.14	5.36	1.51
	8	1E-6	8486	576	0.27	5.12	5.39	1.50
	10	1E-6	9918	574	0.31	5.39	5.70	1.42
1952	ILU(0)		7744	1578	0.16	27.01	27.17	–
	5	1E-6	14580	896	0.44	17.66	18.10	1.50
	8	1E-6	18871	792	0.63	16.62	17.25	1.58
	10	1E-6	21273	781	0.75	17.04	17.69	1.54
	15	1E-6	28807	702	1.09	16.60	17.79	1.53
4560	ILU(0)		34295	1254	0.48	69.00	69.48	–
	1	1E-6	36892	1168	2.46	65.85	68.31	1.02
	4	1E-6	49817	850	3.74	52.70	56.44	1.23
	5	1E-6	54058	824	3.80	50.71	54.51	1.27
	5	1E-4	52541	823	2.31	50.33	52.64	1.30
	10	1E-6	74743	775	5.96	52.45	58.41	1.19
7744	ILU(0)		30848	2843	0.50	215.73	215.23	–
	5	1E-6	59394	1254	1.75	110.27	112.02	1.92
	10	1E-6	88136	957	3.24	88.28	91.52	2.35
	12	1E-6	99614	927	3.76	88.41	92.16	2.34
	12	1E-4	98966	925	2.78	87.22	90.00	2.40
	15	1E-6	116813	880	4.73	88.41	93.14	2.31
	20	1E-6	145410	857	6.48	93.71	100.19	2.14
10593	ILU(0)		41665	5295	0.54	527.40	527.94	–
	5	1E-6	83895	1560	2.23	173.54	175.77	3.04
	10	1E-6	136442	1045	3.65	133.84	137.49	3.82
	12	1E-6	156442	962	4.31	128.96	133.27	3.96
	15	1E-6	188879	906	5.55	131.55	137.10	3.85

Table 1. Performance of the ILUT(p, τ) preconditioner for various values of parameters p and τ in terms of iteration number, total CPU time and fill-in of the U factor (NCOEFU) in the evaluation of the 40 leftmost eigenpairs of the six sample problems. TOL1=10^{-8}

The DACG procedure with different preconditioners has been applied to six sample problems arising form the FE integration of 2-D and 3-D groundwater flow equations with size N= 441, 812, 1952, 4560, 7744 and 10593. The distribution of the 40 leftmost eigenvalues of some of these eigenproblems can be found in [*Gambolati and Bergamaschi, 1992*]. Computations have been performed on an IBM RISC 6000 model 560 workstation in double precision arithmetics. Table 1 yields the overall CPU times, and number of iterations required for the evaluation of the 40 smallest eigenpairs of (1), and the fill-in of the $U = L^T$ factor in the ILUT factorization. In

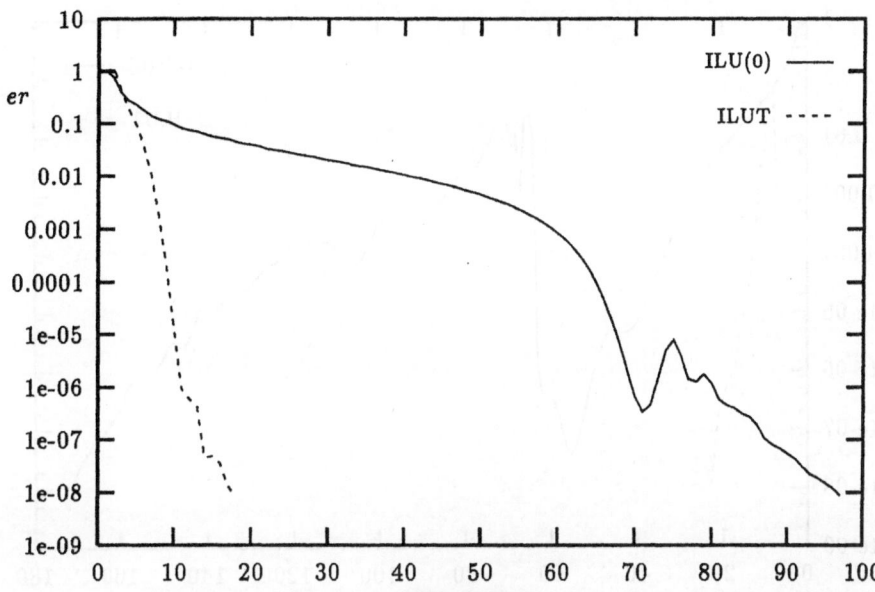

Figure 1. Convergence profiles in the computation of $\lambda_N, \mathbf{u}_N$ for the $N = 10593$ problem. The preconditioners are ILU(0) (solid line) and ILUT with optimal parameters (dashed line).

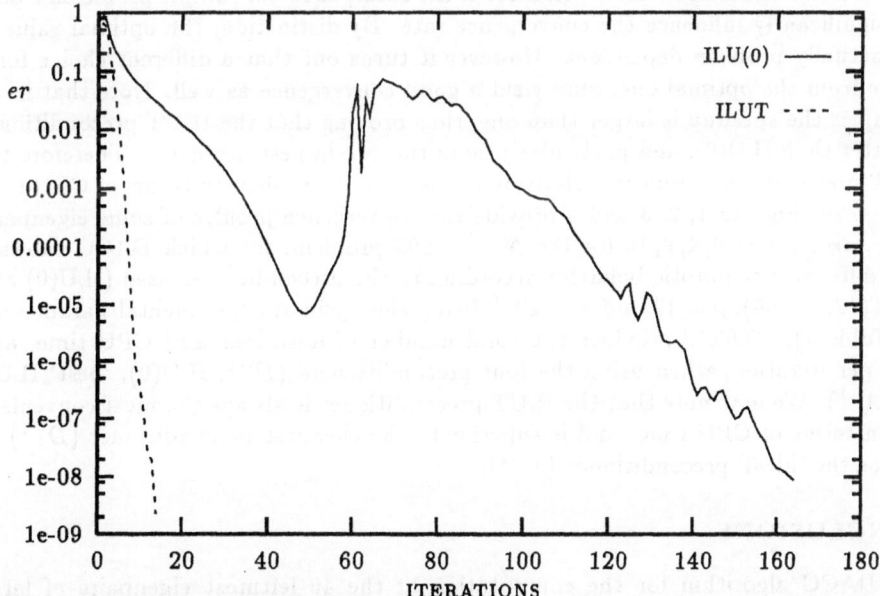

Figure 2. The same as Figure 1 for $\lambda_{N-4}, \mathbf{u}_{N-4}$

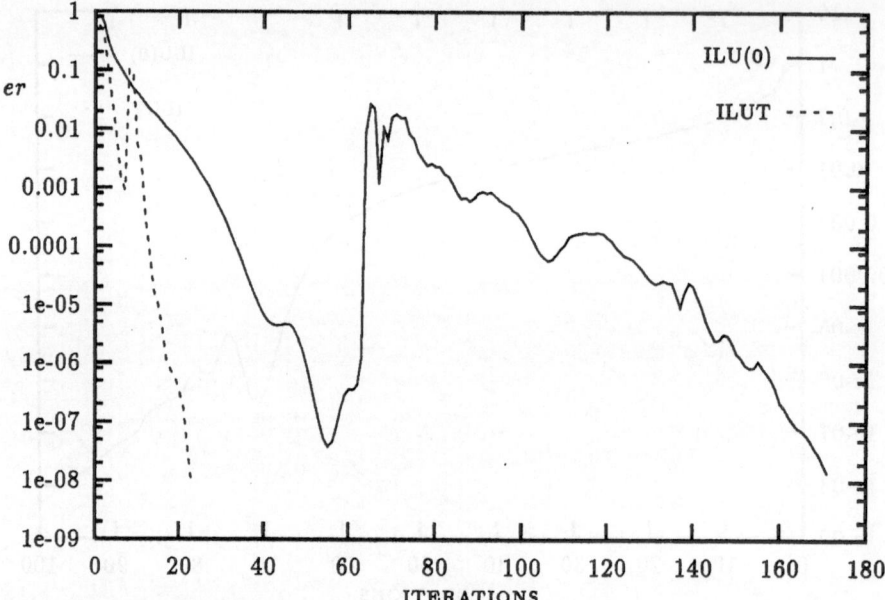

Figure 3. The same as Figure 1 for λ_{N-9}, u_{N-9}

Table 1 the speedup values are also given, with reference to the ILU(0) performance. The τ value has been set to 10^{-6} in most of the cases, as in our sample problems τ does not significantly influence the convergence rate. By distinction, the optimal value of p is actually problem dependent. However it turns out that a different choice for p and τ from the optimal one, may yield a good convergence as well. Note that in all examples the speedup is larger than one, thus proving that the ILUT preconditioner is better than ILU(0), and particularly so in the two largest problems. Therefore the ILUT preconditioner appears relatively robust and convenient in terms of CPU time for $p \leq 20$. Figures 1, 2, 3 and 4 provide the convergence profiles of some eigenpairs $\lambda_{N-j}, u_{N-j}, j = 0, 4, 9, 19$ for the $N = 10593$ problem, for which DACG shows a very different asymptotic behavior according to the preconditioner used (ILU(0) and ILUT(12, 1E-06), $p = 12$ and $\tau = 10^{-6}$ being the optimal experimental parameters, see Table 1). Table 2 provides the total number of iterations and CPU time, and time per iteration, when using the four preconditioners (D^{-1}, ILU(0), 'best' ILUT and A^{-1}). We may note that the ILUT preconditioner is always the most convenient one in terms of CPU time, and is superior to the cheapest preconditioner (D^{-1}) as well as the 'ideal' preconditioner (A^{-1}).

CONCLUSIONS

The DACG algorithm for the computation of the 40 leftmost eigenpairs of large symmetric generalized finite element eigenproblems $Ax = \lambda Bx$ has been accelerated by the use of an incomplete preconditioner (ILUT) whose rate of fill-in is dependent

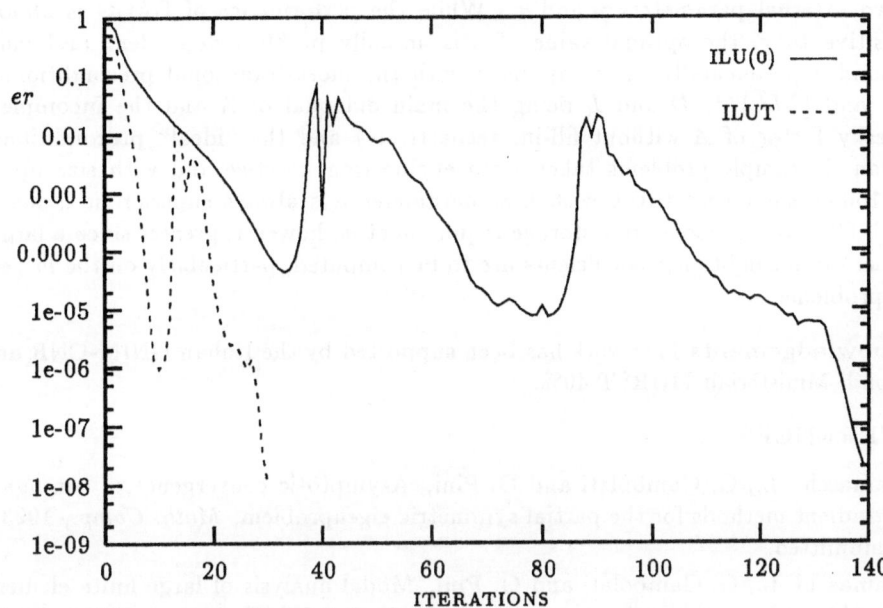

Figure 4. The same as Figure 1 for λ_{N-19}, u_{N-19}

N	K^{-1}	# it	time	t it.
441	D^{-1}	2764	8.87	0.32E−02
	ILU(0)	1019	3.88	0.38E−02
	ILUT	795	3.20	0.40E−02
	A^{-1}	906	11.14	0.12E−01
812	D^{-1}	12031	61.68	0.51E−02
	ILU(0)	1130	8.10	0.72E−02
	ILUT	624	5.36	0.85E−02
	A^{-1}	475	23.21	0.49E−01
1952	D^{-1}	3464	50.16	0.14E−01
	ILU(0)	1578	27.17	0.17E−01
	ILUT	792	17.25	0.22E−01
	A^{-1}	645	85.82	0.13E+00
4560	D^{-1}	15132	631.62	0.42E−01
	ILU(0)	1254	69.48	0.55E−01
	ILUT	823	50.33	0.61E−01
	A^{-1}	718	293.80	0.41E+00
7744	D^{-1}	6606	409.70	0.62E−01
	ILU(0)	2843	215.73	0.76E−01
	ILUT	925	90.00	0.97E−01
	A^{-1}	759	621.48	0.82E+00
10593	D^{-1}	17638	1539.86	0.87E−01
	ILU(0)	5295	527.40	0.10E+00
	ILUT	962	133.27	0.14E+00
	A^{-1}	757	593.79	0.78E+00

Table 2. *Performance* of the various preconditioners in terms of total number of iterations and CPU time, and time per iteration, in the computation of the 40 leftmost eigenpairs of the six sample problems. TOL1=10^{-8}.

on two external parameters p and τ. While the performance of DACG is almost insensitive to τ, the optimal value of p is actually problem dependent and must be found experimentally. A comparison with the most traditional preconditioners (D^{-1} and $(LL^T)^{-1}$, D and L being the main diagonal of A and the incomplete Cholesky factor of A without fill-in, respectively) and the "ideal" preconditioner A^{-1} on six sample problems taken from engineering practice and with size up to over 10000, show that ILUT with best parameter p is always superior in terms of overall CPU times. Computer storage requirement is, however, greater since a larger number of preconditioner coefficients are to be computed, particularly on the largest eigenproblems.

Acknowledgements This work has been supported by the Italian GNIM-CNR and by Fondi Ministeriali MURST 40%.

REFERENCES

Bergamaschi, L., G. Gambolati and G. Pini, Asymptotic convergence of conjugate gradient methods for the partial symmetric eigenproblem, *Math. Comp.*, 1993a. submitted.

Bergamaschi, L., G. Gambolati and G. Pini, Modal analysis of large finite element problems by optimization methods, *J. Shock and Vibr.*, 1993b. submitted.

Gambolati, G., On time integration of groundwater flow equations by spectral methods, *Water Resour. Res.* 29(4), 1257–1267, 1993a.

Gambolati, G., Solution of large eigenvalue problems, In: Papadrakakis, M. (ed.) *Solving Large Scale Problems in Mechanics: The Development and Application of Computational Solution Methods*. John Wiley, New York, pp 125–156, 1993b.

Gambolati, G., F. Sartoretto and P. Florian, An orthogonal accelerated deflation technique for large symmetric eigenproblems, *Comp. Methods App. Mech. Eng.* 94, 13–23, 1992.

Gambolati, G. and L. Bergamaschi, Partial eigensolution for transient groundwater flow equations, In: T. R. Russell et al. (eds.) *ICCMWR IX. Vol. 1: Numerical Methods in Water Resources*. CM Publications, Southampton, pp 175–192, 1992.

Meijerink, J. A. and H. A. van der Vorst, An iterative solution method for linear systems of which the coefficient matrix is a symmetric M-matrix, *Math. Comp.* 31, 148–162, 1977.

Perdon, A. M. and G. Gambolati, Extreme eigenvalues of large sparse matrices by Rayleigh quotient and modified conjugate gradients, *Comp. Methods App. Mech. Eng.* 56, 251–264, 1986.

Polak, E., *Computational Methods in Optimization: A Unified Approach*. Academic Press, New York, 1971.

Saad, Y., ILUT: a dual strategy accurate incomplete ILU factorization. Technical Report, Minnesota Supercomputer Institute, University of Minnesota, 1991.

PREPROCESSORS FOR THE SIMULATION OF GROUNDWATER FLOW PHENOMENA WITH THE FINITE ELEMENT METHOD

J.F. BOTHA, J. BUYS, J.P. VERWEY AND I. VAN DER VOORT
Institute for Groundwater Studies
University of the Orange Free State
Bloemfontein
Republic of South Africa

ABSTRACT

This paper describes two methods that can be used to reduce the errors associated with the numerical simulation of groundwater flow phenomena. The first is a finite element mesh generator, based on the Laplacian map of a rectangular domain onto an irregular domain, and the second an interpolation method for the estimation of initial conditions at nodes where no values have been observed. The methods are particularly useful when the initial conditions are subject to steep gradients, or if the domain contains holes. The paper is concluded with a brief discussion of the numerical model, developed for the marble aquifer at Otjiwarongo in Namibia.

INTRODUCTION

Numerical models of the groundwater flow and hydrodynamic dispersion equations have contributed significantly towards a better understanding of flow and mass transport in the subsurface of the earth. It must be kept in mind though that these numerical solutions are only approximations to the exact solution of the differential equation. The success of such a numerical model, therefore, hinges on the ability of the modeller to keep the error, associated with any numerical method, within acceptable bounds.

The main sources of errors in the numerical solution of a parabolic partial differential equation, such as the groundwater flow and hydrodynamic dispersion equations, are the approximation of: (a) the flow domain and its associated boundary, and (b) the initial water levels and concentrations. These errors can be reduced considerably, if one uses a suitably refined finite element mesh (Van Sandwyk *et al.*, 1992), and an accurate method for the estimation of the initial water levels and concentrations at nodes where they are not known (Van Tonder *et al.*, 1990).

A. Peters et al. (eds.), Computational Methods in Water Resources X, 19–26.
© 1994 *Kluwer Academic Publishers. Printed in the Netherlands.*

This paper describes two methods that the authors found to be very useful in reducing errors, associated with the numerical simulation of groundwater flow phenomena. The first is a finite element mesh generator, and the second an interpolation method for the estimation of initial conditions at nodes where no values have been observed. The advantage of the mesh generator, which is based on the Laplacian map of a rectangular domain onto an irregular domain, is that it can generate a mesh with the same accuracy, required for the solution of the differential equation, while the interpolation method is particularly useful when the initial conditions are subject to steep gradients, or domains with holes. The success of the methods can be judged from the discussion of the numerical model, developed for the marble aquifer at Otjiwarongo in Namibia.

MESH GENERATOR

Although it is possible to put various options into a mesh generator, the present generator was limited to the following basic objectives.

(a) Number the elements and nodes automatically, the latter in such a way that the band width for the coefficient matrix in the finite element implementation is as small as possible, but without the use of elaborate search techniques.

(b) Generate all the nodal coordinates from the coordinates of a limited number of 'boundary coordinates', which need not be restricted to the actual boundary of the domain of interest.

(c) Distinguish between domains with different characteristic properties.

The first two objectives can be easily achieved on a rectangular mesh, or a mesh consisting of the union of rectangular blocks. This type of mesh is, unfortunately, rarely encountered in practical problems. Nevertheless, the construction and optimal numbering of nodes for a such a mesh can be easily automated. Moreover, a rectangular mesh can, in principle, be mapped onto any domain of interest. It is for this reason that the present generator, generates the mesh in two phases.

Phase 1   The generation of a rectangular mesh in a space, Γ, with coordinates $\xi = (\xi_1, \xi_2)$, henceforth referred to as the *cardinal finite element mesh*.

Phase 2   Mapping of the cardinal mesh from Γ onto the global domain, Ω, with coordinates $x = (x_1, x_2)$, to obtain the *global finite element mesh*.

There are a number of methods that can be used to map the cardinal mesh in Γ onto the global domain, Ω. The present generator uses a vector-valued Laplacian map defined by the Dirichlet problem

$$\nabla^2 x(\xi) = 0$$

(1)

with boundary condition $\xi_0 = x_0$ on the boundary $\partial\Gamma$ of the cardinal domain Γ.

The simplest domain to use with the transformation in Equation (1), is a square cardinal domain. However, this may not always yield an acceptable map for a highly deformed global domain Ω. Moreover, there exist multiple-connected domains, which cannot be represented by a simply connected rectangular domain, without introducing cuts. The present generator therefore allows one to divide the global domain (Ω) into a number of subdomains, and then represent Ω as the union of a set of simply connected rectangular blocks in the domain Γ. This approach has two advantages. It allows one to: (a) handle multiple connected domains, as well as domains with different characteristic properties, and (b) vary the mesh size, over the global domain, from one rectangular block, to another (Botha *et al.*, 1993).

The final step in the generation of a global finite element mesh is to solve Equation (1) over the cardinal finite element mesh, subject to prescribed boundary coordinates. If Equation (1) was to be solved analytically, the boundary coordinates have to lie on the global boundary $\partial\Omega$. However, any point can serve as a boundary point, when Equation (1) is solved numerically. This implies that one can fix the positions of points, even points within Ω, by merely including them in the set of boundary points x_0. The advantage of this freedom in the choice of boundary points is that one can force suitable points, the positions of boreholes for example, to coincide with element nodes, at will.

Another advantage of the Laplace map is that one can generate the global mesh with the same basis functions, as those used in solving the differential equation. A finite element mesh, generated with such a Laplace map, therefore will have the same accuracy as the solution of the differential equation. This can reduce the errors in the solution of the differential equation considerably (Botha *et al.*, 1993).

The versatility of the present mesh generator is perhaps best illustrated by the finite element mesh of the marble aquifer, at Otjiwarongo in Figure 1; a mesh that two commercial mesh generators could not generate.

INVERSE DISTANCE WEIGHTED GRADIENT INTERPOLATION

The variables, associated with initial conditions in subsurface flow phenomena, are conventionally represented by regionalized variables. Kriging would thus be the ideal method to use for the estimation of the initial conditions in simulations of subsurface flow phenomena. However, experience has shown that the estimates are not very reliable, if the regionalized variable displays a trend, or steep gradients (Van Tonder *et al.*, 1990). In such cases, one must use either universal Kriging, or Bayes estimation (Van Sandwyk *et al.*, 1992). These methods yield reliable estimates, if the regionalized variable varies smoothly, but not if it is C^0-continuous, or contains steep gradients. For such variables the inverse distance weighted gradient (IDWG) interpolation, yields better results.

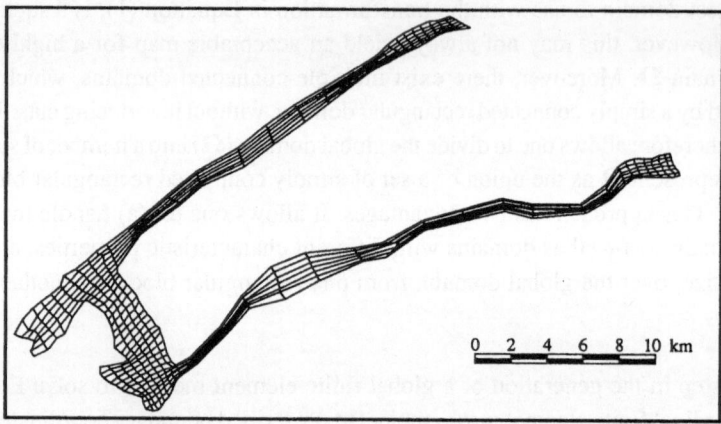

Figure 1 Finite element mesh of the marble aquifer at Otjiwarongo generated with the Laplacian map, defined in Equation 1.

One method to derive an IDWG-approximation for a given regionalized variable, $z(\mathbf{x})$ say, is to triangulate the domain of $z(\mathbf{x})$ with Delaunay triangles and then interpolate $z(\mathbf{x})$ over this triangulation This will yield a set of triangles with vertices defined by the values of the variable at three neighbouring points, as shown in Figure 2 (Buys $et\ al.$, 1992). The plane, $T(x,y,z)$, of the triangle in Figure 2, may thus be regarded as an approximation of the gradient of $z(\mathbf{x})$, at any point $\mathbf{x}$ within the triangle.

One difficulty with the previous approach is that each point, $z(\mathbf{x}_i)$, in a set of triangulated observations $\{z(\mathbf{x}_i) \mid i = 1, \dots, n\}$, is surrounded by a subset of triangles (the Delaunay

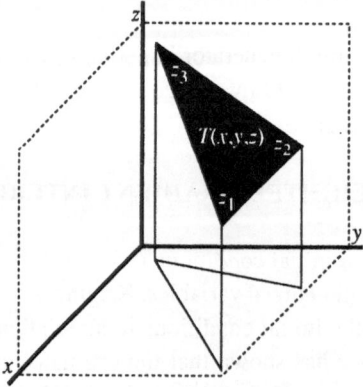

Figure 2 Triangular plane formed by the values of a regionalized variable, $z(\mathbf{x})$ at the three observation points ($\mathbf{x}_1$, $\mathbf{x}_2$, $\mathbf{x}_3$).

point group) which share the same vertex. The gradient of $z(\mathbf{x})$, at the point $\mathbf{x}_i$, will therefore depend on which triangle is used to compute $T(x_i, y_i, z_i)$. This observation led Watson and Philip (1985) to introduce what they called the mean gradient plane of the variable $z(\mathbf{x}_i)$, with normal vector

$$N = \sum_{i=1}^{m} \mathbf{n}_k$$

where $\mathbf{n}_k$ ($k = 1, \ldots, m$) are the unit normal vectors of the m triangular planes, which surround the point $z(\mathbf{x}_i)$. Let

$$\mathbf{r} = \mathbf{i}(x - x_i) + \mathbf{j}(y - y_i) + \mathbf{k}(z - z_i)$$

be a vector that passes through the point $[x_i, y_i, z(\mathbf{x}_i)]$. Since the scalar product

$$\mathbf{N} \cdot \mathbf{r} = 0$$

if $\mathbf{r}$ lies within the plane, the plane must satisfy the equation

$$g_i(x,y) \equiv z = \left(N_1/N_3\right)(x_i - x) + \left(N_2/N_3\right)(y_i - y) + z_i$$

These gradient planes and the method of inverse distance weighted (IDW) interpolation (McLain, 1976) can now be used to obtain an estimate for the regionalized variable at a point $\mathbf{x}$, where it is not known, in the form

$$\hat{z}(\mathbf{x}) = \sum_{i=1}^{n} w_i g_i(x,y)$$

with w_i is a suitable set of weights, that are functions of the distance.

To illustrate the IDWG-interpolation technique, consider the water levels in the marble aquifer at Otjiwarongo shown in Figure 3. A very interesting property of these water levels is that the area between the legs of the aquifer is underlain by impermeable rocks (Seimons, 1990). The behaviour of the water levels in one leg therefore do not depend on the behaviour of the water levels in the other leg. Since there are no known methods with which to account for this behaviour in Kriging, kriged estimates of the water levels must be interpreted with care. IDWG-interpolation, on the other hand, can handle the situation quite easily. All that one need to do, is to impose a dividing line, such as AB in Figure 3, between the two legs of the aquifer, as discussed by Buys et al. (1990). The difference between the two interpolation methods is clearly illustrated by the contours of the interpolated water levels, for the finite element mesh of the marble aquifer at Otjiwarongo, see Figure 1, in Figure 4.

The root mean square errors (RMSE) for both methods were estimated with the well-known procedure of (Dudo and Hart, 1973), where one data point is deleted from the set at a time and its value estimated from the rest. This yielded a RMSE of 16 for the Kriging contours in Figure 4(a) and 14 for the IDWG contours in Figure 4(b). Although this difference may seem to be insignificant, the IDWG-contours in Figure 4 (b) agree better with

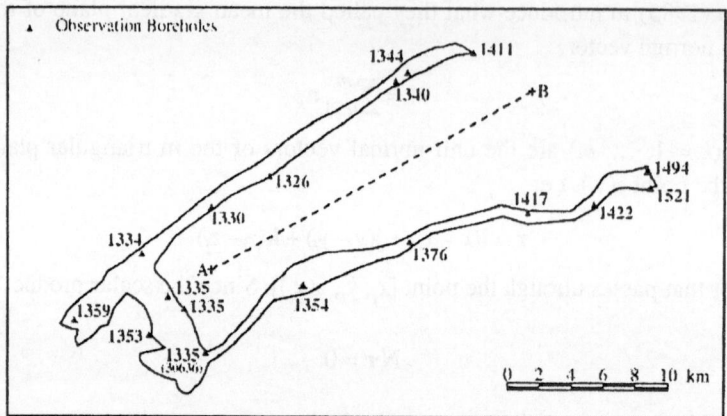

Figure 3   Observed groundwater levels in the marble aquifer at Otjiwarongo. The line, AB, is used to account for the independence of the levels in the two legs of the aquifer.

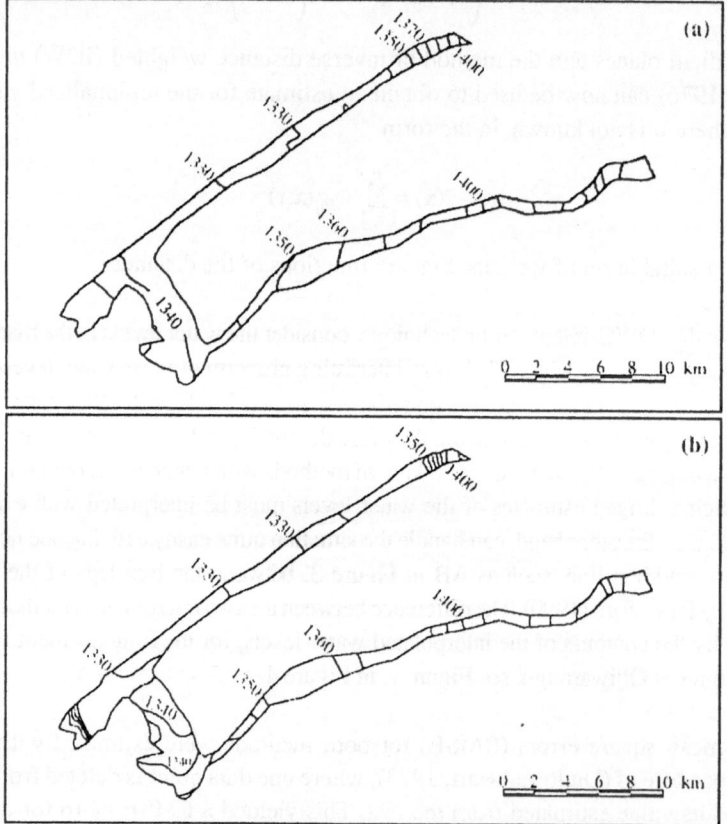

Figure 4   Contours of observed groundwater levels in the marble aquifer at Otjiwarongo (a) derived from Kriging, and (b) from IDWG-interpolation.

the observed groundwater levels than the Kriging contours in Figure 4 (a). This result confirms a well-known principle from approximation theory—it is always worthwhile to consider the behaviour of a variable, that must be approximated, before choosing an approximation method.

SIMULATION OF THE MARBLE AQUIFER

Two finite element computer models of the aquifer were developed for the mesh in Figure 1, one with the initial heads in Figure 4 (a) and the second with the initial heads in Figure 4 (b). Both models were calibrated over a period of 29 months with an automated inverse fitting method, and the observed water levels at fourteen observation boreholes. The results for borehole 30636 are compared with the observed levels in Figure 5. The significance of this graph is that it clearly illustrates the small, but significant influence, that the initial conditions have, on the solution of a parabolic partial differential equation.

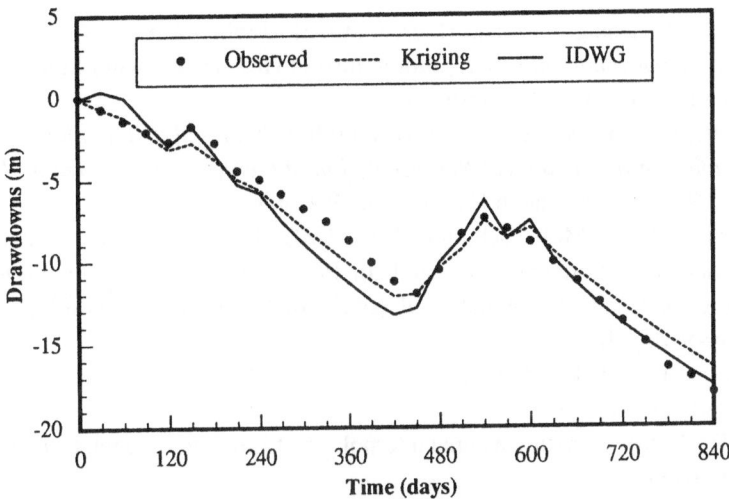

Figure 5   A comparison between the variations in the observed and finite element simulated water-levels in a boreholes in the marble aquifer.

Another result, obtained from the investigation, was that the water levels, simulated with the IDWG interpolated initial heads, approximated the observed water levels consistently better than those derived from Kriging. This is illustrated by the root mean square errors for the water levels at the 14 observation in Table 1.

CONCLUSIONS

The accuracy with which groundwater flow can be simulated numerically, depends on a number of approximations. Two of the more important of these approximations are: the

Table 1   Root mean square error for the drawdowns at the observation boreholes, computed from the intial water levels estimated with Kriging and IDWG-interpolation.

Simulation time	Kriging	IDWG
12 months	4,61	3,54
24 months	3,48	3,07

spatial discretization of the aquifer, and interpolation of the initial water levels. The Laplacian mesh generator and IDWG-interpolation, discussed in this paper, can reduce the influence of both these approximations considerably. The approximations were consequently used to develop full scale preprocessors for the simulation of groundwater levels. Copies of these programs are available from the authors.

REFERENCES

Botha, J. F., Van Rensburg, N. J. and Buys, J. (1993) *Finite Element Mesh Generation, using the Two-dimensional Laplace Equation.* Institute for Groundwater Studies, University of the Orange Free State, Bloemfontein.

Buys, J., Botha, J. F. and Messerschmidt, H. J. (1992) *Triangular Irregular Meshes and Their Application in the Graphical Representation of Geohydrological Data.* WRC Report No 271/2/92. Water Research Commission, Pretoria.

Buys, J., Botha, J. F. and Messerschmidt, H. J. (1990) Triangular finite element meshes and their application in groundwater research. In: Proceedings of the *Computational Methods in Water Resources.* G. Gambolati *et al.* (eds.) Vol 2. Subsurface Hydrology, pp. 115-122. Springer-Verlag, Berlin.

Dudo, R. O. and Hart, P. E. *Pattern Classification and Scene Analysis.* John Wiley & Sons. New York, N.Y. (1973)

McLain, D. H. (1976) Two dimensional interpolation from random data. *Computer Journal.* **19** (2), 178-181.

Seimons, W. S. (1990) *Geohydrological Characteristics of the Otjiwarongo Marble Aquifer, Namibia.* Unpublished MSc. Thesis, Geohydrology of , Geohydrology, University of the Orange Free State P.O. Box 339, Bloemfontein 9300, Republic of South Africa.

Van Sandwyk, L., Van Tonder, G. J., De Waal, D. J. and Botha, J. F. (1992) *A Comparison of Spatial Bayesian Estimation and Classical Kriging Procedure.* WRC Report No 271/3/92. Water Research Commission, Pretoria.

Van Tonder, G. J., Botha, J. F. and De Waal, D. J. (1990) Bayesian estimation of water levels. In: Proceedings of the *Conference on the Calibration and Reliability in Groundwater Modelling.* K. Kovar (eds.) The Hague. IAHS Publication No. 195.

Watson, D. F. and Philip, G. M. (1985) A refinement of inverse distance weighted interpolation. *Geo-Processing.* **2**, 315-327.

A SUBDOMAIN COLLOCATION APPROACH IN TETRAHEDRAL FINITE ELEMENTS

C. CORDES[*] and W. KINZELBACH[**]
[*] Department of Civil Engineering, Kassel University, 34109 Kassel
[**] Institute of Environmental Physics, Heidelberg University, 69120 Heidelberg

In finite difference groundwater models the direction of discrete fluxes always corresponds with the physical flow direction. This is not the case for finite elements: the stiffness coefficient between two neighbouring nodes may become positive, so that a hydraulic gradient with heads decreasing from node A towards node B leads to a flux from B towards A.

Forsyth (1991) shows for the two-dimensional case that this unphysical behaviour can only be avoided if linear triangular finite elements are used and certain discretization requirements are satisfied. In an aquifer with a constant transmissivity and arbitrarily distributed nodal points, for example, elements are uniquely defined by the condition that the circumcircle of each element contains no other node apart from the three element nodes in its interior (empty circle criterion). The analogy to the three-dimensional case does not hold, i.e. in tetrahedral finite elements that satisfy the empty sphere criterion, positive stiffness coefficients between neighbouring nodes can not be avoided, thus leading to unphysical fluxes (Letniowski and Forsyth, 1991).

In linear elements (triangles and tetrahedra) the weighted volume procedure (e.g. Galerkin method) can always be interpreted as an integration over subdomain boundaries (Narasimhan and Witherspoon, 1976). Unlike in two dimensions, subdomains are not conforming in the 3D-case and thus the weighted volume procedure is physically not sound. A new subdomain collocation approach is presented, which is the only possible numerical approach for arbitrarily distributed nodal points in space, for which the stiffness coefficient between two different nodes is never positive and the stiffness matrix becomes an M-matrix. The procedure is easily implemented in the element stiffness matrices of existing finite element codes which are based on tetrahedra or triangular prisms, involving no additional computer costs and leading to much better results. The latter point is illustrated by two examples.

Orthogonal subdomain collocation (OSC)

When a mesh of linear finite elements satisfies the empty circle criterion, each element edge belongs to one face of the corresponding Thiessen-polygons (Figure

A. Peters et al. (eds.), Computational Methods in Water Resources X, 27–34.

1). The face is always mid-perpendicular to the element edge. Narasimhan and Witherspoon (1976) proved that the integrated finite difference method (IFD), with cells in the shape of Thiessen-polygons, is identical to the Galerkin approach in linear elements. Using IFD, the stiffness coefficient between two neighbouring nodes is obtained as easily as in finite differences, i.e. by conductivity k times the area of the common cell face divided by the nodal distance.

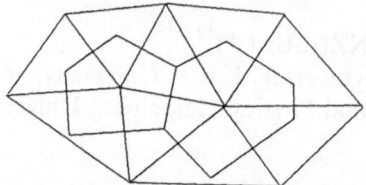

Fig. 1 Linear elements and corresponding Thiessen-polygons (= IFD-cells)

Generally, three-dimensional IFD-cells take the shape of "Thiessen-polyhedra", with each face being mid-perpendicular to the connection line between the two nodes. This orthogonality requirement leads to very complex geometries and to our knowledge no program code for groundwater flow exists applying exact cell geometries, exept for layers with constant thickness.

Yet, the solution to this problem is quite simple. Instead of starting with the shape of polyhedra and fitting the nodes afterwards, arbitrarily distributed nodal points can be defined first, because polyhedra cells are uniquely given by the orthogonality conditions and the geometry can be computed automatically just from the nodal coordinates.

In space, the connection lines between cell nodes form tetrahedra which exactly satisfsy the empty sphere criterion. This allows to consider the geometry of polyhedra tetrahedron-wise. Each tetrahedron is divided into four subdomains by six polyhedron faces (Figure 2).

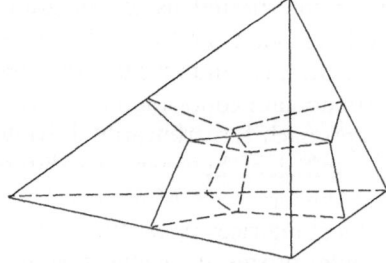

Fig. 2 Tetrahedral finite element with orthogonal subdomains

Now the stiffness between two nodes can be determined element-wise. The general subdomain collocation formulation reads (Huyakorn and Pinder, 1983):

$$S_{ij} = k \int_{\Gamma_j} \nabla N_i d\vec{n}_j \tag{1}$$

where $d\vec{n}_j$ denotes the normal vector on the surface Γ_j of the subdomain belonging to node j. N_i, the interpolation function, is equal to 1 at node i, and 0 at the other nodes in the element. In a tetrahedron ∇N_i becomes a constant vector. Using ortho - gonal subdomains it can be shown (Cordes, 1994) that eq. 1 is nothing but: k times the area of the subdomain face between nodes i and j divided by their distance.

It may surprise that under certain conditions IFD and Galerkin-FE lead to identical results in 2-D, but not in 3-D. For example, let us compare the stiffness coefficients in a prism which is discretized into three tetrahedra (Figure 3).

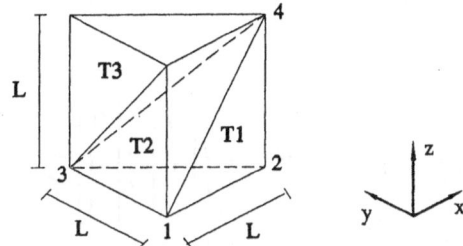

Fig. 3 Prism, discretized into three tetrahedra

Each of the four faces of a tetrahedron can be uniquely defined by an inward oriented normal vector, the length of which represents the area of the face. As index we use the number of the node opposite to the face, e.g. for the face 1-2-3 in tetrahedron T1:

$$\vec{A}_4 = \frac{1}{2}\vec{r}_{12} \times \vec{r}_{13} \tag{2}$$

$\vec{r}_{ij}$ denotes the vector from node i towards node j. For a unique definition the four nodes must always be numbered in such a way that $\vec{r}_{12}$, $\vec{r}_{13}$, and $\vec{r}_{14}$ form a right-handed-system. Then the volume of the tetrahedron becomes

$$V = \frac{1}{6}(\vec{r}_{12} \times \vec{r}_{13})\vec{r}_{14} \tag{3}$$

Since the value of N_i is 1 at node i and 0 at the three remaining nodes, the gradient is perpendicular to the tetrahedron face opposite to node i, leading to:

$$\nabla N_i = \frac{\vec{A}_i}{3V} \tag{4}$$

Thus, the Galerkin approach

$$S_{ij} = k \int_V \nabla N_i \nabla N_j dV \tag{5}$$

for the stiffness coefficients between nodes 1 and 2 and nodes 1 and 3 in tetrahedron T1 yields:

$$S_{12}^{T1} = k \frac{1}{L} \begin{bmatrix} 1 \\ 0 \\ -1 \end{bmatrix} \cdot \frac{1}{L} \begin{bmatrix} -1 \\ -1 \\ 0 \end{bmatrix} \frac{L^3}{6} = -k\frac{L}{6} \qquad S_{13}^{T1} = k \frac{1}{L} \begin{bmatrix} 0 \\ 1 \\ 0 \end{bmatrix} \cdot \frac{1}{L} \begin{bmatrix} -1 \\ -1 \\ 0 \end{bmatrix} \frac{L^3}{6} = -k\frac{L}{6} \tag{6}$$

and in tetrahedron T2:

$$S_{12}^{T2} = 0 \qquad S_{13}^{T2} = k \frac{1}{L} \begin{bmatrix} 0 \\ 1 \\ 0 \end{bmatrix} \cdot \frac{1}{L} \begin{bmatrix} 0 \\ -1 \\ -1 \end{bmatrix} \frac{L^3}{6} = -k\frac{L}{6} \tag{7}$$

The prism stiffness coefficients are obtained by summing up the tetrahedra stiffness coefficients:

$$S_{12} = -k\frac{L}{6} \qquad S_{13} = -k\frac{L}{3} \tag{8}$$

Because of the prism's symmetry, S_{12} and S_{13} ought to be identical, i.e. the Galerkin weighting in tetrahedra, in contrast to triangles, leads to false results. The exact stiffness coefficients can only be obtained by using the OSC method: in tetrahedron T1 the area of the face between nodes 1 and 2 is $L^2/4$, and $L^2/8$ between nodes 1 and 3 (Figure 4).

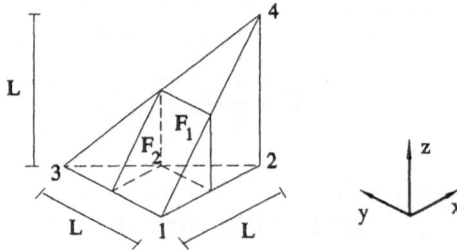

Fig. 4 Faces F_1 and F_2 in tetrahedron T1

Since F_2 is contained in tetrahedron T2 too, the exact coefficients are:

$$S_{12} = -k\frac{L}{4} \qquad S_{13} = -k\frac{L}{4} \tag{9}$$

The stiffness coefficient between nodes 1 and 2 in an arbitrarily shaped tetrahedron is obtained from geometry (Cordes, 1994):

$$S_{12} = -\frac{k}{48V}\left[2(\vec{r}_{13}\vec{r}_{23})(\vec{r}_{14}\vec{r}_{24}) + \vec{A}_3\vec{A}_4\left(\frac{(\vec{r}_{13}\vec{r}_{23})^2}{\vec{A}_4\vec{A}_4} + \frac{(\vec{r}_{14}\vec{r}_{24})^2}{\vec{A}_3\vec{A}_3}\right)\right] \tag{10}$$

The coefficients between the other nodes are determined by permutation of indices or by renumbering the tetrahedron, and can easily be implemented in the element stiffness matrix of existing computer codes.

Usually, when three-dimensional groundwater flow is simulated, prism elements are used, with identical horizontal discretization in all layers. Often prisms are already divided into tetrahedra. When tetrahedra are supposed to satisfy the empty sphere criterion, first the horizontal mesh has to be discretized in triangles that fulfill the empty circle criterion. Spatially, this leads to triangular prism elements, which have to be divided into three tetrahedra each. This is uniquely achieved by the definition of the three diagonals on the three lateral faces. The empty sphere criterion is satisfied, when the shorter diagonal is chosen on each of the lateral faces (Figure 5).

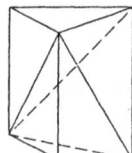

Fig. 5 Discretization of tetrahedra in a triangular prism

Anisotropy

In an anisotropic medium a procedure in analogy to the Galerkin method is not possible, because in general

$$\int_{\Gamma_j}\nabla N_i K\,d\vec{n}_j \neq \int_{\Gamma_i}\nabla N_j K\,d\vec{n}_i \tag{11}$$

K is the conductivity matrix, which we assume to be diagonal. Instead, an average conductivity $k = (k_{xx}k_{yy}k_{zz})^{1/3}$ has to be used and the tetrahedron has to be distorted in its spatial directions by factors $(k/k_{xx})^{1/2}$, $(k/k_{yy})^{1/2}$, and $(k/k_{zz})^{1/2}$ respectively (e.g. Bear, 1972).

The mass matrix

Huyakorn and Pinder (1983) give the mass matrix for the subdomain collocation in the distributed and lumped formulation. However, the determination of the mass matrix is independent from the computation of the stiffness matrix, i.e. the OSC-stiffness-matrix can be combined with a Galerkin mass matrix without any problems. Both the distributed and the lumped OSC-formulation are given in (Cordes, 1994).

Examples

In the first example OSC is compared to the Galerkin approach in a cubic model with edgelength L. The model consists of one element with trilinear interpolation functions N_i (e.g. Zienkiewicz) and eight nodes. Because of the geometry, the Galerkin stiffness matrix (eq. 5) can be calculated analytically without numerical integration:

$$
\begin{aligned}
S_{ij} &= k\frac{L}{3} \quad for\ i{=}j \\
&= 0 \quad when\ i\ and\ j\ share\ the\ same\ edge \\
&= -k\frac{L}{12} \quad else
\end{aligned}
\tag{12}
$$

Using OSC, the element is divided into eight cubic subelements with edgelength L/2, yielding subdomain faces with area $L^2/4$. Similarly to finite differences, two nodes only have an influence upon each other when they share the same edge. Coefficients on the diagonale of the stiffness matrix, S_{ii}, stem from the condition, that the sum of each column and row is zero:

$$
\begin{aligned}
S_{ij} &= \frac{3}{4}kL \quad for\ i{=}j \\
&= -k\frac{L}{4} \quad when\ i\ and\ j\ share\ the\ same\ edge \\
&= 0 \quad else
\end{aligned}
\tag{13}
$$

In contrast to eq. 13, direct neighbours in eq. 12 have no influence upon each other. How can it be explained that such different stiffness matrices lead to similar results as is known from experience?

As long as there is parallel flow in any spatial direction, both approaches, FE and FD/OSC, lead to identical heads. Differences occur in truly 3D-flow. In our model this is simulated by an inflow $Q = kL^2$ at one node and a fixed head $h = 0$ at another node. Since the remaining nodes have no outer fluxes, the outflow will appear at the fixed head node. The solution of the system of 8 equations, $[S]\{h\} = \{Q\}$, is shown in Figure 6.

In all three cases a), b), and c) the Galerkin weighting leads to absolutely unrealistic

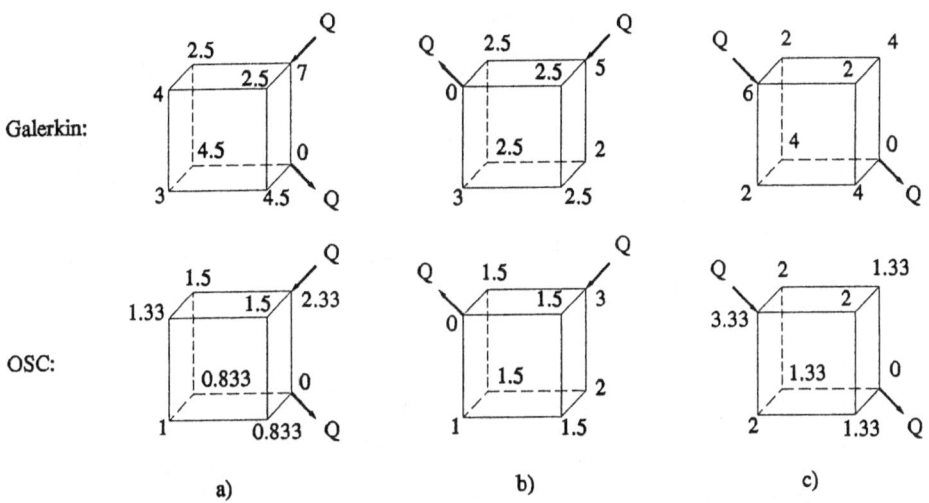

Fig. 6 Comparison of heads [in multiples of edgelength L] in a cubic FE-model with 8 nodes

head distributions, where the head decrease along the edges is opposite to the induced flow. Only the average gradient is exact. For example in case a) the average head of the four upper nodes is 4L and of the four lower nodes 3L. This explains why mesh refinement improves the results - because in that case the gradient in a single element would become nearly constant. If the model was discretized in a large number of elements, in essence the same error would occur in the corner elements, but the contribution to the absolute error would be reduced.

This shows the main problem of the weighted volume procedure. From the mathematical point of view finite elements (specially of higher order) allow a good approximation to the exact solution - but only under the condition of a sufficiently smooth distribution.

The second model consists of 20 cubic elements and 60 nodes (Figure 7). The boundary condition on the right hand side is a fixed head while the remaining five faces are impervious. The flow towards the fixed head boundary originates from a point source which is placed at the bottom of the left corner. As in the first example, the trilinear interpolation leads to heads which do not correspond to the flow direction. On the top surface the maximum head does not show in the corner but at the two neighbouring nodes lying on the edges. At the bottom the maximum head actually occurs at the point source in the corner, but the two neighbours on the edges do not show the second-largest heads, as they should.

The use of Galerkin tetrahedra distinctively improves the head distribution at the bottom, but merely little on the top. Only the OSC approach, which ensures that the stiffness coefficient between two neighbouring nodes is always negative, yields a solution which is very similar to the analytical solution (Cordes, 1994).

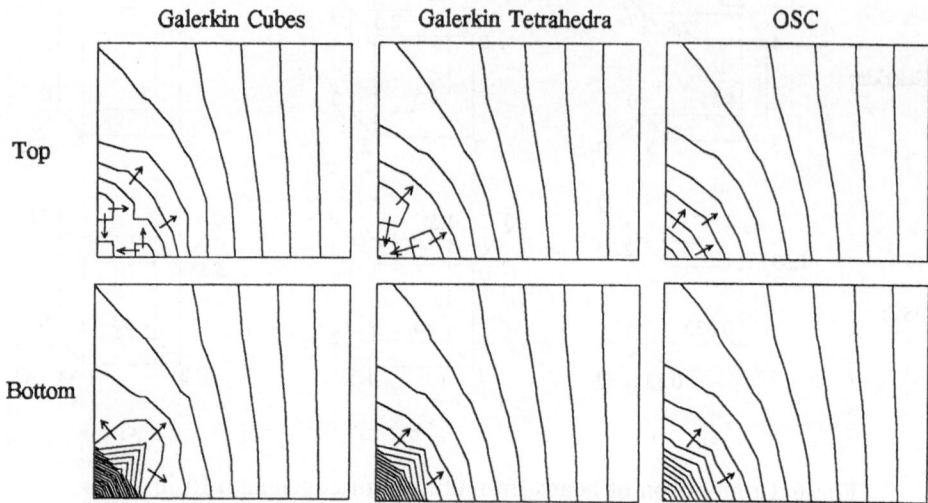

Fig. 7 Heads and direction of head decrease in a flow model with 20 cubic elements (length*width*depth = 5*4*1), or 120 tetrahedra respectively

If trilinear elements are not cubic or the hydraulic conductivity is not isotropic, the Galerkin stiffness coefficient between two nodes which share the same edge (eq. 12) becomes even positive. In that case the error is not limited to elements in the vicinity of a singular point. By means of a layered groundwater model with a small vertical permeability Cordes (1994) shows that the horizontal components of the head gradient can be inverted in a whole layer. This is avoided when OSC is used.

References

Bear, Jacob (1972) Dynamics of Fluids in Porous Media, Dover Publications Inc., New York.

Cordes, C. (1994) "Bahnkurven in Potentialströmungen", Dissertation, Fachgebiet Technische Hydraulik, FB 14, Universität - Gesamthochschule Kassel, 34109 Kassel.

Forsyth,P.A. (1991) "A Control Volume Finite Element Approach to NAPL Groundwater Contamination", SIAM J. Sci. Stat. Comput., Vol. 12, No. 5, 1029-1057.

Huyakorn, Peter S. and Pinder, George F. (1983) Computational Methods in Sub surface Flow, Academic Press, New York.

Letniowski, F.W. and Forsyth, P.A. (1991) "A Control Volume Finite Element Method for Three-dimensional NAPL Groundwater Contamination", Int. J. Num. Methods in Fluids, Vol. 13, 955-970.

Narasimhan, T.N. and Witherspoon, P.A. (1976) "An Integrated Finite Difference Method for Analyzing Fluid Flow in Porous Media", WRR, Vol.12, No. 1, 57-64.

BEM TO CALCULATE GROUNDWATER-SUPPLY-SYSTEMS CONSISTING OF WELLS, COLLECTOR WELLS AND INFILTRATION TUNNELS

I. DAVID *) and H. GERDES
Technische Hochschule Darmstadt,
Institut für Numerische Verfahren und Informatik im Bauwesen und Institut für Wasserbau, Petersenstr. 13, 64287 Darmstadt, Germany
*) On leave from Department of Hydraulic, Technical University Timisoara, Romania

A mathematical model is presented to incorporate local three-dimensional flow patterns into the plane Boundary Element Method (BEM). The flow may be generated by wells with laterals, infiltration tunnels, horizontal drainage pipes, shallow creeks or rivers. Close to those components of the groundwater-supply-system the flow has an actual three-dimensional character.

The proposed method is based on the introduction of an equivalent plane flow-system. The local streamline constriction in the vertical plane is considered as an additional head loss. The head losses of turbulent flow inside the laterals, tunnels and drains are computed as well.

Incorporating this procedure into BEM leads to a more realistic calculation of groundwater supply systems. Tests of the mathematical model and several examples are presented.

INTRODUCTION

In spite of vast improvements in modelling the transport of toxic substances, (Kinzelbach 1987-1993, Kobus 1992) groundwater-supply-systems are still calculated very roughly with the use of obsolete equations.

The area-based methods like the Finite Element Method (FEM) or the Finite Difference Method (FDM) are not useful to analyse local effects close to wells, horizontal drainage pipes etc., which act like mathematical singularities in the flow field.

This paper introduces a new method of transforming three - dimensional flow pattern into a two - dimensional representation by using a local equivalence technique (David,

A. Peters et al. (eds.), Computational Methods in Water Resources X, 35–42.

1977, 1978, 1987, 1992) and solving the two - dimensional problem (including singularities) by the BEM as a new variant to the common BEM.

MATHEMATICAL REPRESENTATION

Figure 1 shows the sketch of a groundwater-supply-system consisting of different components. The flow - domain is limited by different boundary conditions including three - dimensional (partially - penetrating) effects as well.

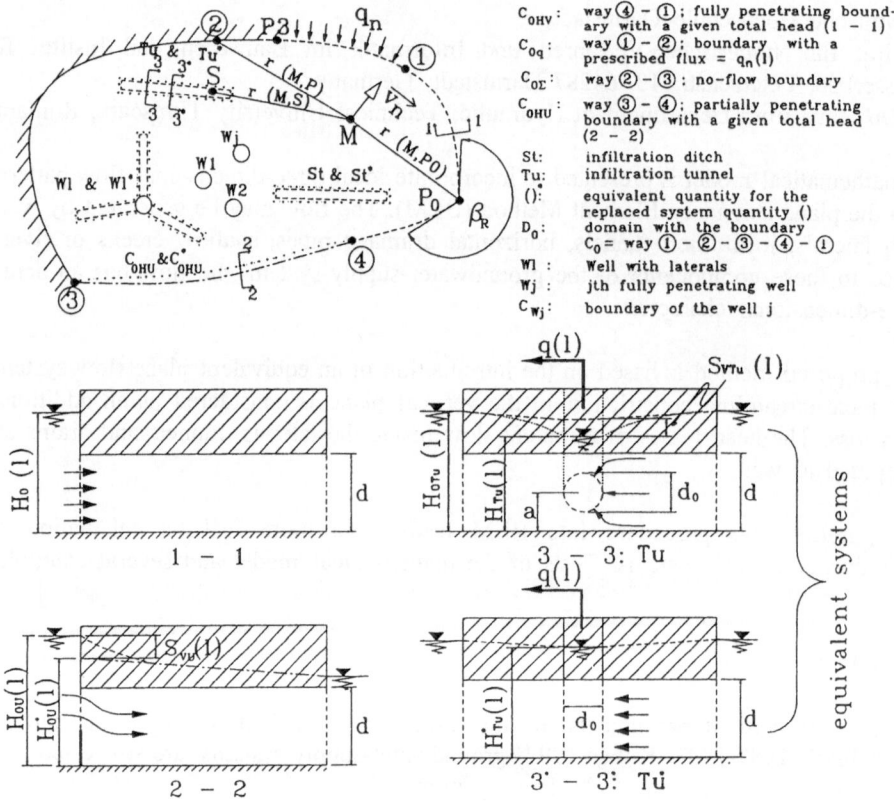

Fig.1: Sketch of a groundwater - supply - system consisting of several interacting components.

Replacing the three-dimensional effects close to partially-penetrating boundaries or a component of the supply-system by a vertical-plane and horizontal-plane system the head loss in the vertical plane is dependent on the distribution of flux into e. g. the tunnel or other components respectively along the boundary C_{OHU} (David, 1991):

$$S_{v(.)} = S_{v(.)} (l) = \frac{q (l)}{T} E_{(.)} (l) \quad , \quad T \equiv k_f d \quad , \quad (.) = Tu , St , Wl , U \quad ; \tag{1}$$

with the head loss of the component (.) $S_{V(.)}$ (see figure 1), the influx q, the coefficient of permeability k_f, the transmissivity T, the depth d of the aquifer and the geometric factor of imperfectness E along the component (.). The loss of hydraulic head must be covered partly by the horizontal and partly by the vertical flow, so that they satisfy the real boundary conditions along the imperfect ways:

$$H_{0(.)} (l) - h_{V(.)} (l) = H_{0(.)}^* (l) + S_{V(.)} (l) \quad , \tag{2}$$

with the given total head $H_{0(.)}$ along the component (.), the equivalent total head in the replaced component $H_{0(.)}^*$ [1], the head loss inside the component $h_{V(.)}$ and the head loss $S_{V(.)}$ along a partially - penetrating component (.). With this scheme we transform the real imperfect three - dimensional flow pattern into an equivalent perfect plane flow pattern. Thus, we obtain a plane boundary problem with implicit boundary conditions (see figure 1 for the indexes, ε is groundwater recharge and δ is Diracs'delta function):

$$\frac{\partial^2 h}{\partial x^2} + \frac{\partial^2 h}{\partial y^2} = - \frac{\varepsilon}{T} - \sum \frac{Q_j}{T} \delta (M , W_j)$$

$$h|_{C_{0HV}} = H_{0V} (l)$$

$$h|_{C_{0HU}^*} = H_{0U}^* (l) = H_{0U} (l) + S_{VU} (l)$$

$$h|_{Tu^*} = H_{0Tu}^* (l) = H_{0Tu} (l) + S_{VTu} (l) - h_{VTu} (l) \tag{3}$$

$$h|_{St^*} = H_{0St}^* (l) = H_{0St} (l) + S_{VSt} (l)$$

$$h|_{Wl^*} = H_{0Wl}^* (l) = H_{0Wl} (l) + S_{VWl} (l) - h_{VWl} (l)$$

$$h|_{Wj} = H_{0j}$$

$$\frac{\partial h}{\partial n}|_{C_{0\varepsilon}} = 0 \quad ; \quad - T \frac{\partial h}{\partial n}|_{C_{0q}} = q_n \quad ,$$

where $S_{V(.)}$ and $h_{V(.)}$ are dependent on the distribution of influx along the boundary C_{0HV} or the component (.). For an infiltration tunnel or a lateral of diameter d_0 we have:

$$E_{Tu} (l) = E_{Wl} (l) = \frac{d}{\pi} \ln \left[\frac{d}{\pi d_o} \frac{1}{\sin \left[\frac{\pi a}{d} \right]} \right] \quad . \tag{4}$$

Furthermore, for a partially penetrating boundary C_{H0U} or a partially penetrating ditch we have:

---

[1]   the star (.)* will from now on be used to sign all equivalent quantities of the replaced system quantity (.).

$$E_{St} \, (l) = E_{CU} \, (l) = \frac{2 \, d}{\pi} \, \ln \, \left[\cos \, \left[\frac{\pi \, p}{2 \, d} \right] \right]^{-1} \quad , \qquad (5)$$

with the penetrating depth p. The inner head losses are considered to be:

$$\frac{dh_v \, (l)}{dl} = \frac{8}{\pi \, g \, d_0^5} \, \left[\lambda - 2 \, d_0 \, \frac{q \, (l)}{Q \, (l)} \right] \, Q^2 \, (l) \quad , \qquad (6)$$

with the friction coefficient (Prandl-Colebrook) λ, the distribution of influx q along the drain and the internal flow Q through the drain-pipe, which is a function of the distribution of influx.

BOUNDARY-INTEGRAL-REPRESENTATION OF THE PROBLEM

The total head h (M) at any point M in the domain D_0^* (M $\in$ D_0^*) of the replaced horizontal system is given by the indirect boundary integral representation (David, 1992):

$$h \, (M) = \frac{1}{2 \, \pi \, T} \, \left[\int_{C_0^*} \psi \, (P) \, G \, (M \, , \, P) \, dl + \int_{D_{0*}^*} \varepsilon \, (N) \, G \, (M \, , \, N) \, dA \right.$$

$$\left. + \int_{Tu^* \, \cup \, St^* \, \cup \, Wl^*} q \, (S) \, G \, (M \, , \, S) \, dl + \sum_{j=1}^{N_w} Q_j \, G \, (M \, , \, W_j) \right] + C \quad . \qquad (7)$$

The velocity at this point M $\in$ D_0^* in a given direction n_M is the derivation of (7):

$$\frac{v_n \, (M)}{k_f} = - \frac{1}{2 \, \pi \, T} \, \left[\int_{C_0^*} \psi \, (P) \, F \, (M \, , \, P) \, dl + \int_{D_{0*}^*} \varepsilon \, (N) \, F \, (M \, , \, N) \, dA \right.$$

$$\left. + \int_{Tu^* \, \cup \, St^* \, \cup \, Wl^*} q \, (S) \, F \, (M \, , \, S) \, dl + \sum_{j=1}^{N_w} Q_j \, F \, (M \, , \, W_j) \right] \quad , \qquad (8)$$

where ψ denotes the indirect and unknown density distribution along the boundary C_0^* and:

$$G \, (M \, , \, P) = \ln r \, (M \, , \, P) \quad ; \quad F \, (M \, , \, P) = \frac{\partial \ln r \, (M \, , \, P)}{\partial n_M} \quad . \qquad (9)$$

With the limit of the integral representations (7) and (8)

$$M \rightarrow P_0 \in C_0^* \, \cup \, Tu^* \, \cup \, Wl^* \, \cup \, C_{Wj} \quad , \qquad (10)$$

and concerning the implicit boundary conditions (3) we obtain the system of boundary-integral-equations:

$$\frac{1}{2 \pi T} \left[\int_{C_o^\cdot} \psi(P) \, G(P_0, P) \, dl + \int_{D_{o\epsilon}^\cdot} \epsilon(N) \, G(P_0, N) \, dA \right.$$

$$\left. + \int_{Tu^\cdot \cup St^\cdot \cup Wl^\cdot} q(S) \, G(P_0, S) \, dl + \sum_{j=1}^{N_w} Q_j \, G(P_0, W_j) \right] + C$$

$$= \begin{cases} H_{0V}(P_0) \;, \; P_0 \in C_{0HV} \\ H_{0Tu} + \dfrac{q(P_0)}{T} E_{Tu}(P_0) - h_{VTu}(P_0) \;, \; P_0 \in Tu^* \\ H_{0St} + \dfrac{q(P_0)}{T} E_{St}(P_0) \;, \; P_0 \in St^* \\ H_{0Wl} + \dfrac{q(P_0)}{T} E_{Wl}(P_0) - h_{VWl}(P_0) \;, \; P_0 \in Wl^* \\ H_{0j} \;, \; P_0 \in C_{Wj} \end{cases} , \qquad (11)$$

$$\frac{1}{2 \pi T} \left[\int_{C_o^\cdot} \psi(P) \, L(P_1, P) \, dl + \int_{D_{o\epsilon}^\cdot} \epsilon(N) \, L(P_0, N) \, dA \right.$$

$$\left. + \int_{Tu^\cdot \cup St^\cdot \cup Wl^\cdot} q(S) \, L(P_0, S) \, dl + \sum_{j=1}^{N_w} Q_j \, L(P_0, W_j) \right] + C \qquad (12)$$

$$= H_{0U}(P_0) + \frac{\beta_R(P_1)}{2 \pi T} \Psi(P_0) \, E_{CU}(P_0) \;, \; P_0 \in C_{0HU}^* \;\;,$$

$$\frac{1}{2 \pi T} \left[\int_{C_o^\cdot} \psi(P) \, F(P_0, P) \, dl + \int_{D_{o\epsilon}^\cdot} \epsilon(N) \, F(P_0, N) \, dA - \frac{\beta_R(P_0)}{2 \pi T} \Psi(P_0) \right.$$

$$\left. + \int_{Tu^\cdot \cup St^\cdot \cup Wl^\cdot} q_r(S) \, F(P_0, S) \, dl \right]$$

$$= \begin{cases} \dfrac{q_n(P_0)}{T} \;, \; P_0 \in C_{0q} \\ 0 \;, \; P_0 \in C_{0\Sigma} \end{cases} \qquad (13)$$

and the equation of balance:

$$\int_{C_o^\cdot} \psi(P) \, dl + \int_{D_{o\epsilon}^\cdot} \epsilon(N) \, dA + \int_{Tu^\cdot \cup St^\cdot \cup Wl^\cdot} q(S) \, dl = 0 \;\;, \qquad (14)$$

with

$$L\ (P_0\ ,\ P)\ =\ G\ (P_0\ ,\ P)\ -\ E\ (P_0)\ F\ (P_0\ ,\ P)\ ,\ \ P_0 \in C^*_{0HU}$$

Equations (11) ÷ (14) describe a system of integral equations with the unknowns: ψ (P), q (P) and Q_j or H_{0j}. If we consider the inner head losses inside the components of the groundwater-supply-system, we see that this system of equations is nonlinear.

NUMERICAL SOLUTION - EXAMPLES

To solve the problem numerically, we subdivide each component of the groundwater-supply-system with true three - dimensional - flow - pattern into subelements. Thus, we obtain an algebraic system of equations that we solve numerically with BEM 3 (Boundary Element Program). First, we tested BEM 3 with any component of the groundwater-supply-system seperately (no interactions). Figure 2 shows' the calculated distribution of influx along an infiltration tunnel that we obtained using BEM 3. These results have been experimentally verified (Gerdes, 1985).

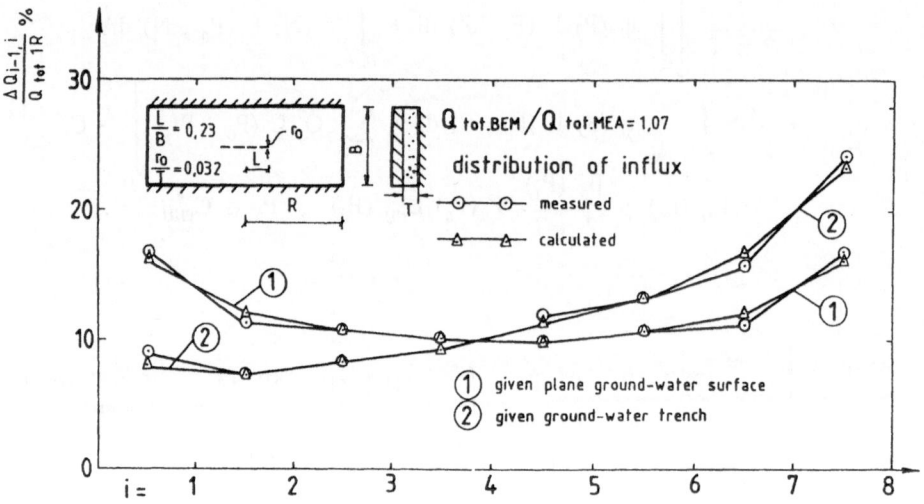

Fig. 2: Comparison of experimental data and numerical results.

Figure 3 shows another example of a well with laterals. Note that there is a great deviation of the results, whether or not considering inner head losses in the laterals:

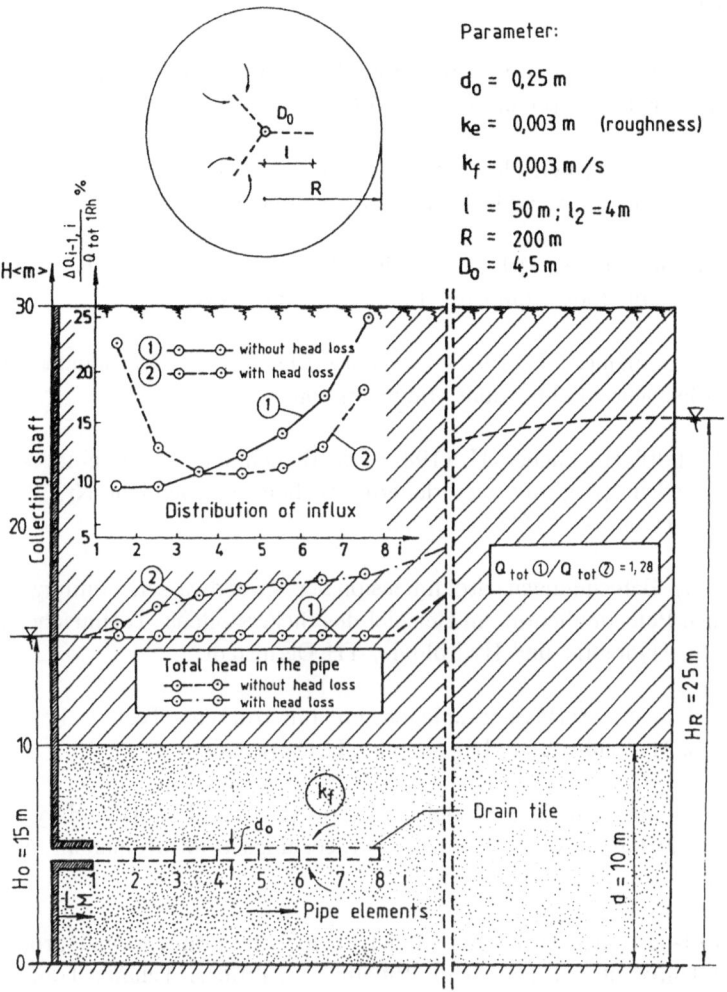

Fig. 3: Numerical results of a well with laterals.

Next, we modelled and checked a whole groundwater-supply-system (interactions) under various boundary conditions with the help of BEM 3.

CONCLUSIONS

We have proven that the introduced method using generalized boundary integral equations and BEM 3 that is based on this method can model a highly interacting groundwater-supply-system. Furthermore, if we consider the geometric parameters of each component of the groundwater-supply-system, we have a new powerful tool to solve important groundwater problems that couldn't be solved so far.

REFERENCES

Kobus, H. (1992) Wärme und Stofftransport im Grundwasser, Band 1, Deutsche Forschungsgemeinschaft, Forschungsbericht, VCH.

Kinzelbach, W. (1987) Numerische Methoden zur Modellierung des Transports von Schadstoffen im Grundwasser, Oldenburg Verlag, München u. Wien.

Banarjee, P.K., Butterfield, R. (1981) Boundary Element Methods in the Engineering Science, McGrraw Hill Book Company, London a. New-York.

Brebbia, C. A., Dominquez, J. (1989) Boundary Element. An Introduction Course, Computational Mechanics Publications, Southampton.

Söhngen, B., Bischoff, H., Lacher, H. (1982) Die Berechnung von Drainagesystemen in horizontal ausgedehnten Grundwasserleitern, Technischer Bereicht Nr. 29, Institut für Hydraulik und Hydrologie, TH Darmstadt.

Gerdes, H. (1985) Berechnungen dreidimensionaler Grundwasserströmung mit den Mitteln der ebenen Potentialtheorie am Beispiel des Sickerstollens, Technischer Bericht Nr. 34, Institut für Hydraulik und Hydrologie, TH Darmstadt.

David, I. (1977) Grundwasserfassungsanlagen mit Filterrohren, Technischer Bereicht Nr. 19, Institut für Hydraulik und Hydrologie, TH Darmstadt.

David, I. (1987) A mathematical Model based on the modified Boundary Element Method for artificial Groundwater Recharge, Proceedings ofthe Rumanian Academiy of Sciences, Serie Mec. Appl. tom 32, nr. 3.

David I. (1992) Plane Boundary Elemet Model for local tree-dimensional groundwater flow, Technical University of Timisoara, Buletin Scientific si Tehnic, Tom 37 (51), Hidrotehnica.

Haitjema, M.H. (1985) Modelling Three-Dimensional Flow in Confined Aquifers by Superposition of Both Two- and Three-Dimensional Anylytic Functions, Water Recources Research, Vol. 2.1, No. 10.

Radoikovic, M., Pecaric, J. (1984) Three-Dimensional Boundary Element Model of Groundwater Flow to Ranney Wells, Hydrosoft 84, Proceedings of the international Conference, Yugoslavia.

IMPLEMENTATION OF INTERACTIVE NUMERICAL MODELING AND HYDRAULIC TESTING IN THE DESIGN AND QUALITY CONTROL OF LOW PERMEABILITY BARRIER SYSTEMS

Th. EIBEN, G. GUDERJAHN and J. SIEBERT
Golder Associates GmbH
Vorbruch 3, 29227 Celle
Germany

The procedure described demonstrates an interactive approach for the design and post-construction in situ testing of low permeability barrier systems. The design stage involves the development of a three dimensional groundwater model. Determination of groundwater flow patterns is used to optimize the performance of the barrier. To plan the in situ test, the model is expanded to include test and observation wells. With the resulting pressure response it is possible to optimize pumping rates and periods as well as number and locations of the observation wells. After completion of the test the model is used for detailed test evaluation. It is shown how the model can be used to estimate the actual in situ conductivity of the barrier and to locate zones of higher permeability within the system.

INTRODUCTION

Our current understanding is that the only way to guarantee the protection of groundwater from contamination is through a combination of geological and artificial barriers. This applies both to new waste deposits and to contaminated sites. The difference is that a multi-component sealing system can be incorporated into a new landfill, but this is impossible for an existing contaminated site, at least as far as a base seal is concerned. The preferred practical option, in combination with other measures, is to cut off the source of contamination from the environment with a vertical barrier system. Proof of the effectiveness of such cut off systems is usually only available later when continued transport of the contaminant is observed. Now however, through a combination of hydraulic tests and numerical simulation, an immediate check of the quality of the system can be obtained.

HYDRAULIC BARRIERS

Hydraulic barriers (cutoff walls) are installed in the ground to control horizontal groundwater flow and contaminant transport by having a lower hydraulic conductivity than the adjacent strata. This can be either for water management or for remediation purposes.
In the environmental field hydraulic barriers are typically designed as circumferential barriers, i.e. they surround the site. However another approach is to use up-gradient installations to reduce the influx of uncontaminated groundwater.

43

A. Peters et al. (eds.), Computational Methods in Water Resources X, 43–50.

For construction of the barrier, different methods including slurry trenching method, grout injection and jet grouting can be used. Installation of steel sheet piles is also possible.

The model used here contains the properties of the most common kind of barrier, the slurry wall. For its construction a vertical trench is excavated and a slurry, typically containing bentonite and water, is used to maintain trench stability. Having completed the excavation the slurry is either left in place for hardening (1-step-method) or it is replaced by a low permeability backfill (2-step-method). The backfill material can be, for example, soil bentonite, cement bentonite, plastic concrete or reinforced concrete. It is also possible to install plastic membranes within the trench backfill.

The conductivity values which can be achieved for the backfill material vary between 10^{-08} and to 10^{-09} m/s for the 1-step-method, to as low as 10^{-11} m/s with the 2-step-method /1/.

CODE DESCRIPTION

The simulation program used consists of a powerful finite difference, 3D, multiphase, fully implicit simulation code and several pre- and post-processing utilities. One of the pre-processors, for example, is an interactive program for generating 3-D simulation grids based on geological models. Conventional block center geometry, as well as radial or complex corner point geometry, may be used. In addition, special features like local grid refinement or coarsening are available.

Since flow in porous media is analyzed the simulation code is based on partial differential equations derived from DARCY's equation of flow, the continuity equation and an equation of state. The simulation program handles simultaneous flow of two liquid phases and one gas phase. The system of partial differential equations describing the flow for each block is transformed into finite difference equations and introduced into a matrix system which is solved with the use of modern algorithms.

The simulation code not only features multiphase flow, also a wide range of other modeling facilities including special options like dual porosity/permeability, diffusion, adsorption and decay is available.

BARRIER DESIGN

For design of the hydraulic barrier system a three dimensional model was built containing an imaginary waste pit. The petrophysical parameters, i.e. permeability and porosity, were taken as homogenous throughout the model, including the waste pit location. The vertical permeability was reduced by a factor of 10. Furthermore, it was assumed that all pores were saturated with water, which means that only single phase flow was considered.

Tab. 1 Main input parameters for simulation

model dimensions	238.6 m x 440 m x 48 m
number of cells	25 x 15 x 12 (= 4400)
porosity	0.15
hydraulic conductivity	$5 \cdot 10^{-06}$ m/s
hydraulic gradient	0.038
seepage velocity	≈ 40 m/a
water saturation	1
waste pit area	10 m x 10 m
depth of waste pit	12 m

Constant pressure boundaries were used on the right and left side of the model (DIRICHLET boundary condition). This results in a steady horizontal groundwater flow through the modeled area. With a hydraulic gradient of 0.038 the groundwater velocity becomes approximately 40 m/a. All other edges of the model were treated as no flow boundaries. The main input parameters for the simulations are summarized in Tab. 1.

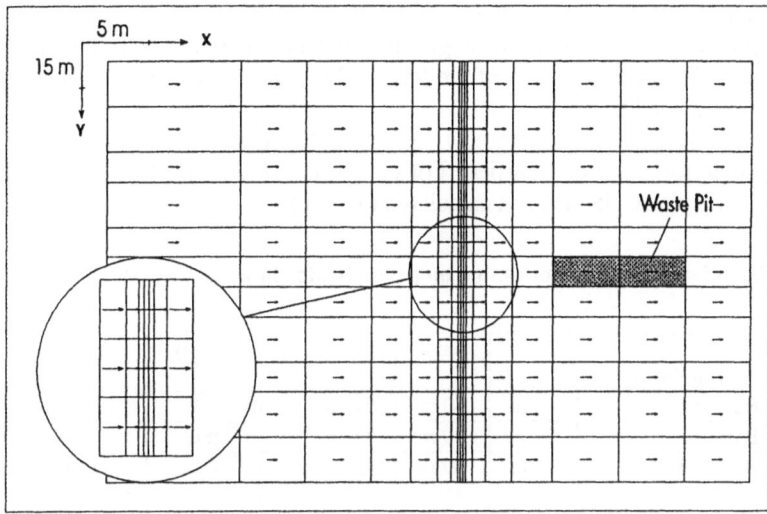

Fig.1   Simulation model showing initial groundwater flow

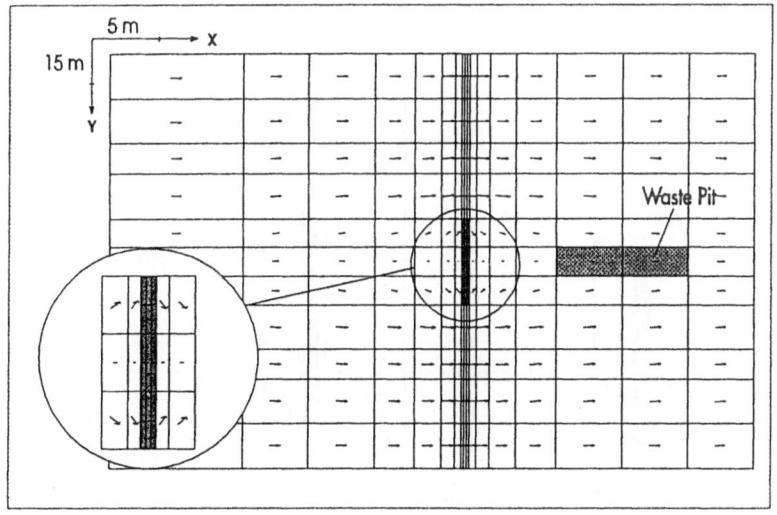

Fig.2   Groundwater flow pattern with barrier system 1

Fig. 1 shows the inner part of the simulation model. The total modeled area is much larger to prevent the results being influenced by boundary effects.

Also included in Fig. 1 is the initial groundwater flow pattern. It should be mentioned here that the arrows mainly show the direction of flow but that their length is also a measure of the seepage velocity.

In the next step a hydraulic barrier with a hydraulic conductivity of 10^{-08} m/s was added to the model. The vertical extension of the barrier was 16 m, its horizontal length was set to 30 m and the width was set to 0.6 m. The influx of the barrier on groundwater movement can be seen in Fig.2.

By building in the hydraulic barrier the groundwater flow is redirected. However the flow still penetrates the waste pit. Part of the flow is through the low permeability barrier, part of it from around the barrier.

Four different hydraulic barrier systems were modeled to show how different designs can influence the performance of a barrier. For comparison the seepage velocity in the middle of the imaginary waste pit has been calculated (Tab. 2).

Tab.2   Effect of different barrier design on groundwater flow

	depth	length	width	conductivity	seepage velocity
without Barrier					≈ 40 m/a
Barrier 1	16 m	30 m	0.6 m	10^{-08} m/s	≈ 30 m/a
Barrier 2	24 m	30 m	0.6 m	10^{-08} m/s	≈ 28 m/a
Barrier 3	24 m	80 m	0.6 m	10^{-08} m/s	≈ 20 m/a
Barrier 4	24 m	80 m	0.6 m	10^{-11} m/s	≈ 15 m/a

Without a hydraulic barrier the groundwater velocity was approximately 40 m/a. For a wall with a depth of 16 m, a length of 30 m and a hydraulic conductivity of 10^{-08} m/s (barrier 1) it was possible to reduce it to about 30 m/a.

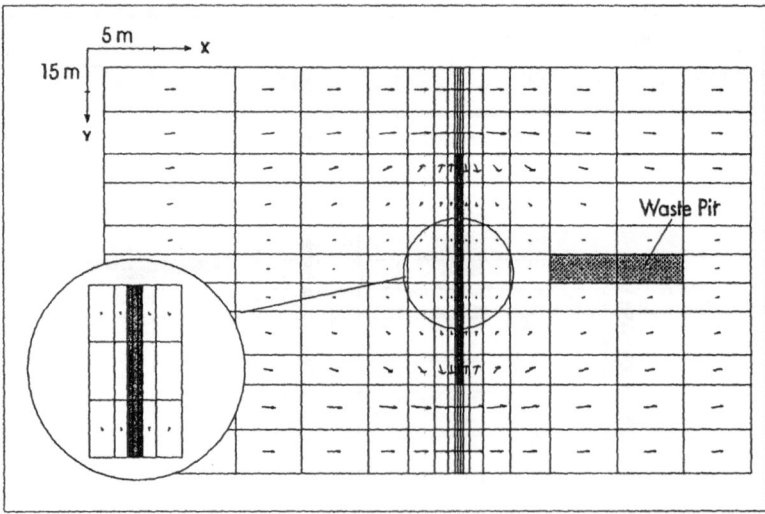

Fig. 3   Groundwater flow pattern with barrier system 4

By extending the barrier to a depth of 24 m (barrier 2) no significant further reduction was achieved. Extending the length of the 24 m deep wall to 80 m (barrier 3) reduced the flow to (20 m/a), 2/3 of the previous value.

Reducing the permeability of the extended barrier from 10^{-08} to 10^{-11} m/s (barrier 4) it was possible to decrease the groundwater velocity further, to approximately 15 m/a. The resulting groundwater flow pattern (now with only a small part of the flow through the waste pit) is shown in Fig.3.

TEST DESIGN

The next step within the methodology is to control the quality of construction of the barrier system designed above. Differences between planned and actual permeability can occur due to variations in material quality (e.g. grain size), improper backfill mixing, slurry entrapment during displacement and failure to remove sediment from the base of the trench. Instead of relying on laboratory measurements the permeability of the barrier is measured in situ. For this purpose an interference test is carried out by injecting water into a test well on one side of the barrier and measuring the resulting pressure response in an observation well on the other side.

Fig. 4 shows how an example test set up consisting of a test well (I) and three observation wells (O) was built into the existing model. All wells were modeled with a local radial grid refinement to minimize the effects of discretization. The depth of the boreholes was set to 24 m with the injection and observation intervals between 8 and 12 m below ground surface. The approach described in this paper uses more than one observation well to be able to examine lateral differences in permeability within the barrier.

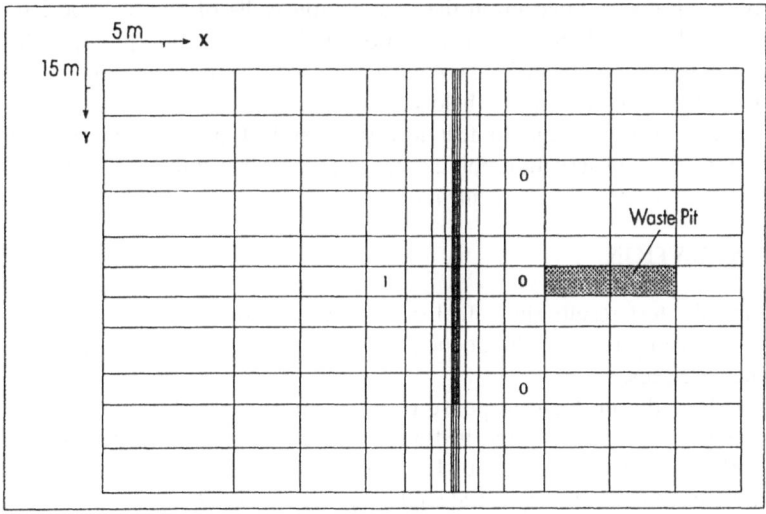

Fig.4   Location of test and observation wells

For the first test design case a hydraulic barrier with a depth of 24 m , a length of 80 m and a conductivity of 10^{-08} m/s (barrier 3) was taken. A pressure pulse was applied to the test well by injecting water at a rate of 35 l/min. for a period of 12 hours. Fig. 5 shows the pressure behavior vs. time in the injection well and in the three observation wells. With the pumping rates and periods mentioned above the pressure difference in the injection well reaches a maximum of 1.6 bar above the hydrostatic value. In the observation wells the maximum pressure response was 0.033 bar in the central well and 0.025 bar in the offset wells.

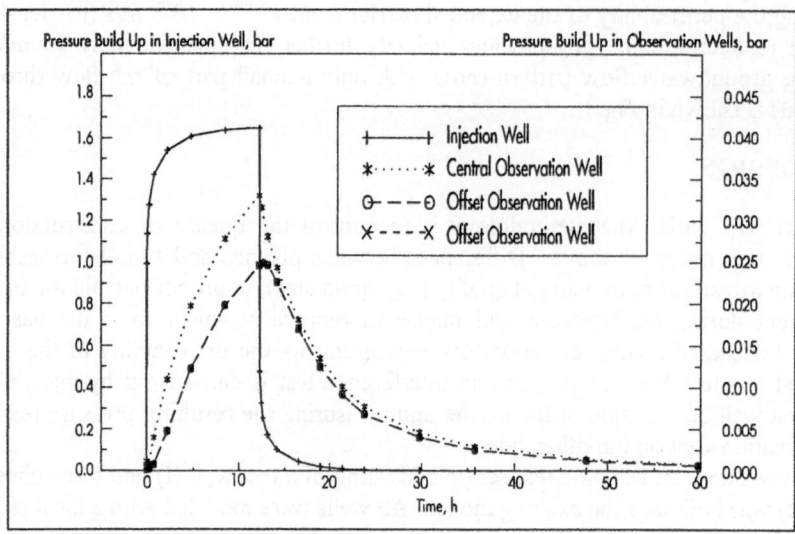

Fig. 5   Pressure build up in test and observation wells

In addition, a comparison of the pressure responses for three different test designs was made. For an injection rate reduced by a factor of 5, the pressure response was much smaller. With a pressure build up in the observation wells of about 0.006 bar it may be impossible to make reliable measurements. For the lower injection pressure the length of the injection period was changed. Though the pressure response was higher it was still too small for reliable measurements (below 0.01 bar).

In a similar way it is possible to test a variety of test parameters, such as injection intervals, depth of the observation points, etc. to assess their influence on the reliability of conclusions drawn from the measurements.

TEST EVALUATION

By evaluating the test results the actual conductivity of the barrier is to be 'found out'. This means estimating the overall conductivity and its distribution, i.e. the location of any more permeable zones.

Fig. 6 shows how the pressure response in the central observation well differs for three different conductivity values of the hydraulic barrier. The highest pressure response is observed with a barrier conductivity of 10^{-07} m/s. If the conductivity of the barrier is 10^{-08} m/s the pressure response in the central well is reduced by 0.09 bar.

The evaluation approach is to successively adjust the wall's conductivity in the model until the observed pressure build up is reproduced. This process will result in the determination of the in situ conductivity of the hydraulic barrier. However, the method is limited by resolution of the measurement equipment. As can be seen in Fig. 6 the pressure response decreases significantly if the barrier's conductivity is getting lower than 10^{-09} m/s.

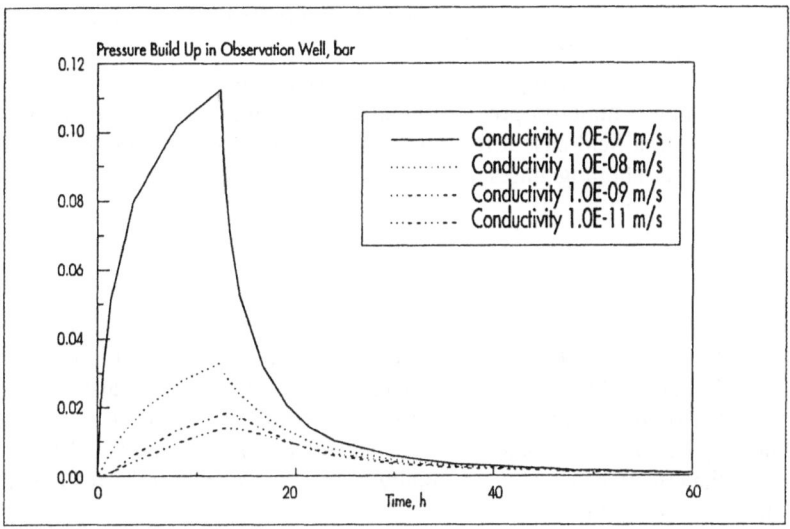

Fig. 6 Pressure build up for different barrier conductivity values

The second aspect of test evaluation is the location of defective zones within the wall. For this purpose a local grid refinement was made in the cells representing the wall. These cells were defined to be 2 m wide, 2 m deep and 0.6 m thick.

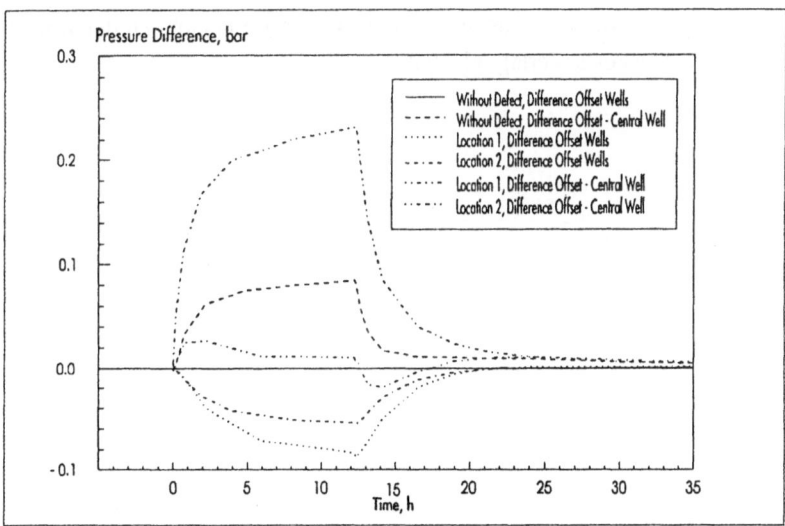

Fig. 7 Pressure build up for different locations of high permeability zones

Fig. 7 shows how two different high conductivity (increased 100 times) cell locations seperated by 15 m influence the pressure response in the observation wells.

Comparison of the pressure response in the offset wells, and of the difference between the response in the offset and the central well enables an initial determination to be made of the location of the defect. For example a large difference between offset and central well responses would indicate a defect near the center of the wall.

By matching the pressure build up in the wells the location of the high permeability zone, or zones, can be defined more closely.

For the described model a homogenous distribution of parameters was assumed. Therefore it is clear that a non symmetric pressure build up is an effect of conductivity variations within the barrier. In a field case a test could be made before construction of the wall to enable the effects of inhomogenities in the ground to be accounted for.

CONCLUSIONS

Installation of low permeability barriers is a well known method for controlling groundwater movement. In this paper it is shown that numerical simulation is a useful tool for the design stage optimization of such control systems. It is also shown that a combination of hydraulic testing and numerical simulation can be used to evaluate the in situ conductivity of the barrier and to detect defective zones within it. This enables an assessment of the effectiveness of the groundwater control system to be made immediately after installation is completed.

REFERENCES

/1/ Daniel, D.E. (1993) Geotechnical Practice for Waste Disposal, Chapman & Hall, London

/2/ Franzius, V., Stegmann, R., Wolf, W., Brandt, E. (1993) Handbuch der Altlasten-sanierung, R. v. Decker's Verlag, Heidelberg

NUMERICAL IMPLEMENTATION OF A THREE-PHASE HYSTERETIC ALGORITHM

P.J. VAN GEEL and J.F. SYKES
Department of Civil Engineering
University of Waterloo
Waterloo, Ontario N2L 3G1
Canada

A fully hysteretic model for the relative permeability-saturation-capillary pressure relationships was implemented in a finite difference multiphase flow and transport code. The code uses an iterative Newton-Raphson scheme to solve the non-linear governing equations which can be formulated using either a fully implicit or an implicit in pressure and explicit in saturation (IMPES) temporal discretization. The fully hysteretic algorithm accounts for fluid entrapment, drainage-imbibition hysteresis in the capillary pressure-saturation relationships, and hysteresis in the relative permeability terms. Two fluid entrapment algorithms were evaluated in terms of their numerical convergence and their effect on the resulting saturations. Criteria were established to determine when a saturation reversal occurs and when fluid entrapment should commence. The effects of updating the hysteretic parameters implicitly at each iteration level or explicitly after each timestep were evaluated.

INTRODUCTION

Hysteresis in k-S-P relationships is often ignored in numerical modelling of three-phase flow. However, several researchers found that significant errors were encountered when modelling non-monotonic multiphase flow without including hysteresis effects. Van Geel and Sykes (1993) demonstrated the importance of hysteresis when modelling a two-dimensional laboratory spill of a lighter-than-water, non-aqueous phase liquid (LNAPL). Essaid et al. (1993) modelled a non-aqueous phase liquid (NAPL) spill in the northern United States. A comparison of the hysteretic and non-hysteretic model results, clearly illustrated the importance of hysteresis in determining the large scale characteristics of the NAPL distribution.

A fully hysteretic model requires the definition of the primary drainage and imbibition(wetting) curves and the entrapped fluid saturations. The primary drainage curve represents the capillary pressure-saturation path taken during monotonic drainage from an initially saturated state and the primary imbibition curve represents the capillary pressure-saturation path followed during monotonic wetting from an initial saturation equal to the irreducible fluid saturation. In addition to the primary curves, a method of developing secondary scanning curves for paths inside the primary curves is required. Based on the saturation history, the volumes of entrapped fluid are determined.

A. Peters et al. (eds.), Computational Methods in Water Resources X, 51–58.
© 1994 *Kluwer Academic Publishers. Printed in the Netherlands.*

Lenhard (1990) and Lenhard and Parker (1987) present a fully hysteretic three-phase model which scales the two-phase relationships to develop the relationships for a three-phase system. They assume that in the presence of three phases (water, NAPL, air) the most preferentially wetting fluid covers the soil particles and is in contact with the second most wetting fluid, which, in turn, is in contact with the least wetting fluid. Hence, two fluid-fluid interfaces exist. Based on the two interfaces, Lenhard (1990) defined a hysteretic model in terms of an apparent water and an apparent total liquid saturation which are defined as

$$\overline{S}_w = S_w + S_{ot} + S_{atw} \tag{1}$$

$$\overline{S}_t = S_w + S_o + S_{at} \tag{2}$$

respectively, where S_w is the effective water saturation; S_{ot} is the saturation of NAPL trapped in the water phase; S_{atw} is the saturation of the gas trapped in the water phase; S_o is the free and trapped NAPL saturation; and S_{at} is the total trapped gas saturation in the water and NAPL phases. The volumes of the trapped phases were scaled between zero and the residual saturation based on the saturation at which the saturation path changed from the primary drainage curve to an imbibition scanning curve. This point is referred to as the saturation reversal point, $^{\Delta}S_j^{ij}$. The maximum entrapped non-wetting fluid saturation, S_{ir}^{ij}, based on the saturation reversal point is defined as

$$S_{ir}^{ij} = \left(1 - {}^{\Delta}S_j^{ij}\right) / \left[1 + R_{ij}\left(1 - {}^{\Delta}S_j^{ij}\right)\right] \tag{3}$$

where

$$R_{ij} = \left(1 \, / \, {}^{I}S_{ir}^{ij}\right) - 1 \tag{4}$$

and $^{I}S_{ir}^{ij}$ is the maximum entrapped saturation for the primary imbibition curve. Based on the current saturation, S_j^{ij}, Lenhard (1990) linearly interpolated between zero and the maximum entrapped saturation, S_{ir}^{ij}, using the following equation.

$$S_{it}^{ij} = S_{ir}^{ij} \left[\left(\overline{S}_j^{ij} - {}^{\Delta}S_j^{ij} \right) / \left(1 - {}^{\Delta}S_j^{ij} \right) \right], \quad \overline{S}_j^{ij} > {}^{\Delta}S_j^{ij} \tag{5}$$

Kaluarachchi and Parker (1992) found that a linear interpolation resulted in numerical problems and modified equation (5) giving

$$S_{it}^{ij} = \left[\frac{1 - {}^{\Delta}S_j^{ij}}{1 + R_{ij}\left(1 - {}^{\Delta}S_j^{ij}\right)} - \frac{1 - \overline{S}_j^{ij}}{1 + R_{ij}\left(1 - \overline{S}_j^{ij}\right)} \right], \quad \overline{S}_j^{ij} > {}^{\Delta}S_j^{ij} \tag{6}$$

The fluid entrapment algorithms based on equations (5) and (6) will hereinafter be referred to as the LP and modified LP hysteretic models, respectively.

Numerical Model

The numerical model used in this study was initially developed by Sleep and Sykes (1993). It is a three-dimensional, three-phase, finite difference flow and transport code. The code was modified to include a fully hysteretic solution based on the LP and modified LP hysteretic models. The hysteretic models are based on the van Genuchten (1980) capillary pressure-saturation relationship. In addition to the van Genuchten parameters the maximum entrapped phase saturations and the two-phase scaling parameters of Lenhard and Parker (1987) are required as input to the hysteretic model. Certain rules were established within the code to determine if a saturation reversal occurred. A minimum change in saturation, S_{rmin}, is required before a reversal point is recognized. Hence, small saturation changes below S_{rmin}, due to numerical fluctuations are not recognized as a saturation reversal point. The volumes of the trapped phases are determined based on the water and total liquid saturation histories.

Validation of the Hysteretic Model

In order to validate the implementation of the fully hysteretic algorithm, the model results were compared to the one-dimensional results presented by White and Lenhard (1993). White and Lenhard (1993) modelled a one-dimensional three-phase flow experiment presented by Lenhard et al. (1993). A 70 cm sand column was instrumented with hydrophobic and hydrophillic tensiometers to measure the water and NAPL pressures. The water and NAPL saturations were measured using a collinear dual-energy γ-radiation system. The column was initially water saturated. The prescribed water pressure at the base of the column was lowered and drainage of the water phase commenced. A LNAPL spill was initiated 12.2 hours after the water pressure was lowered. A total of 250 ml of LNAPL entered the column over a 46 minute period. The water table was then raised, lowered and raised again at 17.4, 19.4 and 21.1 hours, respectively, to simulate a fluctuating water table. White and Lenhard (1993) modelled the experiment using a multiphase flow code based on the LP hysteretic model. The agreement between the model developed in this study and the results presented by White and Lenhard (1993) indicates that the three-phase hysteric algorithm used in for this study is implemented correctly.

LP and Modified LP Hysteretic Model Comparison

The model results for the three-phase experiment presented by Lenhard et al. (1993) were used to illustrate the difference in the LP and modified LP hysteretic algorithms. The water and LNAPL saturations predicted by the LP and modified LP models are presented in Figure 1. The 67, 37 and 17 cm locations refer to the elevations above the base of the sand column. The reason for the differences in the results is discussed in the following paragraphs.

The equations for the maximum phase entrapment based on the saturation reversal point (equations 3 and 4), were developed by Land (1968) based on a series of plots relating the initial air phase saturation to the maximum air phase entrapment after the wetting phase saturation increased to a value of 1.0. Land (1968) found that the difference in the reciprocals of the initial air phase saturation and the maximum entrapped air phase saturation was approximately constant for a given sand. Equation (3) is plotted in Figure 2 for an air-water system in which the maximum entrapped air saturation is 0.2.

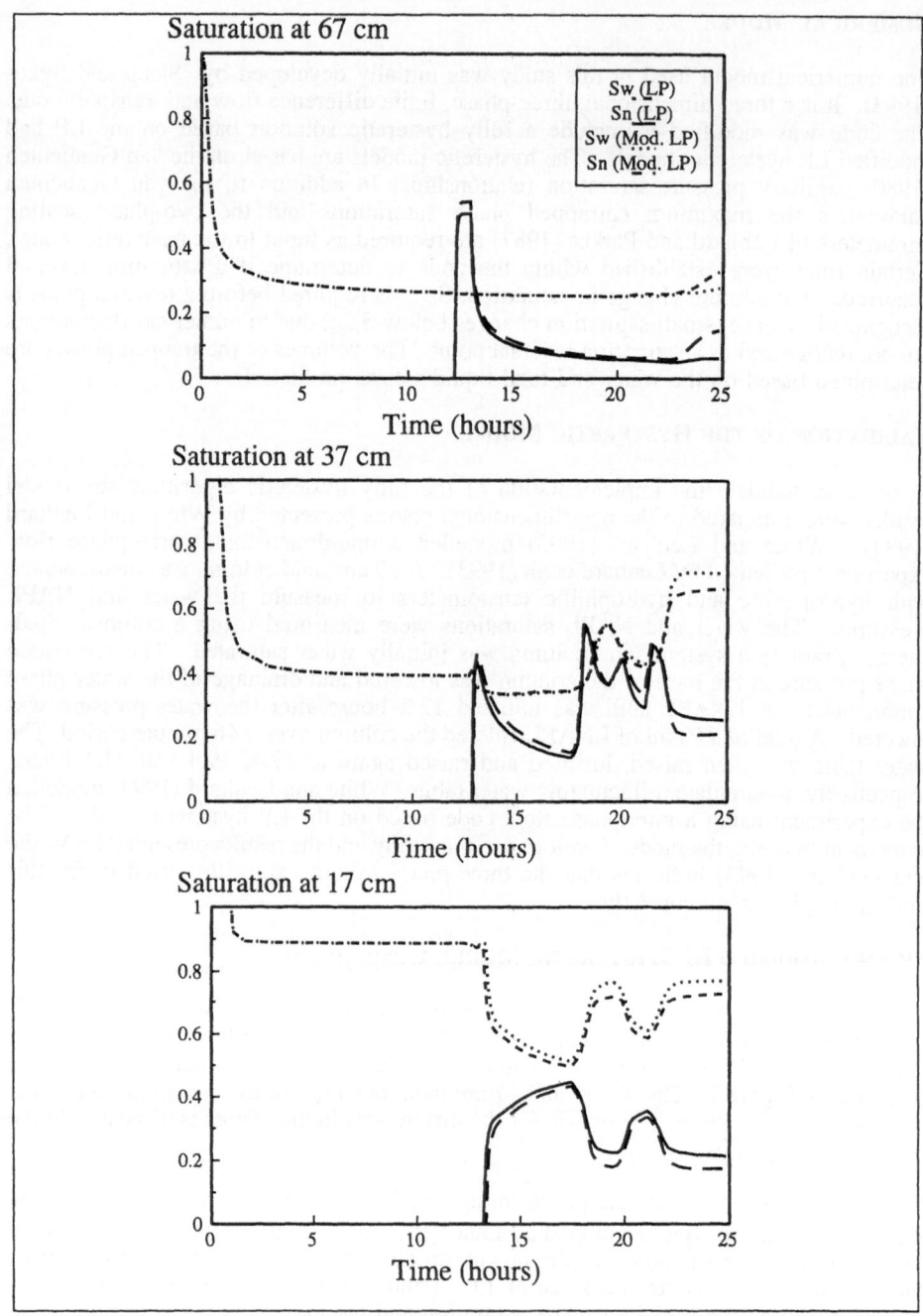

Figure 1 LP and Modified LP Model Results for Three-Phase Experiment

The water saturation reversal point is plotted along the x-axis and reflects the initial air saturation when imbibition commenced. Figure 2 illustrates that for different saturation reversal points, the percent entrapment of the initial air phase saturation differs. For example, a saturation reversal at 0.3 results in a maximum entrapped saturation of 0.184 which is 26.3% of the initial mobile air saturation (0.70) at the reversal point. In comparison, a saturation reversal at 0.9 results in a maximum entrapped saturation of 0.0667 which is 66.67% of initial saturation (0.10) at the reversal point. The derivative of the maximum entrapped saturation with respect to the saturation reversal point gives the percent entrapment of the initial air phase saturation. The derivative is also plotted as a fraction in Figure 2.

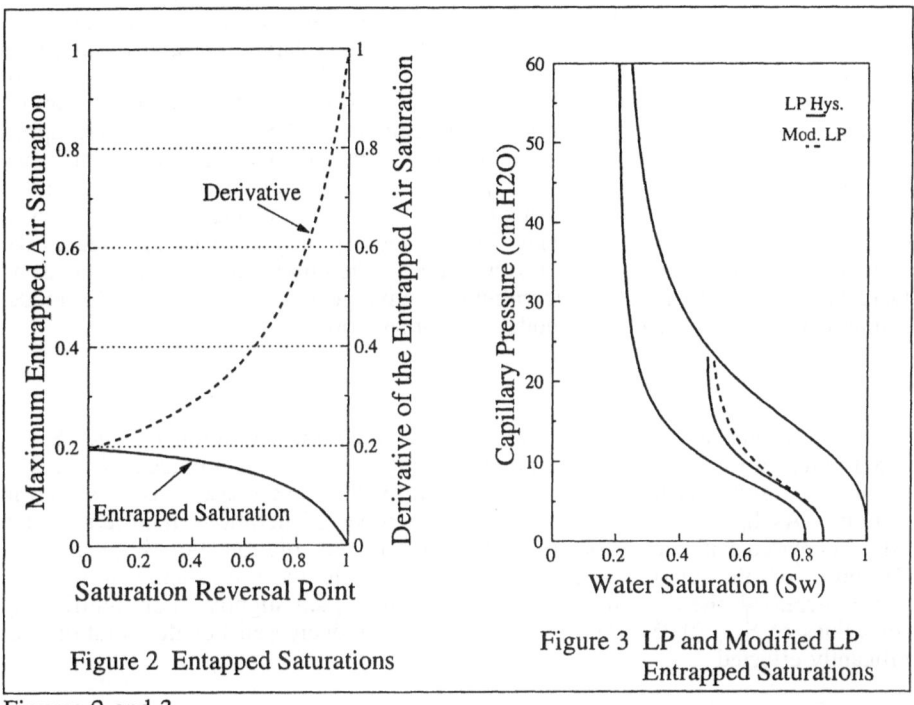

Figure 2 Entapped Saturations

Figure 3 LP and Modified LP Entrapped Saturations

Figures 2 and 3

Both the LP and modified LP hysteretic models use equations (3) and (4) to determine the maximum entrapped saturation based on a saturation reversal point. The LP algorithm calculates the entrapped saturation for a saturation between the reversal point and 1.0 using linear interpolation (equation 5). The modified LP algorithm of Kaluarachchi and Parker (1992) calculates the entrapped saturation as the maximum entrapped saturation based on the saturation reversal point, minus the remaining potential entrapment based on the current saturation (equation 6). As a result, the entrapped volume calculated using the LP model is greater than the entrapped volume calculated using the modified LP model for the same saturation value. Figure 3 illustrates the different saturation paths based on the LP and modified LP entrapment algorithms for a saturation reversal at 0.40. If the saturation reversal for the water-air system presented in Figure 2 is 0.4 and the saturation is increased to 0.6, the entrapped air phase saturations based on the LP and modified LP models are 0.059 and 0.023, respectively. The larger initial entrapped volumes predicted by the LP model result in reduced relative permeabilities which in turn result in different minimum and maximum saturation values as the water table is raised and lowered during the three-phase experiment (Figure 1).

Kaluarachchi and Parker (1992) found that the LP model lead to convergence problems at capillary pressures near zero, with this being attributed to the large changes in the volume of entrapped NAPL with small changes in capillary head. For the model runs performed for this study, the LP model resulted in superior convergence compared to the modified LP model. This is partially due to the larger entrapped saturation derivatives of the modified LP model in comparison the LP model. The derivatives of the entrapped saturations with respect to the apparent water and total liquid saturations are required for the Newton-Raphson system of equations. The derivative of entrapped saturation with respect to the current saturation for the LP model is a constant value. The derivative of the modified LP entrapped saturation is dependent on the saturation and approaches 1.0 as the saturation approaches 1.0. Large changes in the entrapped saturations as the saturation approaches 1.0 and the capillary pressure approaches 0.0 are inherent in the modified LP formulation. Therefore, the reasons stated by Kaluarachchi and Parker (1992) concerning the convergence problems of the LP model in comparison to the modified LP model are not supported.

MODEL SENSITIVITY TO S_{rmin}

The model sensitivity to the magnitude of the minimum saturation change, S_{rmin}, required before a saturation reversal point is recognized, was also evaluated. A S_{rmin} value of 0.01 resulted in the fastest convergence for the three-phase runs presented in this study. As S_{rmin} was decreased, the total number of iterations required for the solution increased without any significant change to the solution. For S_{rmin} values of 0.001 and 0.002, the solution diverged as the water table was raised and lowered. As S_{rmin} was increased, the total number of iterations increased slightly. For significantly larger values of S_{rmin} (0.05), the number of iterations decreased but the solution was significantly effected.

IMPLICIT-EXPLICIT UPDATING OF THE HYSTERETIC PARAMETERS

The hysteretic parameters and the saturation history can be updated explicitly after each timestep or implicitly after each iteration. Three implicit-explicit solution schemes were evaluated: a fully implicit scheme, a fully explicit scheme, and a combined implicit-explicit scheme in which the hysteretic parameters are updated explicitly except for cases where the saturation exceeds the limits of the current scanning curve. If the limits of scanning curve are exceeded, the saturation path falls back onto the previous scanning curve, imbibition or drainage and the saturation history is updated at the iteration level. If the limits of the scanning curve are exceeded in the fully explicit scheme, the correct capillary pressure and its derivatives are calculated based on the correct scanning curve, but the saturation history is not updated until the end of the timestep. The three implicit-explicit solution schemes were evaluated for the LP and modified LP hysteretic models for the three-phase experiment (Lenhard et al.,1993). The results are summarized in Table 1. The simulations consisted of 10234 timesteps which varied in size from 0.01 to 20.0 seconds. The smaller timesteps were required immediately after the prescribed water pressure at the lower boundary was instantaneously increased or decreased. Included in Table 1 are the total number of iterations required for convergence, the number of timesteps for which 8 iterations were required and the number of saturation reversal points recorded during each solution. The fully implicit model recorded a significantly larger number of saturation reversals as a result of updating the saturation history at each iteration level. The total number of iterations required for the LP model runs did not vary significantly with the implicit-explicit updating scheme that was used.

Table 1 LP and Modified LP Convergence Data

		# iterations	# times 8 iterations were exceeded	total # of sat. reversal points
fully implicit	LP Model	27975	161	3734
	Modified LP	diverged	-	-
fully explicit	LP Model	27132	28	658
	Modified LP	35873	533	650
implicit-explicit	LP Model	27105	22	785
	Modified LP	36496	635	817

CONCLUSIONS

A three-phase hysteretic algorithm was implemented in a multiphase flow code. Two entrapment algorithms were evaluated: the LP model presented by Lenhard (1990) and the modified LP model presented by Kaluarachchi and Parker (1992). The LP model uses linear interpolation to scale the entrapped saturations between zero and the maximum entrapped saturation. The modified LP model uses a non-linear equation which increases the percent entrapment as the saturation approaches 1.0. The initial entrapped saturations at the beginning of an imbibition scanning curve predicted by the LP model are slightly larger than the entrapped saturations calculated using the modified LP model. The larger entrapped volumes result in lower relative permeabilities. To illustrate the effects of the two algorithms, the one-dimensional three-phase flow experiment presented by Lenhard et al. (1993) was modelled using each algorithm. The saturations predicted by the LP and modified LP models followed similar saturation paths in response to the water table fluctuations but the saturation values differed by as much as 0.10.

The entrapped saturations are determined based on the water and total liquid saturation histories. To establish when a new saturation scanning curve is required, a minimum saturation reversal term, S_{rmin}, was established. A S_{rmin} value of 0.01 proved to be sufficiently large to improve convergence and avoid the storage of small saturation changes due to numerical fluctuations, and was sufficiently small that it did not significantly effect the solution. Three implicit-explicit updating schemes for the hysteretic parameters and saturation histories were evaluated. The model results appeared to be relatively insensitive to the implicit or explicit evaluation of the hysteretic parameters for the one-dimensional runs performed for this study. This may be partially attributed to the small timesteps used in the simulations. The effect of timestep size on the implicit or explicit updating of the hysteretic parameters was not evaluated.

ACKNOWLEDGEMENTS

This research was funded by a Natural Sciences and Engineering Research Council of Canada Scholarship awarded to P.J. Van Geel and the Ontario Ministry of the Environment and Energy, Environmental Research Program. The authors would like to acknowledge Dr. Brent Sleep of the University of Toronto for his efforts in the development of the multiphase code which was modified for this research.

REFERENCES

Essaid, H.I., W.N. Herkelrath and K.M. Hess, (1993), "Simulation of Fluid Distributions Observed at a Crude Oil Spill Site Incorporating Hysteresis, Oil Entrapment, and Spatial Variability of the Hydraulic Properties", Water Resour. Res., 29(6), pages 1753-1770.

Kaluarachchi, J. J. and J. C. Parker, (1992), "Multiphase Flow with a Simplified Model for Oil Entrapment", Transport in Porous Media, (7), pages 1-14.

Land, C.S., (1968), "Calculation of Imbibition Relative Permeability for Two- and Three-Phase Flow From Rock Properties", Society of Petroleum Engineers Journal, June, pages 149-156.

Lenhard, R. J. and J. C. Parker, (1987), "A Model for Hysteretic Constitutive Relations Governing Multiphase Flow 2. Permeability-Saturation Relations", Water Resour. Res., 23(21), pages 2197-2206.

Lenhard, R.J., (1992), "Measurement and Modeling of Three-Phase Saturation-Pressure Hysteresis", J. of Contaminant Hydrology, 9, pages 243-269.

Lenhard, R.J., T.G. Johnson and J.C. Parker, (1993), "Experimental Observations of Nonaqueous-Phase Liquid Subsurface Movement", Journal of Contaminant Hydrology, (12), pages 79-101.

Sleep, B.E. and J.F. Sykes, (1993), "Compositional Simulation of Groundwater Contamination by Organic Compounds 1. Model Development and Verification", Water Resour. Res., 29(6), pages 1697-1708.

Van Geel, P.J. and J.F. Sykes, 1993, "Laboratory and Numerical Simulations of a LNAPL Spill in a Variably Saturated Sand Medium", submitted to the Journal of Contaminant Hydrology, submitted October 1993.

van Genuchten, M.Th., 1980, " A Closed-Form Equation for Predicting the Hydraulic Conductivity of Unsaturated Soils", Soil Sci. Soc. Am. J., 44, pages 892-898.

White, M.D. and R.J. Lenhard, (1993), "Numerical Analysis of a Three-Phase System with a Fluctuating Water Table", Proceedings of the Thirteenth Annual American Geophysical Union Hydrology Days, March 30 - April 2, 1993, Colorado State University, Fort Collins, Colorado, pages 219-236.

COMBINED USE OF FE-MODELS FOR PEVENTION OF ECOLOGICAL DETERIORATION OF AREAS NEXT TO A RIVER HYDROPOWER COMPLEX

R. KOHANE and R. WELZ
Lahmeyer International GmbH
Lyoner Straße 22
60528 Frankfurt/Main
Germany

The construction of the Pielweichs hydropower scheme will alter the ecological conditions of the areas next to the reservoir. In order to minimize the ecological impact on these areas different compensation measures are planned. The required investigations for the hydraulic design of these measures are carried out using mathematical models of surface water and groundwater flow.

INTRODUCTION

The Pielweichs hydropower scheme is situated at the lower course of the river Isar near to the city of Plattling in Bavaria in the southern part of Germany. The project is part of a rehabilitation program aimed to stop the deepening of the river-bed resulting from erosion induced by existing hydropower schemes situated upstream along the river. The construction of the Pielweichs weir will cause the water surface elevation in the river to raise approx. 6 m above the present level. The backwater will reach the downstream section of the next weir lying upstream along the river. Sealing of the embankments parallel to the reservoir prevents the river water from interacting with the groundwater flow. As a consequence the groundwater dynamics in the ecological sensitive forest areas behind the embankments are supressed. Flooding of the reservoir will also have an impact on the fauna settled on the floodplains of the river.

In order to compensate these effects two mayor corrective measures are planned. The first one considers diverting part of the river's discharge through natural compensation channels located on both sides of the reservoir behind the embankments. This will provide a connection between the downstream part of the river and the reservoir allowing fishes to migrate upstream along the river. Additionally new habitat will be provided along the banks of the new channels for the fauna living in the floodplain areas of the river. The second measure considers simulating artificially the natural fluctuations of the groundwater table in the forest areas by infiltrating river water into the aquifer through buried infiltration pipes running parallel to the lateral dams.

The investigations for planning and designing the compensation measures are carried out with mathematical models. A two dimensional, depth-avereged river model is used for computing the surface flow along the natural compensation channel situated on the left side of the river. Groundwater flow is computed using a two-dimensional depth-avereged model based on Darcy's flow equations. Both models are coupled permiting an accurate computation of mass exchange between surface and subsurface flow.

A. Peters et al. (eds.), Computational Methods in Water Resources X, 59–66.
© 1994 *Kluwer Academic Publishers. Printed in the Netherlands.*

DESCRIPTION OF THE AREA OF STUDY

Figure 1 shows the location of the compensation channel on the left side of the river Isar. The channel flows in its middle and lower sections along two existing natural streams. In its upper portion the channel flows through a system of natural interconnected drainage channels. The discharge of the channel is planned to range between 3 and 9 m³/s. The water presently flowing through the channel (3 m³/s) will be deviated through an artificial interception ditch running parallel to the border separating the forest from the agricultural fields. The interception ditch will prevent the artificial infiltration from altering the groundwater levels in the agricultural fields.

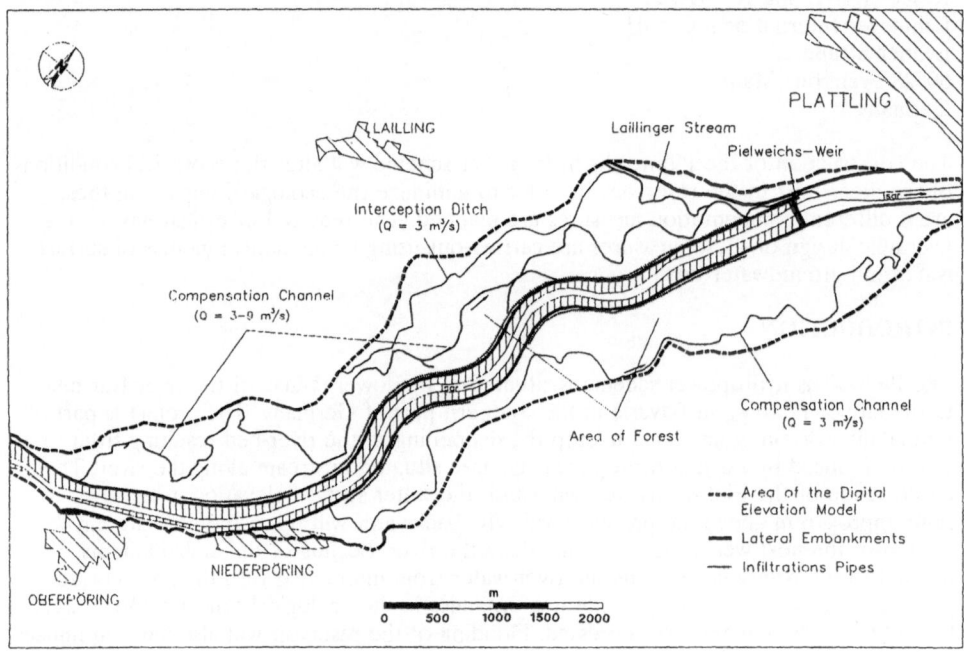

Figure 1:   Map showing the area of study

Figure 1 also shows the location of the infiltration pipes running parallel to the retention embankments. The pipes are made of porous concrete, have a diameter of 0,8 m and are burried approx. 0,5 m underneath the ground surface. River water is pumped into the pipes at different points distributed along the reach and infiltrates through the porous walls into the groundwater aquifer. The infiltration system is designed to carry different flow discharges depending on the flood event of the river Isar which wants to be simulated.

The hydrogeological caracteristics of the groundwater aquifer in the area of study are relative simple. The terrace of the river Isar in the estuary region consists of vast quarternary gravel with an hydraulic conductivity ranging from 5×10^{-2} to 5×10^{-3} m/s. The aquifer has a thickness which varies between 3 and 12 m and is limited underneath by an impervious tertiary stratum. The depth of the water table corresponding to mean hydrological conditions ranges between 1 and 3 m.

MODELING OF SURFACE WATER FLOW

Model description

The core of the model used for simulating the surface water flow is a workstation implementation of the modeling system FESWMS-2DH (Lee and Froehlich, 1989). The model solves the two-dimensional surface water flow equations in the horizontal plane using the finite element method. The flow is described by three nonlinear partial-differential equations, two for conservation of momentum (expressed in conservative form) and one for conservation of mass. The two-dimensional depth-avereged flow equations account for energy losses through bottom friction and turbulent stresses. Bottom friction is considered by using Manning's formula extended to two dimensions and turbulent stresses are computed using the Boussinesq eddy-viscosity concept. Quadratic basis functions are used to interpolate velocity components, and linear basis functions are used to interpolate depths on triangular, six-node, isoparametric elements. The nodal values of the velocity components and depth are solved using Galerkin's method of weighted residuals, a Newton-Raphson iteration scheme and a frontal solution algorithm

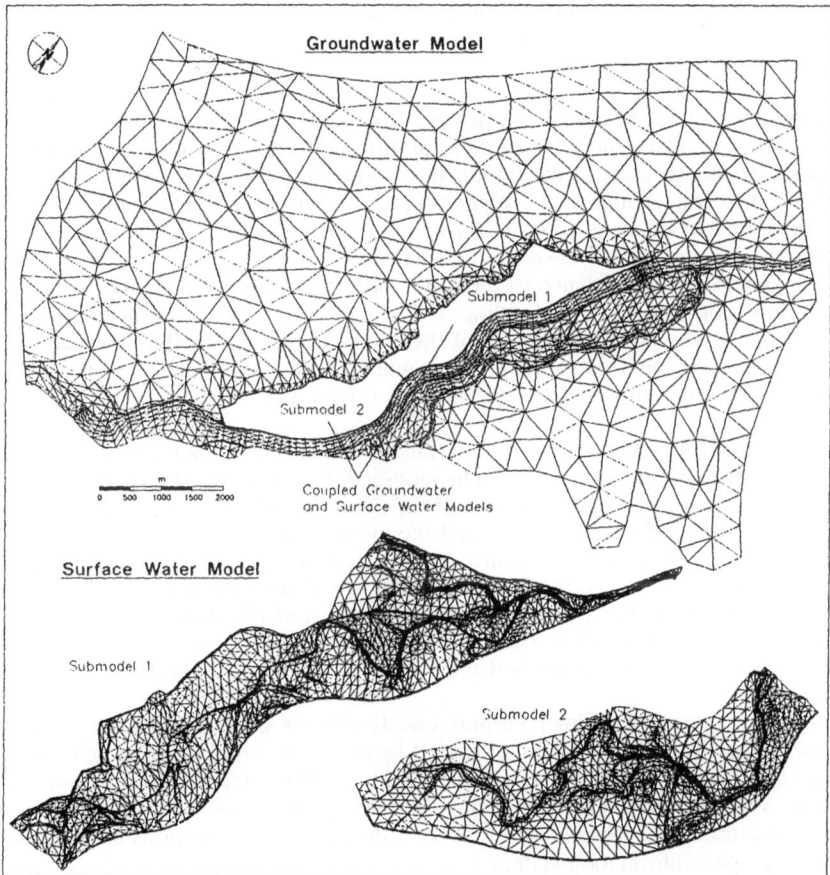

Figure 2:   Finite element grid of the surface water and groundwater models

Preprocessing of the finite element network and associated data and postprocessing of model results are performed using a preprocessing and postprocessing system based on a self-developed AutoCAD application using the ADS programming environment (Vogt et al., 1992).

The region considered by the surface water model and the finite element mesh used in the computations are shown in Fig. 2. The model covers the whole reach of the compensation channel including the lateral floodplains. The interception ditch and the lower part of the Laillinger stream are also included in the model. The model is subdivided into two smaller submodels in order to reduce the computational effort required to carry out the simulations. Each submodel covers approximately half of the total model area. Existing profiles of the flow cross-section of the streams distributed at regular distances along the reach and data included in a digital elevation model are used to design the finite element grid.

For both submodels boundary conditions are defined by specifying flow components at inflow boundary nodes and specifying water-surface elevations at outflow boundary nodes. Along all other boundaries zero normal flow is assumed.

Results of the computational studies

Initially a calibration of the Manning coefficient of the main channel was carried out using available data of water surface elevations at different cros-sections along the reach. Next the roughness coefficients of the floodplains were estimated based on information available about the caracteristics of the vegetation of the floodplains. The calibrated model was then used to simulate different flow situations in which the discharge of the channel was varied in the range between 3 and 9 m³/s and the downstream boundary condition was specified in accordance to the water surface elevations of the river Isar. As example the results of one simulation with a discharge of 6 m³/s are shown in Figure 3. The overbank flow in the downstream part of the channel is caused by the backwater of the river Isar. A surface water elevation corresponding to a 5-year discharge of the river was specified as downstream boundary condition for this simulation.

The results of the investigations carried out so far show that the geometry of the cross-section of the channel will have to be partially modified. For instance, the channel-bed will have to be lowered in the upstream part of the reach in order to reduce the difference between the surface water head and the mean groundwater head. In addition the downstream end of the channel will have to be moved to a point further downstream along the river in order to reduce the flooding of the forest areas caused by the backwater of the river Isar. The geometry and the elevation of the channel-bed will have to be changed at some sections to account for the higher discharges flowing through the channel and to account for the higher bed shear stresses resulting from them.

An important aspect presently being considered deals with the ecological impact on the vegetation of the areas beside the channel caused by the redefinition of the geometry of channel's cross-section. The ecological damage is evaluated by computing the size of each of the affected vegetation zones and balancing the corresponding areas according to the ecological significance of the vegetation present on them. The computations are carried out using a geographic information system.

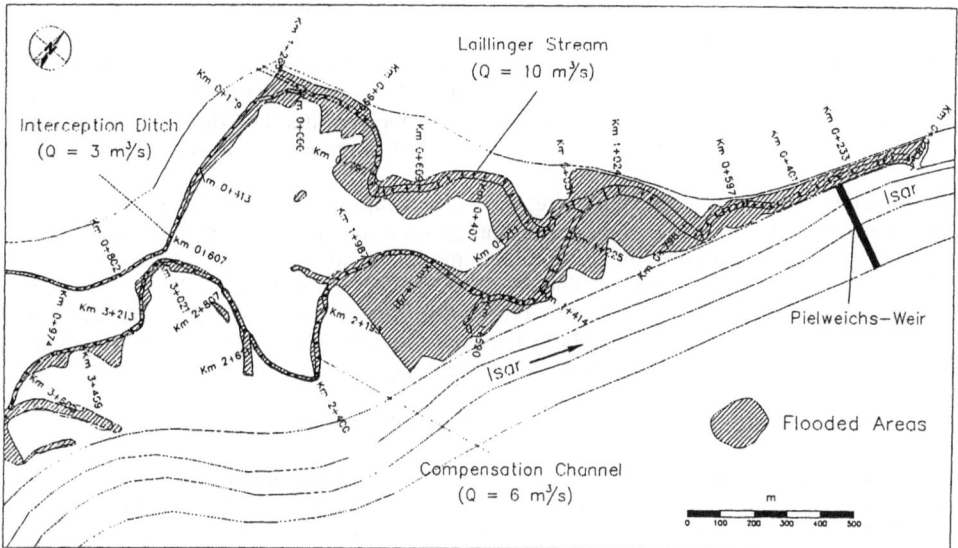

Figure 3:   Flooded areas along the compensation channel for a 5-year discharge of
             the river Isar

MODELING OF GROUNDWATER FLOW

Model description

The program used for computing the groundwater flow is a two-dimensional depth-
avereged groundwater model based on the Galerkin finite element method. The model
solves Darcy's flow equation in the horizontal plane using linear basis functions to
interpolate groundwater heads on triangular elements. Both steady and unsteady flow
situations can be computed and a great variety of point, line and area boundary
conditions can be specified.

The area of the model covering approximately 74 km^2 is shown in Figure 2. The western
border of the model runs almost perpendicular to the main direction of the groundwater
flow. Along this boundary Dirichlet boundary conditions are specified. Dirichlet boundary
conditions are also specified along the eastern boundary of the model. To the south, the
aquifer is limited by a raised tertiary terrace considered in the model to be a natural
boundary. The northern border of the model lies along a natural watershed.

The river Isar and all important streams and natural channels present in the area of study
are considered in the grid discretisation. They are specified in the model using a leakage
boundary condition. The sealing walls along the embankments of the reservoir are
modeled using a row of narrow elements having low hydraulic conductivity. Behind the
sealing walls a row of nodes is used to model the infiltration pipes. In the area of forest
the grid of the groundwater model is chosen to be almost identical to the grid of the
surface water model.

Coupling of the groundwater model with the surface water model

The coupling of the groundwater model with the surface water model is necessary in order to permit an accurate description of the influence of the surface water flow on the groundwater situation. Especially for simulation of flood events the effect of the backwater of the river Isar in the downstream part of the model is of great importance.

The models are coupled in the following way. The computations with the surface water model are carried out without considering the groundwater flow. The computed surface water heads are then used in the groundwater model for specifying point leakage boundary conditions at the nodes coinciding with the surface water flow. The assumption of neglecting the interflow in the computations with the surface water model is admissible considering that the mass exchange between surface water and groundwater is less than approximately 15 % of the total suface flow discharge.

Approach for computing the groundwater recharge due to artificial infiltration

The recharge along the infiltration pipes is considered in the groundwater model by using two approaches. For steady flow computations a point leakage boundary condition is specified at each node of the pipe. The piezometric heads along the pipes, needed for specifying the leakage boundary condition are calculated using a one-dimensional finite element pipe-network model. Because the amount of water flowing through the porous walls of the pipes depends on the difference of the piezometric head inside the pipe and the groundwater head outside the pipe an iterative computational procedure is needed.

The approach described above could also be applied to unsteady flow computations however at a cost of spending much computational effort. An easier approach is to assume constant discharges along the pipes and adjust the required pressure distribution in the pipes by operating the control valves. The boundary condition to be specified in this case for each pipe-node of the groundwater model is of the Neumann type.

Model calibration

After having designed the model grid the aquifer parameters in the model were determined. First, the model hydraulic conductivities and the leakage coefficients were calibrated assuming steady flow conditions. Historical records of hydrological and hydrogeological data were used to define the mean values of groundwater recharge, groundwater heads and surface water elevations used as reference for the calibration. Following the steady state calibration unsteady flow conditions were considered in order to determine the storage coefficients of the model. The validity of the model parameters obtained from the steady state case was examined under unsteady flow conditions and in some cases corrections were made to the corresponding values.

Results of the computational studies

The groundwater model was used to investigate several different hydraulic aspects within the project. Only a brief description of the main results of the model will be given here. In the first stage of the project the sealing walls along the embankments of the reservoir had to be calculated. The hydraulic conductivity required for the sealing walls and the depth needed to penetrate the less permeable terciary stratum lying underneath were calculated with the model.

The hydraulic design of the underground pipe-system running parallel to the embankments was also investigated with the groundwater model. Originally the pipes were designed as drainage system for catching the seepage water passing through the reservoir embankments. Later it was decided that they could also be used as infiltration pipes for simulating artificial groundwater recharge in the areas of the forest. The diameters and the lengths of the pipes and the depth underneath the ground surface at which the pipes are burried were calculated with the model.

For the hydraulic design of the compensation channels on both sides of the reservoir the interaction between surface water and groundwater had to be considered. For each of the studied surface flow cases a coupled computation using the surface flow and the ground-water models was performed in the manner described above. In each case the influence of the surface water flow on the groundwater heads in the sensitive areas of the forest was studied. Surface flow cases causing changes in the groundwater levels were analized and solutions for minimizing these changes were proposed. These included redefining the geometry and the layout of some surface flow cross-sections.

A very important aspect of the model studies was to determine the manner in which the infiltration system should be operated. In order to be able to judge the functioning and effectiveness of the infiltration system a reference event for comparison was defined. The flow situation corresponding to an historical 1-year flood event of the river Isar was used for this purpose. For the considered flood event a great amount of data was available so that accurate boundary conditions for the model could be specified. The simulation with the model was chosen to cover the whole period of the flood starting 2 weeks before the flood peak was reached and ending 4 weeks later.

An adecuate operational scheme of the infiltration system that would simulate the movement of the groundwater table caused by the rise and fall of the water level in the river Isar was investigated. A series of computational runs were carried out with the model varying the amount of water pumped into the aquifer, the distribution of the infiltration discharges along the pipes and the duration of the infiltration process. The infiltration was simulated in the model using the approach decribed above.

Figure 4 shows the results of the operational scheme of the infiltration system found to give the best results for the flow event being considered. A section perpendicular to the reservoir on the left side of river Isar is considered. The time dependent groundwater heads are represented at six different points distributed along this section. The upper diagram corresponds to the natural flood event and the lower diagram to the simulation with artificial infiltration. The infiltration discharges, differing along each pipe, are assumed to remain constant during the infiltration process which has a duration of 6 days. The results show a good agreement between computed and observed groundwater heads. Important for the vegetation in the areas of forest are the peaks of the groundwater levels reached during the flood event. These are well simulated with the artificial infiltration. Figure 4 also shows that the influence of the infiltration decreases when moving away from the reservoir and practically disappears behind the interception ditch.

Computational runs with artificial infiltration were also carried out for a 5-year flood event of the river Isar. The infiltration discharges for each pipe and the total duration of the infiltration process were determined. Finally an operating scheme for the infiltration system was developed coupled with the water level of the river Isar at a gauging station downstream from the reservoir.

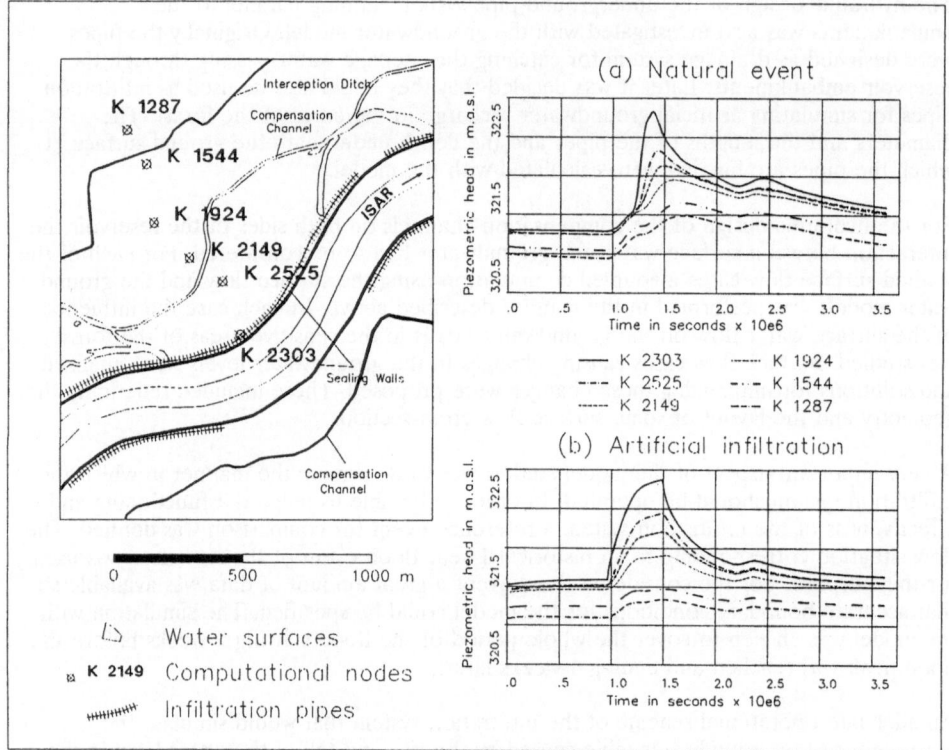

Figure 4:   Groundwater heads along a section perpendicular to the reservoir for a 1-year flood event of the river Isar.

SUMMARY

A practical application on the use of mathematical models for planning and designing different hydraulic aspects of projects in which solutions to environmental issues are demanded is presented. In the model investigations described above two main compensation measures for the ecological sensitive forest areas on the sides of the Pielweichs reservoir are considered using a combination of two standard, two-dimensional, finite element models for computing combined groundwater and surface water flow.

References

Lee, J. K. and Froehlich, D. C. (1989) "Two-dimensional finite-element hydraulic modeling of bridge crossings: Research report", U.S. Geological Survey, Publication No. FHWA-RD-88-146.

Vogt, M. et al. (1992) "Pre- und Postprocessing von Finite Elemente Grundwassermodellen mit Hilfe neuster CAD-Programmier, Animations- und Präsentationsverfahren", XXI. Internationaler Finite Elemente Kongreß, Baden-Baden.

Sedlmair, G. and Asal, P. (1992) "Isar-Stützkraftstufe Pielweichs", Wasserwirtschaft 82, 11.

SOLUTION OF SECOND ORDER EQUATIONS BY GALERKIN LEAST SQUARES PENALTY METHOD ON ORTHOGONAL MESHES

Jeffrey P. Laible
Department of Civil and Environmental Engineering, University of Vermont, Burlington, Vermont , 05405 USA

ABSTRACT

A Galerkin Least Squares Penalty Method (GLSP) formulation for solving secondorder differential equations defined over irregular domains is proposed and tested on a model equation. This approach has the advantage that any irregularly shaped region can be modeled by using a completely orthogonal computational grid mesh. Generation of the computational mesh and any refinement is a straight forward and simple process compared to the traditional task of mesh generation requiredin standard finite element methods. A geometric mesh is required but it need only describe the boundary of the domain. Previous studies have employed a completely least squares formulation using collocation (LESCO) [1,2]. As is well known, the least squares method in general requires the use of higher order basis function than may be used with Galerkin weak formulations. In the solution of the diffusion advection equation for example, the cubic Hermetian polynomials have been used[1]. In this paper the domain equations are cast in the weak form, while the boundary equations are treated by least squares. This combination allows for the use of geometrically orthogonal linear C^o finite elements.

INTRODUCTION

This paper describes a numerical approach that greatly reduces the effort required for solving problems of irregular geometry, while retaining the accuracy of the solution. In this approach one combines a completely orthogonal mesh with a general description of the problem domain geometry. This is achieved by defining two separate meshes. The "computational mesh" is simply an orthogonal set of elements (see Figure 1) and the "geometric" mesh is a connected list of boundary points which need not correspond to the element sides of the computational mesh. The nodes at the corners of the orthogonal elements are referred to as the computational nodes and the points on the geometric boundaries are referred to as the geometric nodes or points. When refinement is used the nodes at the mid points of adjacent elements are treated as slave nodes. In this arrangement, the

A. Peters et al. (eds.), Computational Methods in Water Resources X, 67–74.
© 1994 Kluwer Academic Publishers. Printed in the Netherlands.

computational mesh bears a resemblance to a finite difference mesh and the geometric list bears a resemblance to the boundary mesh used in the boundary element method. As will be shown, the distinguishing features of the proposed method are that linear finite elements are still used to formulate a weak statement of the problem and the interior domain is still discertized, but with a very simple configuration of orthogonal elements. In effect, this approach incorporates some of the advantages of the finite difference and boundary element methods into a finite element formulation.

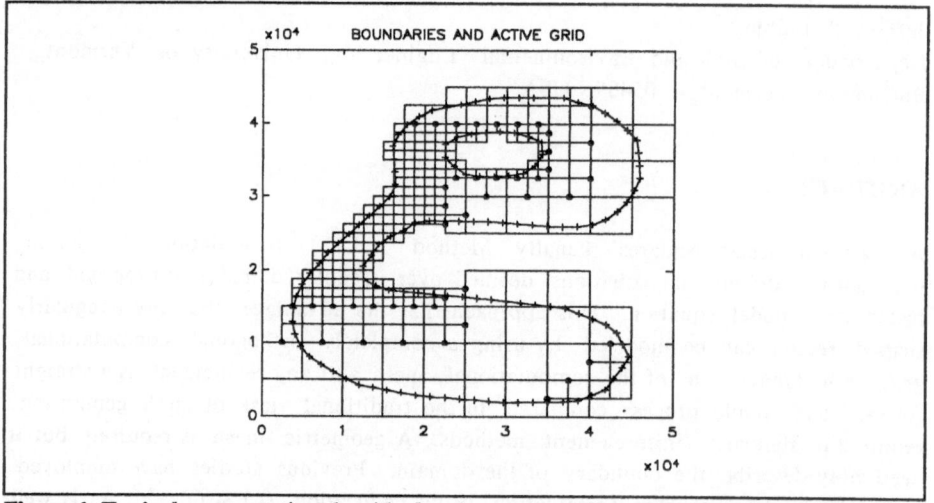

Figure 1 **Typical computational grid showing active elements, and slave nodes from refinement (indicated by o). All other nodes are free nodes. Boundary points are indicated by +.**

This dual mesh approach has been previously applied to the scalar second order diffusion-advection equation[1] , using the least squares collocation method (LESCO). More recently LESCO has been applied to vector first order equations[2] , namely the shallow water equations. In both instances the test functions are essentially the variations of the differential operator residuals and the associated boundary condition operator residuals. For the second order equation, LESCO requires higher order finite elements (in the single equation formulation). Cubic Hermites were used for the diffusion advection problem. For the first order equation, bilinear elements were readily used. Other relevant studies [3,4,5,6,7] are listed in the reference section.

Here, we are concerned with formulating a Galerkin-least squares method for the purpose of imposing boundary conditions on irregular boundaries defined on an orthogonal mesh. A secondary goal is to utilize lower order basis functions than normally required in the least squares formulations. For convenience we will refer to this as the Galerkin Least Squares Penalty Method (GLSP). The method has been verified for a scalar second order equation and verification will be discussed in detail.

MODEL EQUATION

In this paper we will consider the solution of the second order diffusion equation and show computational results for both Dirichlet and Neumann boundary conditions. For completeness however, the theoretical formulation is described for the complete transport equation.

The differential equation and associated boundary conditions are stated as:

$$\frac{\partial c}{\partial t} + \vec{\nabla}^T \vec{q} - f = 0 \qquad on \quad \Omega \tag{1}$$

$$-q_n + h = 0 \qquad on \quad \Gamma_2 \tag{2}$$

$$c - g = 0 \qquad on \quad \Gamma_1 \tag{3}$$

$$\vec{q} = -\ddot{D}\vec{\epsilon} + \vec{v}c \tag{4}$$

$$\vec{\epsilon} = \vec{\nabla}c \tag{5}$$

where: $\vec{v} = \begin{Bmatrix} u \\ v \end{Bmatrix}$ $\vec{\nabla} = \begin{Bmatrix} \frac{\partial}{\partial x} \\ \frac{\partial}{\partial y} \end{Bmatrix}$ $\vec{n} = \begin{Bmatrix} n_x \\ n_y \end{Bmatrix}$ $\ddot{D} = \begin{bmatrix} D_{xx} & D_{xy} \\ D_{yx} & D_{yy} \end{bmatrix}$ $\vec{q} = \begin{Bmatrix} q_x \\ q_y \end{Bmatrix}$ $q_n = \vec{n}^T\vec{q}$

and: c is the field variable; f (source/sink) , h (applied flux) and g (applied c) are functions of x and y; n_x and n_y are the direction cosines of a unit normal directed away from the domain; u and v are the x and y velocities of the host medium; [D] is a symmetric tensor of material properties (eg. dispersion, permeability, diffusion) and t is time.

Weighted Residual Statement In a weighted residual statement we assume that c may be expressed as a summation of functions of x and y and some undetermined coefficient i.e. $c = \vec{\phi}\,\vec{C}$. A weighted residual functional, with appended least squares constraints, is then given as:

Carrying out the differentiation of the last two terms, one obtains:

$$\left\langle w, \left(\frac{\partial c}{\partial t} + \vec{\nabla}^T \vec{q} - f \right) \right\rangle_\Omega + \left\langle w, (-q_n + h) \right\rangle_{\Gamma_2} + \left\langle \frac{\partial}{\partial \vec{C}} \left[(-q_n + h) \left(\frac{\beta_2}{2} \right) (-q_n + h) \right] \right\rangle_{\Gamma_2} +$$

$$\left\langle \frac{\partial}{\partial \vec{C}} \left[(c - g) \left(\frac{\beta_1}{2} \right) (c - g) \right] \right\rangle_{\Gamma_1} = 0$$

$$\left\langle w, \left(\frac{\partial c}{\partial t} + \vec{\nabla}^T \vec{q} - f \right) \right\rangle_\Omega + \left\langle (w + w_2), (-q_n + h) \right\rangle_{\Gamma_2} + \left\langle w_1, (c - g) \right\rangle_{\Gamma_1} = 0 \qquad (6)$$

$$w_2 = -\frac{\partial q_n}{\partial \vec{C}} \beta_2 \qquad w_1 = \frac{\partial c}{\partial \vec{C}} \beta_1 \qquad (7)$$

where: w is a test function, β_1 *and* β_2 are penalty weights, $\langle *, * \rangle$ indicates integration over the appropriate domain or boundary. The discretized equations are then given as:

$$\vec{M} \frac{\partial \vec{C}}{\partial t} + \vec{K} \vec{C} = \vec{F} \qquad (8)$$

$$\vec{M} = \left\langle \vec{\phi}^T \vec{\phi} \right\rangle_\Omega \qquad (9)$$

$$\vec{K} = -\left\langle (\vec{\nabla}\vec{\phi})^T \vec{d} \vec{\phi} \right\rangle_\Omega + \left\langle (\vec{n}^T \vec{d} \vec{\phi})^T \beta_2 (\vec{n}^T \vec{d} \vec{\phi}) \right\rangle_{\Gamma_2} + \left\langle \vec{\phi}^T \beta_1 \vec{\phi} \right\rangle_{\Gamma_1} \qquad (10)$$

$$\vec{F} = \left\langle \vec{\phi}^T f \right\rangle_\Omega + \left\langle (-\vec{\phi} + \beta_2 \vec{n}^T \vec{d} \vec{\phi})^T h \right\rangle_{\Gamma_2} + \left\langle \vec{\phi}^T \beta_1 g \right\rangle_{\Gamma_1} \qquad (11)$$

$$w = \vec{\phi}^T, \quad (\vec{\nabla}w)^T = (\vec{\nabla}\vec{\phi})^T$$

$$\vec{q} = (-\vec{D}\vec{\nabla}c + \vec{v}c) = \vec{d} c = \vec{d} \vec{\phi} \vec{C}$$

$$\vec{d} = -\vec{D}\vec{\nabla} + \vec{v}, \quad q_n = \vec{n}^T \vec{q} = \vec{n}^T \vec{d} \vec{\phi} \vec{C}$$

$$\frac{\partial q_n}{\partial \vec{C}} = (\vec{n}^T \vec{d} \vec{\phi})^T, \quad \frac{\partial c}{\partial \vec{C}} = \vec{\phi}^T$$

EXAMPLES

Convergence To study convergence a steady state diffusion problem with an analytic solution was examined. The circular domain of Figure 2 had an imposed boundary concentration over the arc from 45 to 135 degrees (measured positive from the +x axis). On this portion of the boundary a cubic hill is defined with the concentration and its tangential derivative zero at the extremes and a value of 1 at the center (at 90 degrees). The tangential slope at 90 degrees is also zero.

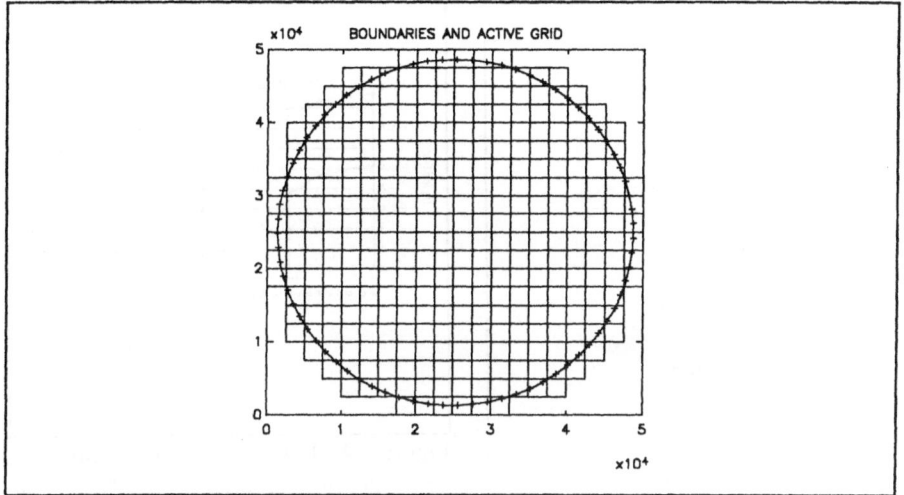

Figure 2 Computational and Geometric Mesh

This problem has an analytical solution given by:

$$c_e = \frac{1}{2\pi} \int_0^{2\pi} f(\acute{\theta}) \frac{a^2 - r^2}{a^2 - 2arCos(\acute{\theta} - \theta) + r^2} \, d\acute{\theta} \qquad (12)$$

where: a is the outer radius, r and θ are the polar coordinates of a point and f() is the equation which describes the imposed c on the boundary. Figure 3 illustrates the resulting concentration surface. Two separate error norms were computed, one by using the boundary points as sampling points and the other by using the nodes of the computational grid that fall inside the domain. Theses norms were computed for nx=ny = 6,8,10,12,14,16,18,20,22,24,32. The slope of the log of the norm vs. the log of the normalized grid spacing (see Figure 4) provide an estimate of the convergence order. In all of the analyses the spacing of the boundary points was 3000 m except for the last analysis with nx=ny=32. The boundary spacing for this run was reduced to 2400 m. This was necessary because nx=32 resulted in numerous computational elements on the boundary that did not have boundary points, ie. $\Delta S > 2\Delta X$. The slope of these curves indicate that the convergence rate is between 1 and 2. Second order accuracy appears to be achievable if the spacing of the boundary points is adjusted as noted above.

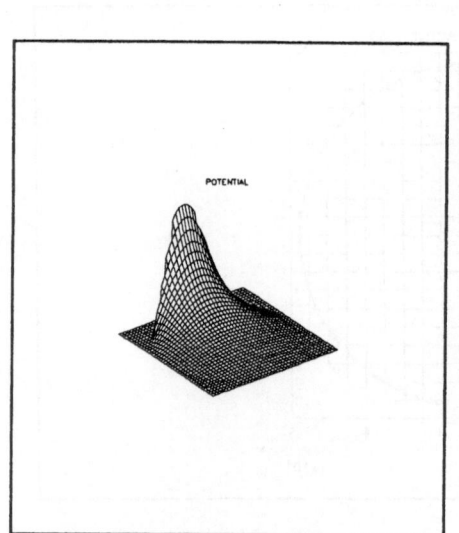

Figure 3 Concentration Surface

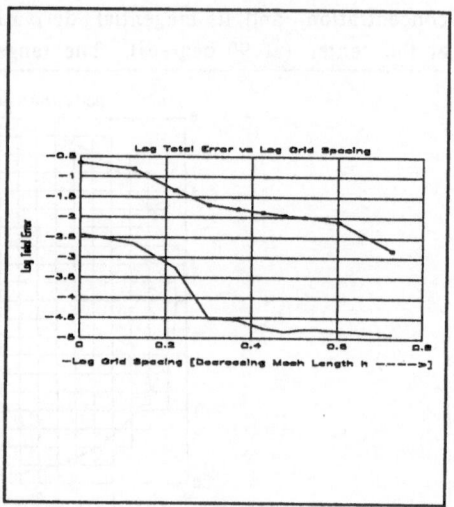

Figure 4 Log plot for computing convergence order. o = Norm at interior computational nodes, --Norm at boundary points.

Choice of Penalty Factor The choice of the penalty factors determines the degree to which the boundary conditions are enforced. As a general rule we use values that result in contributions to the diagonal of the system matrix that are Kfac times the contributions from the non penalty terms. Kfac itself may vary significantly for different problems. In general the total stiffness matrix is composed of the two parts $K = KG + Kfac*(max(diag(KG)))*KP$, where KG is from the differential equation and KP is from the penalty integrations. In a steady state diffusion problem KG is singular (because no boundary constraints are applied). The sensitivity of KG is such that only very small value of Kfac will yield a total system matrix that is singular (in the first problem of the EXAMPLE section Kfac=.01 was used). As Kfac becomes large it may over power KG. While a solution can be found (ie. KP is nonsingular) it merely satisfies the boundary conditions but not the differential equation. This can yield spurious solutions. The present approach is to use a Kfac that is as small as possible and still obtain acceptable results for the prescription of type 1 boundary conditions.

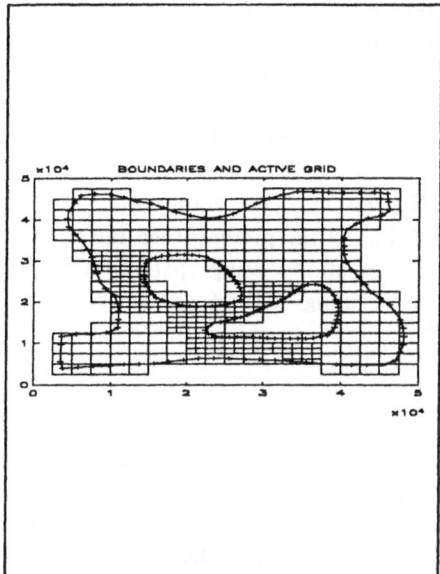

Figure 5 Grid with Refinement

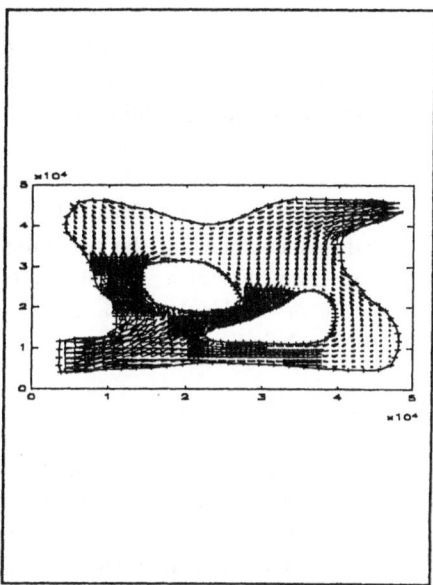

Figure 6 Flux Vectors

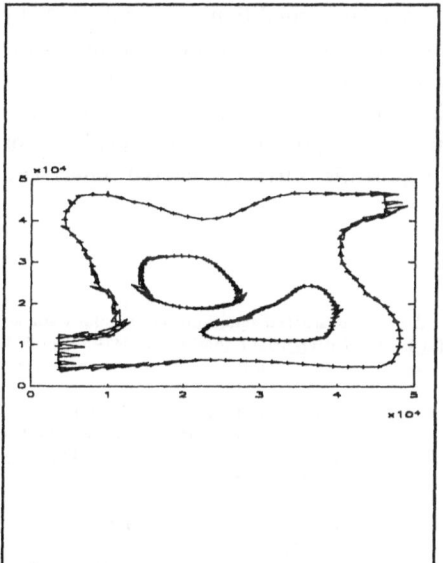

Figure 7 Boundary point flux vectors with no flux penalty.

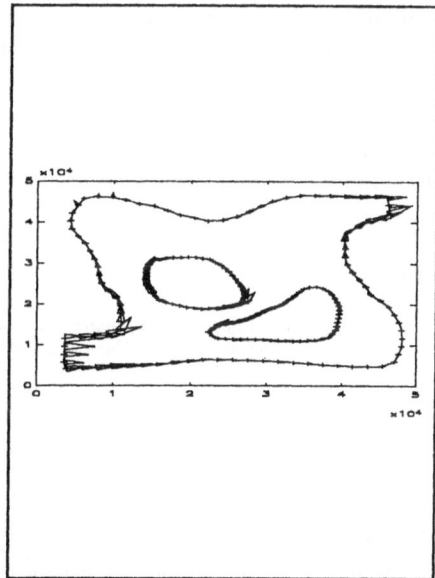

Figure 8 Boundary point flux vectors with flux penalty = 10000*max(DIAG(KG)).

Figures 5 and 6 illustrate the solution of a potential problem with both Dirichet and Neumann boundary conditions. A head of 0 was imposed at the two upper corners and a head of 1 at the lower right entrant. A zero normal flux condition was imposed on the remaining exterior and interior boundaries. Figure 7 shows the boundary flux vectors for $\beta_2 = 0$. At several points where a zero normal flux is imposed the flux vectors are not tangent to the boundary. Figure 8 shows the results for $\beta_2 = 10000*max(DIAG(KG))$. In this case the zero normal flux is achieved without altering the remaining flux field. At the present time the penalty terms are also the same for all points. Experimentation using spatially variable penalty factors and automated procedures for defining Kfac based on condition number are being pursued.

CONCLUSIONS

The solution of a boundary value problem, defined over an irregular shaped domain is solved using a completely orthogonal set of finite elements for the computational mesh. A separate geometric mesh need only describe the boundaries of the domain. The use of completely orthogonal elements simplifies the process of developing grid meshes. The method (GLSP) is a combination of the Galerkin MWR and the least squares penalty method. The Galerkin method is applied to the domain operator and the boundary conditions are enforced by least squares. The method is shown to be accurate and convergent. The proposed method can be applied to more complex problems including time dependent vector equations in two and three spatial dimensional . Further developments are presently under way to examine the method as applied to the equations of elasticity and moving boundary problems (eg. phreatic surface, material interfaces and tidal flat problems).

RERERENCES

1 J.P. Laible and G.F. Pinder, 'Least squares collocation solution of differential equations on irregularly shaped domains using orthogonal meshes', Numerical Methods for Partial Differential Equations, 5, 347–361 (1989).
2 J.P. Laible and G.F. Pinder, 'Solution of the shallow water equations by least squares collocation', Water Resources   Research, 29, 445–455 (1993).
3 J.N. Reddy, 'An introduction to the finite element method', McGraw–Hill, 2nd Ed., 488–494 (1993).
4 J.N. Reddy, 'On the finite element method with penalty for incompressible fluid flow problems', in The Mathematics of Finite   Elements and Applications III., ed. J.R. Whitman, Academic Press, New York, (1979).
5 P.P. Lynn and S.K. Arya, 'Finite element formulation by weighted discrete least squares method', Int. J. Num. Methods in   Eng., 8 71–90 (1974).
6 O.C. Zienkiewicz, D.R.J. Owen and K.N. Lee, 'Least squares finite elements for elasto–static problems–use of reduced   integration', Int. J. Num. Methods in Eng., 8, 341–346 (1974).
7 T.J.R. Hughes, L.P. Franca and B.M. Hulbert, 'A new finite element formulation for computational fluid dynamics: VIII.   The Galerkin/least–squares method for advective–diffusive equations', Computer Methods in Applied Mech. and Engr., 73,   173–189 (1989).

SIMULATION OF ALTERNATE WETTING AND DRYING IN UNSATURATED FLOW ZONES

J. LETHA and K. ELANGO
Department of Civil Engineering
Indian Institute of Technology, Madras - 600 036
India.

ABSTRACT

Alternate wetting and drying of a soil medium under unsaturated conditions is a natural process that occurs in many cases. A finite element model based on Richards equation for the simulation of this process is reported. It provides a bank of alternative formulations for soil moisture characteristics to suit a range of soil types and soil moisture status. The model considers also the hysteretic aspect of soils through main, primary and scanning wetting and drying curves based on the reversal point and the assumption of similarity of these curves to the main wetting and drying curves.

Two specific cases are taken up for studying the application of this model. The first case reported already deals with subirrigation and drainage processes occurring independently. The second case simulates the functioning of a combined drainage-cum-subirrigation pipe system going sequentially through the wetting and drying phases under unsaturated conditions. The simulation results are used to assess the effect of hysteresis in terms of time and space scales.

1   INTRODUCTION

.Renewed   interest has been   evinced in the recent past on the hydrological processes taking place in the vadose zone. The inherent complexity of these processes is mainly due to the fact that the zone is unsaturated. The influence of unsaturated zone is important in many practical cases like aquifer interacting with a stream, surface-irrigation system with evapotranspiration cycles, furrow irrigation through ditches and subsurface drains and in a system of combined drainage and subirrigation.. In all such cases the soil

A. Peters et al. (eds.), Computational Methods in Water Resources X, 75–82.
© 1994 Kluwer Academic Publishers. Printed in the Netherlands.

medium undergoes repeated wetting and drying cycles. The analysis of this process becomes further complicated , if the hysteresis in the behaviour of the soils is significant.

Studies on the concept of drainage and subirrigation under unsaturated conditions have been reported by Fouss et. al [1989], Shirmohammadi et. al[1991], Skaggs[1991]. DRAINMOD [1982] is a popular model to deal with the design of subsurface drainage and controlled drainage. In most of these earlier studies, drainage and subirrigation are considered as separate independent events and hysteresis in soil behaviour is neglected.

The present paper deals with simulation of cyclic wetting and drying process in unsaturated soils. The hysteretic aspect of soils is considered in detail. A finite element model (written in C) based on Richards equation is developed for dealing with the processes. The model is applied to two practical situations. The first one is subirrigation and drainage through parallel ditches. This case has been studied already in detail using a finite difference model by Tang and Skaggs(1977). Their results are used to validate the present model. As a second application , the drainage-cum-subirrigation carried out sequentially through a subsurface pipe is taken up. Results are assessed in terms of time and space scales to bring out the significance of acknowledging the hysteresis of soils.

2   MAIN FEATURES OF THE FINITE ELEMENT MODEL

The basis of finite element model developed in the study is the Richards equation which may be written in the following form.:.

$$C(h) \frac{\partial h}{\partial t} = \frac{\partial}{\partial x}\left[k(h)\frac{\partial h}{\partial x}\right] + \frac{\partial}{\partial z}\left[k(h)\frac{\partial h}{\partial z}\right] + \frac{\partial k(h)}{\partial z} \qquad\qquad ....(1)$$

wherein, h   -   soil water pressure head (L)
         C(h)   - $d\theta/dh$ - soil moisture capacity (1/L)
         θ -   soil moisture content
         k(h)   -   unsaturated hydraulic conductivity (L/T)
         t   -   time (T)
         x,z   -   space coordinates , x measured horizontally towards right,
                            and z measured vertically upwards.(L)

2.1 Soil moisture characteristic equations:

The fluid flow through an unsaturated medium is thus influenced by the capacitance, C, conductivity, k, as well as the moisture content and pressure head fields. Equation sets, one relating soil moisture content, θ and soil water pressure head,.h and the other relating hydraulic conductivity, k to either θ or h are generally termed soil moisture characteristic equations. Several analytical expressions are quoted in the literature. Brooks and Corey[1964] and Van Genuchten[1980] forms are the most popular.

The present numerical model is designed to be quite versatile by providing a choice from four combinations of h-θ and k-θ relationships.. An earlier study has identified combinations of characteristic equations that are better suited to specific soil types and soil moisture ranges. [Letha and Elango 1993]. Based on this, Van Genuchten form for h-θ relation and Brooks and Corey form for k-θ relation are used in the present study. For cases in which data are lacking for k-θ relation , Van Genuchten form of k-θ relation is used.

2.2 Hysteresis in soils:

Many soils exhibit hysteresis in the h-θ characteristic relationship. On the other hand hysteresis in k-θ relation is usually negligible.[Kool and Parker,1987]. Thus the features of an h-θ relation which is not unique are expressed through main wetting , main drying and different scanning curves (primary and higher order).

Numerical simulation acknowledging the hysteretic behaviour of soils requires the specification of these various curves. Kool and Parker have described in detail a procedure to account for this hysteretic aspect. The evolution -path of primary and secondary scanning curves are derived based on the reversal point and the assumption of similarity of these curves to the main curves for the wetting and drying phases. This procedure has been adopted in the present numerical model. The program logic is designed in a way that .whenever the soil medium undergoes a phase change it is capable of picking out the appropriate scanning curve from the family of these primary and secondary curves .

2. 3 Computer code:

The computer code is written in C language . Many special features of C language (dynamic memory allocation and release , acquiring memory from heap etc.) are exploited in this regard.

3 APPLICATION OF THE MODEL

The model is employed in the simulation of situations of practical interest. The first one is a case of drainage and subirrigation through parallel ditches as already reported by Tang and Skaggs[1977]. This case for which results are available is used for validating the present model.

3.1 Validation:

Herein the problem along with the boundary conditions for subirrigation and drainage presented by Tang and Skaggs is adopted. Typical results from simulation of independent drainage and subirrigation phases are compared with those reported by Tang and Skaggs in Fig 1.(a) and 1(b) . As seen from these figures, the match is quite good. The seepage face occurring in the drainage phase, disappears within one hour. Hence along the left boundary (at x =0.0m) the pressure is atmospheric for a height of 0.3m which is the water level in the ditch maintained constant for this phase. . Figure 1(b) is plotted for a subirrigation phase at x=8.5m, which is the middle point between two parallel ditches. Since the control of pressure head is in the ditch ,the pressure distributions at x=8.5m   for various time instants do not have a common fulcrum as in the other case.

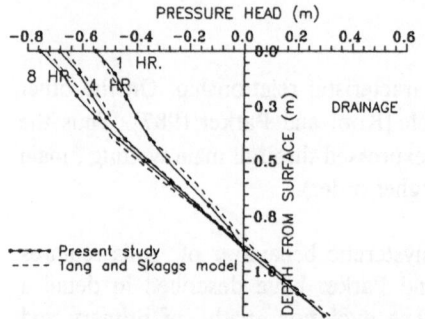

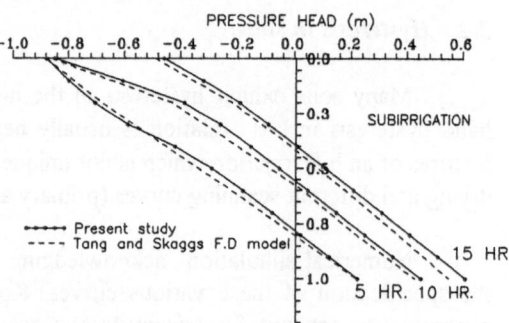

Fig. 1(a). Pressure distribution plots at
x = 0.0 for drainage phase..

Fig. 1(b). Pressure head plots at x= 8.5 m
for subirrigation phase

3.2 Application:

The case of cyclic operation of a subsurface pipe system under alternating drainage and subirrigation modes is considered for application of the numerical model.

Drainage and subirrigation through a subsurface pipe:

. The impermeable and symmetric boundaries for a system serving both drainage and subirrigation through a subsurface pipe shown in the schematic sketch are listed from(2a) to (2e).

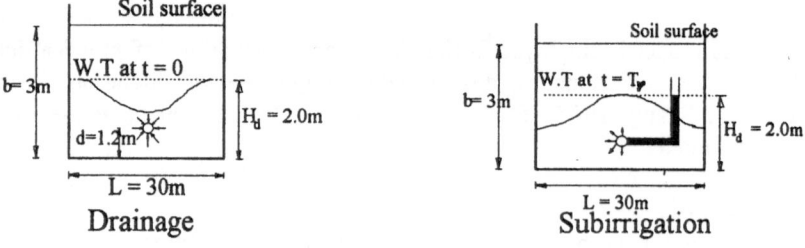

$$H = H_d, \qquad 0 \le x \le L, \qquad 0 \le z \le b, \qquad t = 0 \qquad ...(2a)$$

$$\frac{\partial H}{\partial x} = 0, \qquad x = 0, \qquad 0 \le z \le b, \qquad t > 0 \qquad ...(2b)$$

$$\frac{\partial H}{\partial z} = 0, \qquad 0 \le x \le L, \qquad z = 0, b, \qquad t > 0 \qquad ...(2c)$$

$$H = d + r_e, \qquad x = L/2, \qquad z = d, \qquad T_r > t > 0 \qquad ...(2d)$$
$$\text{for } r = 1, 3, 5 ....$$

$$H = H_d, \qquad x = L/2, \qquad z = d, \qquad T_r > t > 0 \qquad ...(2e)$$
$$\text{for } r = 2, 4, 6....$$

wherein r_e is the effective radius . The boundary conditions (2d) and (2e) represent the drainage and subirrigation modes respectively. These are invoked alternately depending on the values of r which is the mode-change index. In the present system, the change in mode is effected, based on trigger values of moisture content at a specified control point in the medium. The trigger values adopted herein are twenty percent of (θ_s - θ_r) for change over from drainage to subirrigation mode and ninety five percent of the same for reverting back to the drainage mode. Herein, θ_s is the moisture content at saturation and θ_r is the residual moisture content , with the difference indicating the maximum crop available water. The term T_r represents the transition time from one mode to the other. This is a parameter whose value gets decided through the progress of the simulation run. The control point may represent a lumped root zone in a practical irrigation setting. In the present version of formulation evapotransipiration losses are not yet considered.

3.2.1. Simulation:

For this part of the study a suitable soil medium and its characteristics are to be first identified. For this purpose, the sand medium used in our laboratory experiments on unsaturated flow is considered. The main drying curve for this soil medium is determined using a pressure plate apparatus. Since the main wetting curve could not be determined experimentally, the suggestion by Kool and Parker [1987] for generation of main wetting curve in the absence of experimental data is adopted herein. Accordingly, the ratio of the Van Genuchten parameters for the wetting and drying curves (α_w/α_d) is assumed equal to two.

The geometry of the subsurface pipe system adopted here is the one considered by Fipps et. al[1991]. The spacing between the drains, L , the depth of soil, b , the depth of subsurface pipe to the impervious datum, d are assumed values indicated in the definition sketch. The control point representing the lumped root zone as described above is assumed to be at 5m from the drain and 0.6 m above the drain. The effective radius is retained as 0.01m. Initial distribution for moisture contents for the saturated portion were derived using the main wet curve and the unsaturated portion from the main dry curve.

A Van Genuchten h-θ form was fitted to the measured data of the main drying curve and parameters α_d , m, n are optimally evaluated. The parameters for the wetting curve generated are indicated in Table 1. The fitted curves for both phases is shown in Fig 2. The saturated hydraulic conductivity is estimated to be 0.294 m/hr . The Van Genuchten k-θ form is chosen for k-θ relation.

The simulation runs were carried out for three complete cycles, each comprising of one drainage mode and one subirrigation mode. Separate runs were made first neglecting hysteresis and then considering hysteresis. The h-θ curve used for non-hysteretic simulation is the main dry curve. Hysteresis in k-θ relation is neglected in both cases.

Table 1. Optimal parameter values fitted for Van Genuchten h-θ form

Main drying Curve (measured)	α_d = 3.66 /m n = 4.75 m = 0.7894 θ_s = 0.36 θ_r = 0.012
Main wetting curve (generated)	α_d = 7.32 /m n = 4.75 m = 0.7894 θ_s = 0.36 θ_r = 0.012

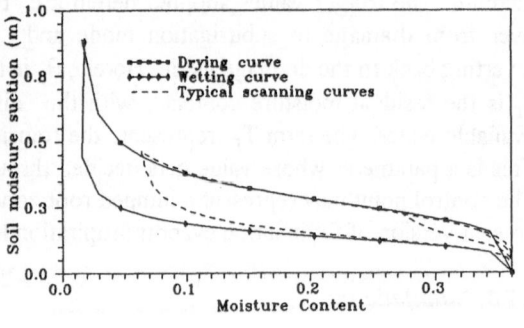

Fig. 2. Fitted drying and generated wetting curve with some typical scanning curves

3.2.2   Results from Simulation:

Typical results are furnished from Fig. 3 to Fig. 5 The role of hysteresis can be assessed by comparing the plots. From the results, a narrow drying front formation is noticed. Many secondary scanning curves tend to merge with the main dry curve for this chosen sand medium. This is seen from Fig. 2, where some typical scanning curves which are used somewhere through simulation are plotted. Spatial plots of pressure head at the pipe and midway between the pipes is shown in Fig 3a and Fig.3b. As is clear from the Fig. 3a, in a hysteretic simulation the extent of unsaturated zone is deeper than the other one with the water table location at the pipe about 1.5m from the surface for the hysteretic case and 1.2m for the non-hysteretic case.

Fig.4 gives the temporal plot of moisture contents at the control point. over time for both cases. The time scales involved in the two cases are different. The hysteretic simulation covered 300hrs to span the three cycles, while it is 278 hrs for the non-hysteretic simulation. Influx and efflux rates also show variation between both cases which is evident from Fig. 5. Hysteretic simulation always shows a larger flow rate than the other throughout the process. Thus the results indicate the effect of hysteresis on the extent of unsaturated zone and the influx and efflux rates in a drainage-cum-subirrigation system can be significant.

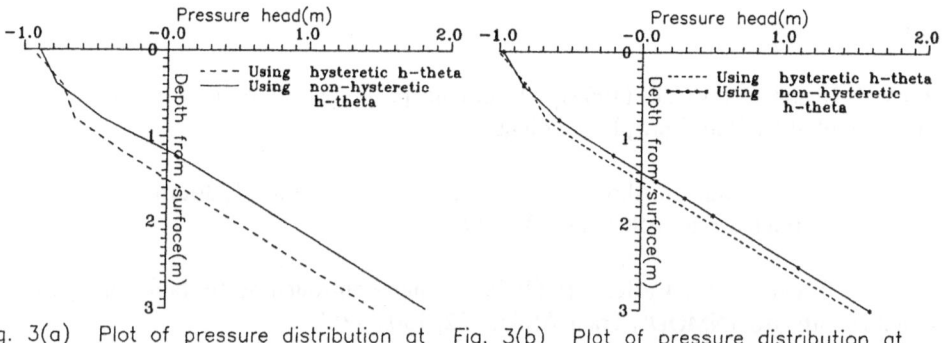

Fig. 3(a) Plot of pressure distribution at the pipe at 150 hr after start of simulation.

Fig. 3(b) Plot of pressure distribution at midway between pipes at 150 hr after start of simulation.

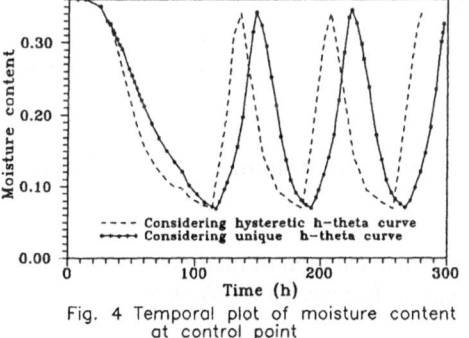

Fig. 4 Temporal plot of moisture content at control point

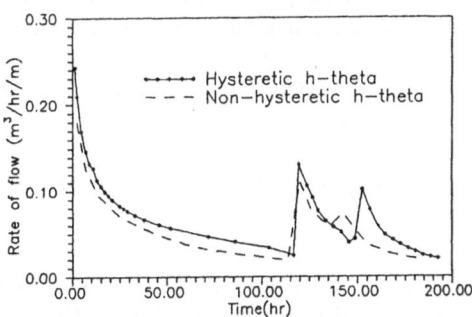

Fig.5 Rate of flow over time

4 CONCLUSION

The numerical model is validated satisfactorily with the available results The model is shown effective in dealing with the case of a cyclic operation of drainage-cum-subirrigation of a subsurface pipe system. It helps in mapping the unsaturated domain with the space-time variation of pressure and moisture content fields. It is found that when the wetting and drying curves differ substantially, space time picture of moisture variation differs significantly due to hysteresis. In general, the efflux, influx rates for a combined system is higher when hysteresis in soil is accounted. Thus it may be necessary to consider hysteresis in problems of this nature.

REFERENCES

Brooks,R.H and Corey,C.T. (1964) " Hydraulic properties of porous media", Hydrol. Paper 3. Colorado State Univ., Fort Collins.

Fipps,G. (1991) " Simple methods for predicting flow to drains ", Journal of Irrigation and Drainage Engineering, ASCE,117, 881-896.

Fouss,J.L, Rogers,J.S and Carter,C.E. (1989) " Sump-controlled water table management predicted with DRAINMOD", Trans.ASAE , 32, 1303-1308.

Kool,J.B and Parker,J.C. (1987) " Development of closed-form expressions for hysteretic soil hydraulic properties", Water Resources Research, 23, 105-114.

Letha,J and Elango,K. (1993) " Simulation of mildly unsaturated flow", Journal of Hydrology(accepted for publication).

Shirmohammadi,A, Thomas,D.L and Smith,M.C. (1991) " Drainage-subirrigation design for pelham loamy sand", Trans.ASAE, 34,73-80.

Skaggs,R.W.(1982) " Field evaluation of a water management simulation model", Trans.ASAE, 25, 666-674.

Skaggs,R.W.(1991) "Modeling water table response to subirrigation and drainage", Trans.ASAE, 34, 169-175.

Tang,Y.K and Skagggs,R.W. (1977) " Experimental evaluation of theoretical solutions for subsurface drainage and irrigation", Water Resources Research,13, 957-965.

Van Genuchten, M.Th.(1980) " A closed form equation for predicting the hydraulic conductivity of unsaturated soils", Soil.Sci.Soc.Am.J.,44,892-898.

COMBINING THE ANALYTICAL AND FINITE ELEMENT MODELS OF THE RIVER-GROUNDWATER INTERACTION

M. NAWALANY
Institute of Environmental Engineering Systems
Warsaw University of Technology
Nowowiejska 20, 00-653 Warsaw
Poland

It is commonly accepted assumption that the interaction between surface waters and groundwater can be represented by a boundary condition of the third type along the line of physical contact between the river and the aquifer. However, for the two-dimensional horizontal approximation of groundwater flow water balance (conservation of mass) can be considerably in error even of order 70%. Although the three-dimensional groundwater flow models are not corrupted by errors of this kind their use on regional scale is impractical from computational point of view. Combining the river flow models with two-dimensional horizontal models for regional groundwater flow by using the three-dimensional analytical elements is expected to improve the operational efficiency of hydrological models considerably. Hence, it is proposed in the paper to construct a buffer zone of such elements between a river bed and the adjacent aquifer in order to maintain the water balance of the whole system.

INTRODUCTION

River-aquifer interaction (r-a-i) always poses problems when modelled as a part of some larger hydrological systems. Especially when incorporated into the two-dimensional horizontal flow models of groundwater systems (r-a-i) may not be mimicked accurately thus resulting in an unacceptable inaccuracy of the global mass balance. Rivers are modelled either as linear source/sink terms of flow equations or as fully penetrating internal boundaries. In the latter case the third type boundary conditions are assumed along the river bed.

It is the intention of this paper to check whether (and if 'yes' how) the (r-a-i) can be applied within a framework of the horizontal flow models. For the simple case of a river that recharges the adjacent aquifer two analytical models - the two-dimensional horizontal and the three-dimensional one - are presented. They are compared with each other in terms of the total seepage within the series of computer experiments. From the calculations important recommendations on the modelling aspects of (r-a-i) are drawn. In particular, hybrid analytical-numerical model of the (r-a-i) is suggested as a remedy for inaccuracies generated by two-dimensional horizontal flow models.

A. Peters et al. (eds.), Computational Methods in Water Resources X, 83–90.
© 1994 Kluwer Academic Publishers. Printed in the Netherlands.

ASSUMPTIONS

Throughout this paper a river bed is assumed to have a rectangular cross-section. It is located in the middle of the rectangular, homogeneous and confined aquifer and it penetrates the aquifer only partly. Also a constant water table position is assumed in the river. At both ends of the aquifer constant piezometric heads - ϕ'_o and ϕ''_o - are specified. They are assumed to be equal to each other and lesser than the piezometric head in the river, i.e. $\phi'_o = \phi''_o = \phi_o < \phi_r$. This implies that interaction between the river and the aquifer is symmetric in space. Physically the interaction is a seepage through the sediments cumulated at the river's bottom and banks. The seepage is proportional to the difference between the piezometric heads in the river and the aquifer and reprocical to the flow resistivity through the sediments - c. The resistivity c is also assumed homogeneous. Because of a geometric symmetry of the case, only one half of the (r-a-i) system needs to be considered - Figure 1.

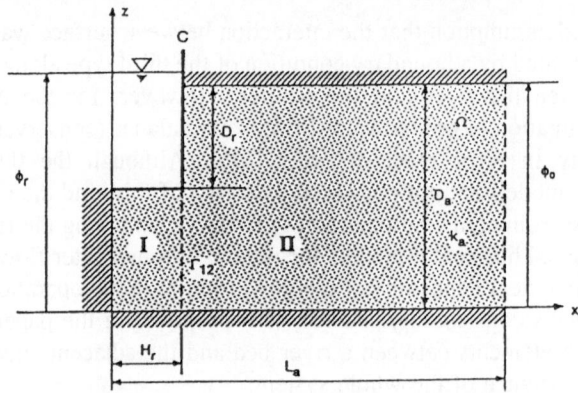

Fig. 1. River-aquifer system (simplified)

TWO-DIMENSIONAL MODEL OF THE RIVER-AQUIFER INTERACTION

For this case a piezometric head in the aquifer is considered depth-averaged. If ϕ^* is a value of piezometric head at $x = H_r$ the total outflow Q_H can be calculated from the following formula:

$$Q_H = T_a \cdot \frac{\phi^* - \phi_o}{L_a - H_r} \tag{1}$$

where $T_a = D_a \cdot k_a$.

Analytical horizontal flow model is derived by considering the 2D-flow equation with the source term (seepage flux) q_s which for $x \in [0, H_r]$ is equal to

$$q_s(x) = \frac{\phi_r - \phi}{c} \tag{2}$$

Steady state flow equation for this region is therefore

$$T_r \frac{\partial \phi}{\partial x^2} + \frac{\phi_r - \phi}{c} = 0 \tag{3}$$

where $T_r = (D_a - D_r) \cdot k_a$. \hfill (4)

Boundary conditions for the region are follows:

$$\frac{\partial \phi}{\partial x}\bigg|_{x=0} = 0, \quad \phi|_{x=H_r} = \phi^*. \tag{5}$$

By introducing notation: $\quad \tilde{\lambda}^2 = T_r \cdot c$ $\tag{6}$

a particular solution to equation (3) can be written as:

$$\phi(x) = \phi_r - (\phi_r - \phi^*) \frac{\cosh(x/\tilde{\lambda})}{\cosh(H_r/\tilde{\lambda})}. \tag{7}$$

where

$$\phi^* = \frac{\phi_o + \phi_r \psi}{1 + \psi}, \qquad \psi = \frac{L_a - H_r}{T_a} \left\{ \frac{T_r}{\tilde{\lambda}} \text{tgh}(H_r/\tilde{\lambda}) + \frac{D_r}{c} \right\} \tag{8-9}$$

$$T_r = (1 - p) T_a, \quad p = D_r/D_a, \quad \lambda^2 = T_a c. \tag{10}$$

Formula (7) together with formulae (8)-(10) define the sought solution to the 2D-horizontal flow equation that include the river-aquifer interaction. For the rest of the aquifer a piezometric head changes linearly.

It can be observed that the total seepage ($=Q_H$) calculated from the 2D-model is always larger than the exact Q which can possibly be calculated for the 3D situations. This can be deduced by even without deriving the formulae for the 3D-model.

THREE-DIMENSIONAL MODEL OF THE RIVER-AQUIFER INTERACTION

In order to solve the groundwater flow problem in three dimensions we subdivide a flow domain Ω into two parts - I and II - (see Figure 1) and solve two separate flow problems for the two parts of Ω. After that we merge the solutions along the common boundary Γ_{12} in terms of piezometric heads and normal fluxes, i.e.:

$$\phi^I(x,z) = \phi^{II}(x,z) \quad and \quad \frac{\partial \phi^I(x,z)}{\partial x} = \frac{\partial \phi^{II}(x,z)}{\partial x} \tag{11-12}$$

for $x = H_r$, for $z \in [0, D_a - D_r]$.

Solution for region I
Piezometric head in region I, $\phi^I(x,z)$, (noted $\phi(x,z)$) must satisfy the flow equation:

$$\frac{\partial^2 \phi}{\partial x^2} + \frac{\partial^2 \phi}{\partial z^2} = 0 \quad for \quad \begin{cases} x \in [0, H_r] \\ z \in [0, D_a - D_r] \end{cases} \tag{13}$$

and the following boundary conditions:

$$
\left\{
\begin{array}{llll}
\Gamma_1: & \left.\dfrac{\partial \phi}{\partial x}\right|_{x=0} = 0 & \text{for } z \in [0, D_a - D_r] \\[3mm]
\Gamma_2: & -k_a \left.\dfrac{\partial \phi}{\partial z}\right|_{z=D_a-D_r} = \left.\dfrac{\phi - \phi_r}{c}\right|_{z=D_a-D_r} & \text{for } x \in [0, H_r] \\[3mm]
\Gamma_{12}: & \phi|_{x=H_r} = \phi^{II}|_{x=H_r} & \text{for } z \in [0, D_a - D_r] \\[3mm]
\Gamma_4: & \left.\dfrac{\partial \phi}{\partial z}\right|_{z=0} = 0 & \text{for } x \in [0, H_r]
\end{array}
\right.
\tag{14}
$$

The solution for region I can be expressed as:

$$
\phi(x,z) = \phi_r + \sum_{k=1}^{\infty} A_k \cosh(\lambda_k x) \cos(\lambda_k z) \quad \text{for } 0 < x \le H_r, \;\; 0 < z \le D_a - D_r.
\tag{15}
$$

where λ_k $(k=1,2,\ldots)$ - satisfy the following equation:

$$
tg[\lambda_k (D_a - D_r)] = 1/k_a \cdot c
\tag{16}
$$

where A_k $(k=1,2,\ldots)$ are unknown constants that can be calculated from merging conditions (11-12).

Solution for region II

Piezometric head $\tilde{\phi}^{II} = \phi^{II} - \phi_r$ (abbr. hereafter as $\tilde{\phi}$) must satisfy the flow equation:

$$
\frac{\partial \tilde{\phi}}{\partial x^2} + \frac{\partial^2 \tilde{\phi}}{\partial z^2} = 0 \quad \text{for }
\left\{
\begin{array}{l}
x \in [H_r, L_a] \\[2mm]
z \in [0, D_a]
\end{array}
\right.
\tag{17}
$$

mering conditions (11-12) and

boundary conditions:

$$
\begin{cases}
\Gamma_{12}'': & \left.\dfrac{\partial\tilde{\phi}}{\partial x}\right|_{x=H_r} = \left.\dfrac{\tilde{\phi}}{\chi^2}\right|_{x=H_r} & \text{for } z\in[D_a-D_r,D_a] \\[3mm]
\Gamma_2: & \left.\dfrac{\partial\tilde{\phi}}{\partial z}\right|_{z=D_a} = 0 & \text{for } x\in[H_r,L_a] \\[3mm]
\Gamma_3: & \tilde{\phi}\big|_{x=L_a} = \phi_o-\phi_r = \tilde{\phi}_o & \text{for } z\in[0,D_a] \\[3mm]
\Gamma_4: & \left.\dfrac{\partial\tilde{\phi}}{\partial z}\right|_{z=0} = 0 & \text{for } x\in[H_r,L_a]
\end{cases}
\tag{18-21}
$$

The solution in region II can be expressed as:

$$
\tilde{\phi}(x,z) = \sum_{l=1}^{\infty} D_l\sinh[\,\mu_l(L_a-x)]\cos\mu_l z + \tilde{\phi}_o - \beta^*(L_a-x)
\tag{22}
$$

where $\mu_l = l\pi/D_a$; $(l=1,2,...)$.

Now merging conditions (11-12) between regions I and II can be written as follows:

$$
\begin{cases}
\displaystyle\sum_{k=1}^{\infty} A_k\cosh(\lambda_k H_r)\cos(\lambda_k z) = \sum_{l=1}^{\infty} D_l\sinh[\mu_l(L_a-H_r)]\cos(\mu_l z)+L(H_r) & (23) \\[4mm]
\displaystyle\sum_{k=1}^{\infty} A_k\lambda_k\sinh(\lambda_k H_r)\cos(\lambda_k z) = -\sum_{l=1}^{\infty} D_l\mu_l\cosh[\mu_l(L_a-H_r)]\cos(\mu_l z)+\beta^* & (24)
\end{cases}
$$

for all $z \in [0,D_a-D_r]$.
where

$$
\beta^* = \sum_{k=1}^{\infty} A_k\frac{\chi^2}{M}\sin[\lambda_k(D_a-D_r)]\sinh[\lambda_{kH_r}] \; - \sum_{l=1}^{\infty} D_l\frac{1}{M}\sinh[\mu_l(L_a-H_r)]c_l + \phi^*
\tag{25}
$$

$$
\phi^* = \frac{\tilde{\phi}_o D_r}{M} \quad M = D_a\chi^2 + (L_a-H_r)D_r. \quad \chi^2 = k_a\cdot c
\tag{26}
$$

$$
c_i = \int_0^{D_a-D_r} \cos\mu_i z\,dz = \frac{1}{\mu_i}\sin[\mu_i(D_a-D_r)], \quad (i=1,2...).
\tag{27}
$$

After substituting (25-27) to (23-24) we obtain the following set of algebraic equations:

$$
\begin{bmatrix}
A_{11} & \vdots & A_{12} \\
\hdashline
A_{21} & \vdots & A_{22}
\end{bmatrix}
\begin{bmatrix}
A_1 \\
A_2 \\
\vdots \\
\hdashline
D_1 \\
D_2 \\
\vdots
\end{bmatrix}
=
\begin{bmatrix}
\underline{b}_1 \\
\hdashline
\underline{b}_2
\end{bmatrix}
\tag{28}
$$

where for $(i,j,k = 1,2,...)$

$$
\begin{cases}
A_{11}(i,k) = \cosh(\lambda_k H_r)\gamma_{ki} + \dfrac{\chi^2(L_a-H_r)}{M}\sinh[\lambda_k H_r]\sin[\lambda_k(D_a-D_r)]c_i \\[2ex]
A_{12}(i,k) = -\sinh[\mu_k(L_a-H_r)]\left\{ verphi_{ki} + \dfrac{(L_a-H_r)}{M}c_k c_i \right\} \\[2ex]
A_{21}(i,k) = \sinh(\lambda_k H_r)\left\{ \lambda_k\varphi_{ki} - \dfrac{\chi^2}{M}\sin[\lambda_k(D_a-D_r)]c_i \right\} \\[2ex]
A_{22}(i,k) = \mu_k\cosh[\mu_k(L_a-H_r)]\varphi_{ki} + \dfrac{1}{M}\sinh[\mu_k(L_a-H_r)]c_k c_i
\end{cases}
\tag{29}
$$

$$
\varphi_{ij} = \frac{\sin[|\mu_i-\mu_j|(D_a-D_r)]}{2|\mu_i-\mu_j|} + \frac{\sin[(\mu_i+\mu_j)(D_a-D_r)]}{2(\mu_i+\mu_j)}
\tag{30}
$$

$$
\gamma_{kj} = \frac{\sin[|\lambda_k-\mu_j|(D_a-D_r)]}{2|\lambda_k-\mu_j|} + \frac{\sin[(\lambda_k+\mu_j)(D_a-D_r)]}{2(\lambda_k+\mu_j)}
\tag{31}
$$

and

$$
\begin{cases}
b_1(i) = [\bar{\phi}_o - \phi^*(L_a-H_r)]c_i \\[1ex]
b_2(i) = \phi^* c_i \ , \quad (i=1,2,...).
\end{cases}
\tag{32}
$$

RESULTS

A number of computer experiments have been carried out to find out how the ratio between the exact total Q_{3D} - formula (26) - and the two-dimensional total flow Q_{2D} -

formula (7) - depends on hydraulic and geometric parameters of the river-aquifer system. The results of experiments have been illustrated on the following fig. 2 and 3.

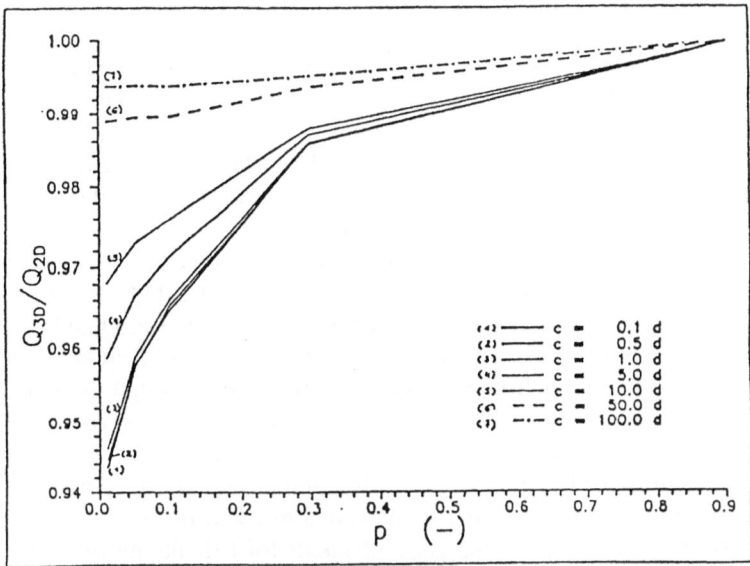

Fig. 2.   Shows Q_{3D}/Q_{2D} as a function of penetration (p) of the river into aquifer for a number of resistivity values c and for $H_r = 5.0m$, $(k_a) = 10.0m/d$

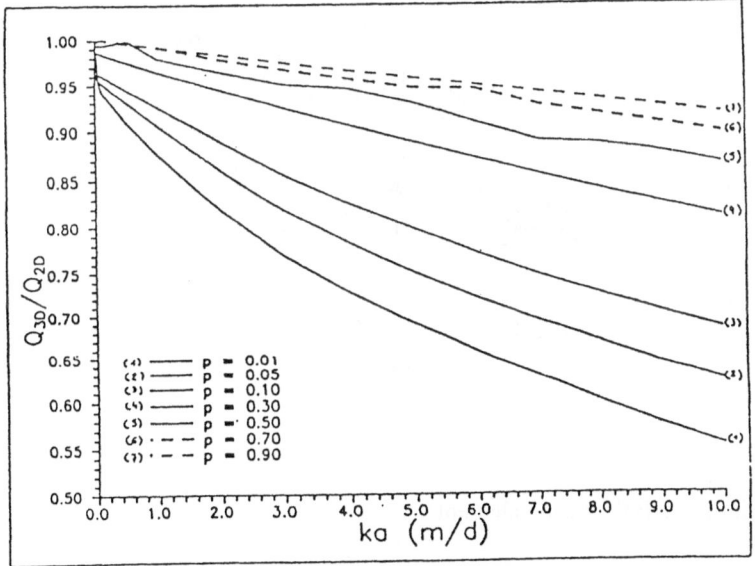

Fig. 3.   Shows Q_{3D}/Q_{2D} as a function of hydraulic conductivity of the aquifer (k_a) for a number of river penetration values (p) and for $H_r = 5.0m$, $c = 1.0d$

CONCLUSIONS/RECOMMENDATIONS

The results show several common features:

- the relationship between the Q_{3D}/Q_{2D} and the system parameters is complex and nonlinear and therefore introducing an artificial resistivity in addition to physical resistivity c is in general case difficult
- ratio Q_{3D}/Q_{2D} is always lesser than 1.0. The 2D-horizontal models of groundwater flow always overestimate the total flow in the aquifer in the presence of the river-aquifer interaction
- all the hydraulic factors (like increasing k_a or decreasing c) that allow water from the river to penetrate deeper below the river's bottom into the aquifer cause a difference between the two models increasing (i.e. Q_{3D}/Q_{2D} decreasing)
- all the geometric factors (like increasing penetration p or decreasing width of the river H_r) cause water from the river to flow mostly through the river bank and then continuing almost horizontally in an aquifer. As the result a difference between the two models decreases (i.e. Q_{3D}/Q_{2D} increases)
- since there is no clear relationship between the ratio Q_{3D}/Q_{2D} and parameters of the river-aquifer system it is recommended to use a buffer zone (see Figure 4) of the three-dimensional analytical elements for calculating flow in the aquifer under a river bed.

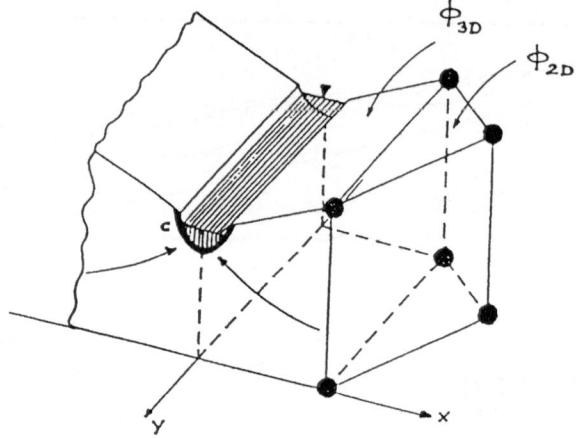

Figure 4. Buffer zone element

- Hybrid 3D-analytical and 2D-finite element models of (r-a-i) need not to incorporate any additional artificial resistivity in order to reproduce the correct water balance.

CONTINUOUS TRANSMISSIVITY TRANSITIONS FOR HORIZONTAL GROUNDWATER FLOW MODELS: FROM UNSATURATED TO CONFINED CONDITIONS

B. ODENWALD[1], J. STAMM and B. HERRLING
Inst. of Hydromechanics
University of Karlsruhe
Kaiserstr. 12, D-76128 Karlsruhe
Germany

In the southern part of germany, shallow and mostly unconfined valley aquifers are widespreaded. The groundwater flow describing differential equation and boundary conditions vary for different flow conditions, due to transient variations of the groundwater table. The authors present a continuous physical-mathematical formulation of two-dimensional horizontal groundwater flow, which allows a simultaneous computation of watersaturated unconfined and partially unsaturated or confined systems. This requires a physically senseful functional relationsship between the transmissivity and the piezometric head. For the numerical approximation of the groundwater flow describing nonlinear parabolic differential equation in space direction the finite element method is used and for the time derivation a finite difference method. The applied Newton-method serves as an efficient tool to solve the large nonlinear algebraic equation system.

INTRODUCTION

For large scale groundwater flow models with horizontal dimensions which are in orders of magnitude of the vertical, a two-dimensional horizontal description of the flow problem, based on the Dupuit assumptions is commonly used (Bear, 1979). In shallow aquifers with phreatic groundwater surface, the dependency of watersaturated aquifer thickness and the piezometric head cannot be neglected. Furthermore, the computation of a transient flow problem yields to a moving boundary problem for these aquifers, illustrated in figure 1. Due to transient changes of the waterlevel, in areas with impermeable overlying layers a transition from unconfined to confined conditions can occur, leading to additional problems in the numerical approximation. Scope of this contribution was to develop an useful mathematical-physical model as well as an appropriate numerical solving scheme, which allows a two-dimensional horizontal computation of groundwater flow for unconfined saturated and partially unsaturated or confined conditions. An extensive description of the presented procedure is given in Odenwald (1994).

---

[1] now at WAT Wasser- und Abfalltechnik Ingenieurgesellschaft mbH, Kriegsstraße 127, D-76135 Karlsruhe, Germany

A. Peters et al. (eds.), Computational Methods in Water Resources X, 91–98.
© 1994 Kluwer Academic Publishers. Printed in the Netherlands.

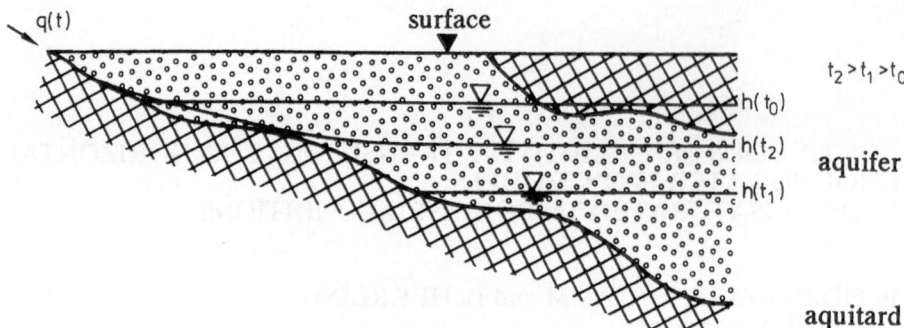

<u>Fig.1:</u>   Vertical profile of an aquifer with groundwater tables for different points
           of time t_0, t_1, and t_2.

BASIC EQUATIONS

The basic equations for the description of groundwater flow for exclusively
confined or unconfined aquifers are presented in extensive literature (e.g. Bear,
1979). Darcy's law embedded in the continuity equation yields for horizontal,
transient groundwater flow in confined and isotropic aquifers to:

$$\frac{\partial}{\partial x}(T\frac{\partial h}{\partial x}) + \frac{\partial}{\partial y}(T\frac{\partial h}{\partial y}) + q^* = S\frac{\partial h}{\partial t} \qquad\qquad onto\ \Omega \times T \qquad (1)$$

with the transmissivity T, the piezometric head h, the space coordinates x and y,
the time t, the sink and source term q^*, and the storage coefficient S. Ω is the
model domain and T the simulation time. The transmissivity is defined as the
product of saturated aquifer thickness m and the mean value of hydraulic
conductivity k. In respect to the scale, the hydraulic conductivity of the aquifer is
assumed to be isotropic.

In unconfined aquifers with a phreatic groundwater surface, the thickness m and
consequently the transmissivity T are functions of the piezometric head h.
Furthermore, the storage coefficient S has to be replaced by the porosity n, which
is flooded or drained because of changing waterlevels. Groundwater flow for
unconfined and saturated conditions can be described by:

$$\frac{\partial}{\partial x}(km\frac{\partial h}{\partial x}) + \frac{\partial}{\partial y}(km\frac{\partial h}{\partial y}) + q^* = n\frac{\partial h}{\partial t} \qquad\qquad onto\ \Omega \times T \qquad (2)$$

In contrast to the linear parabolic differential equation (1), the formal description
of unconfined groundwater flow has a nonlinear character (eq. 2). The numerical
computation of a flow problem with transitions of confined to unconfined
conditions by switching the describing equations (eq. 1 and eq. 2), leads to
convergence problems due to the abrupt transition between the porosity n and
the storage coefficient S.

In the past, several methods were developed to feasible the consideration of desiccant regions with unsaturated flow in a horizontal model. Herrling (1982) presented a moving boundary method, where the estimation of groundwater flow can be restricted on the saturated area, by adapting the boundary at the saturated domain in every calculation time step. Here a continuous transmissivity expression will be used to describe saturated and unsaturated conditions. Herefore, an explicit functional relationship between aquifer thickness m and piezometric head h is defined (eq. 3).

In the saturated zone (h≥s), the aquifer thickness m is equal to the difference between s and the groundwater surface h and the aquifer bottom s, enlarged with the soil dependent capillary fringe r. In the unsaturated zone (h<s), the value of m decreases exponentially to the residual saturated thickness of r_0. The defined relationship between the piezometric head h and the aquifer thickness m causes, that m is positiv in the whole calculation domain and furthermore continuously differentiable with respect to the piezometric head h (eq. 4). The latter is necessary for the application of the Newton-method, which will be seen later.

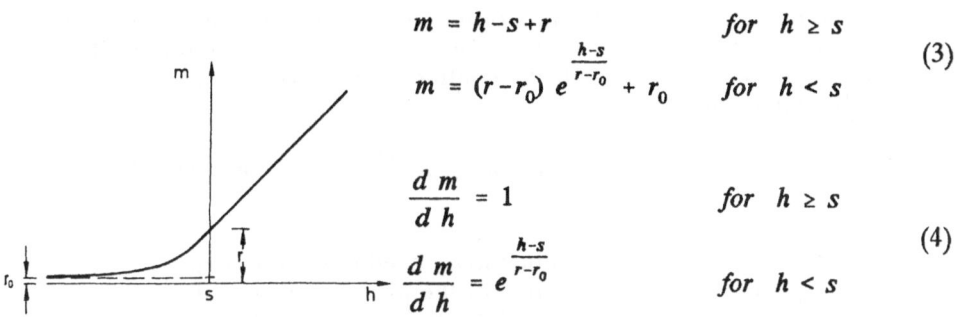

$$m = h - s + r \qquad \text{for } h \geq s$$
$$m = (r - r_0) \, e^{\frac{h-s}{r-r_0}} + r_0 \qquad \text{for } h < s \tag{3}$$

$$\frac{d\,m}{d\,h} = 1 \qquad \text{for } h \geq s$$
$$\frac{d\,m}{d\,h} = e^{\frac{h-s}{r-r_0}} \qquad \text{for } h < s \tag{4}$$

Fig. 2:   Aquifer thickness m as a function of the piezometric head h for unsaturated and saturated unconfined flow conditions

Analogously, a continuous expression for m can be defined in the transition zone from unconfined to confined groundwater flow (eq. 5). The lower surface of the overlying impermeable layer is signed with d, referred to the same datum level as h and s. For confined aquifer conditions m corresponds asymptotically to the thickness of the groundwater conducting layer.

This simplified concept is justified as long as only an insignificant part of the whole modeled domain is confined and so the storage capacity in this regions is negligible in comparision with the unconfined area. Equation 6 fullfills the demand of continuous differentiability, too.

The definition of the storage term as a product of porosity n and the partial derivation with respect to the aquifer thickness by the time $\partial m/\partial t$ and the

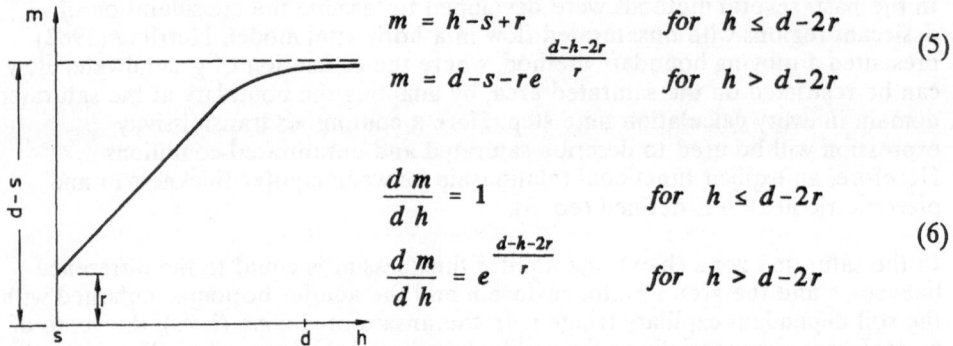

$$m = h - s + r \qquad for \quad h \le d - 2r \tag{5}$$

$$m = d - s - r e^{\frac{d-h-2r}{r}} \qquad for \quad h > d - 2r$$

$$\frac{d\,m}{d\,h} = 1 \qquad for \quad h \le d - 2r \tag{6}$$

$$\frac{d\,m}{d\,h} = e^{\frac{d-h-2r}{r}} \qquad for \quad h > d - 2r$$

Fig. 3:   Aquifer thickness m as a function of piezometric head h for saturated unconfined and confined flow conditions

definition of thickness m as a function of the piezometric head h in consequence of eq. 3 and 5, enables the formal description of horizontal groundwater flow for unsaturated, saturated unconfined, and confined aquifer conditions by a single partial differential equation:

$$\frac{\partial}{\partial x}\left(km\frac{\partial h}{\partial x}\right) + \frac{\partial}{\partial y}\left(km\frac{\partial h}{\partial y}\right) + q^* = n\frac{\partial m}{\partial t} \qquad onto \ \Omega \times T \tag{7}$$

The use of the following initial and boundary conditions (B.C.) allows a plain solution of equation 7. The prescription of the piezometric head h onto the boundary Γ_1 complies with a Dirichlet B.C. and the prescription of the discharge $\bar{q}$ normal to the boundary Γ_2 matchs a Neumann B.C.. At the beginning of the simulation, h_0 defines the initial piezometric head in the whole domain Ω.

$$h - \bar{h} = 0 \qquad auf \ \Gamma_1 \times T \tag{8}$$

$$\left(km\frac{\partial h}{\partial x}\right)n_x + \left(km\frac{\partial h}{\partial y}\right)n_y - \bar{q} = 0 \qquad auf \ \Gamma_2 \times T \tag{9}$$

$$h(t=0) = h_0 \qquad auf \ \Omega \tag{10}$$

NUMERICAL APPROXIMATION

For the numerical approximation in space direction, a finite element method is used. Thereby an ordinary differential equation system (11) is formed. N_e signs the number of finite elements and N the number of knots. The coefficient matrix A_{ELK} depends on the geometry of the aquifer and the hydraulic conductivities; the storage matrix H_{EL} depends on the porosity. The vector q_L incorporates the

$$m_E A_{ELK} h_K + \frac{\partial m_E}{\partial t} H_{EL} = q_L \tag{11}$$

with $E = 1,2,3,...,N_e$ and $L,K = 1,2,3,...,N$

discharges within and normal to the border of the model domain.

For the numerical approximation of equation system 11 in time direction, a finite difference method is employed. The time deviation term of the aquifer thickness $\partial m_E/\partial t$ can be expressed by:

$$\frac{\partial m_E}{\partial t} \approx \frac{m_E^{t+\Delta t} - m_E^t}{\Delta t} \qquad \text{with} \quad E = 1,2,3,...,N_e \tag{12}$$

Additionally, the question arises, at which moment within the time interval Δt the unknown values h_K and m_E should be evaluated. Assuming a linear course within Δt, the piezometric head h_K and the aquifer thickness m_E can be described by the use of the time weighting factor α and α^*, respectively.

$$h_K^{t+\alpha \Delta t} \approx (1-\alpha) h_K^t + \alpha h_K^{t+\Delta t} \qquad \text{for} \quad 0.0 \leq \alpha \leq 1.0 \tag{13}$$

with $K = 1,2,3,...,N$

$$m_E^{t+\alpha^* \Delta t} \approx (1-\alpha^*) m_E^t + \alpha^* m_E^{t+\Delta t} \qquad \text{for} \quad 0.0 \leq \alpha^* \leq 1.0 \tag{14}$$

with $E = 1,2,3,...,N_e$

Several time weighting schemes have been investigated, the fully implicit scheme ($\alpha=\alpha^*=1.0$), the Crank-Nicolson scheme ($\alpha=\alpha^*=0.5$) and a mixed formulation ($\alpha=1.0$, $\alpha^*=0.5$), whereby the latter yields the best results. Investigations by Herrling and Leismann (1984) returned to the same understanding for the numerical treatment of saturated and unsaturated vertical groundwater flow. Substituting the adequate terms in eq. 11 by eq. 13 and 14, and the use of the mixed time weighting formulation leads to the algebraic equation system:

$$\frac{1}{2} \left(m_E^t + *m_E^{t+\Delta t} \right) A_{ELK} \, h_K^{t+\Delta t} + \frac{1}{\Delta t} \left(m_E^{t+\Delta t} - m_E^t \right) H_{EL} = q_L^{\Delta t} \tag{15}$$

with $E = 1,2,3,...,N_e$ and $L,K = 1,2,3,...,N$

The resolution of this nonlinear equation system requires an iterative calculation of the piezometric head h within each time step, starting from a known inital state for t=0. The here presented proceeding uses a generalized Newton algorithm, which is designated by a local quadratic convergence behavior (Gill, 1981). The application of the Newton method onto equation 15 generates the following equation:

$$\left\{ \left(\frac{\partial m_E}{\partial h_K}\right)^{t+\Delta t} \left(\frac{1}{2} A_{ELH} h_H^{t+\Delta t} + \frac{1}{\Delta t} H_{EL}\right) + \frac{1}{2}\left(m_E^t + m_E^{t+\Delta t}\right) A_{ELK} \right\}^{(i)} \Delta h_K^{(i)} =$$

$$-\left\{ \frac{1}{2}\left(m_E^t + m_E^{t+\Delta t}\right) A_{ELK} \, h_K^{t+\Delta t} + \frac{1}{\Delta t}\left(m_E^{t+\Delta t} - m_E^t\right) H_{EL} \right\}^{(i)} + q_L^{\Delta t} \qquad (16)$$

$$\text{with} \quad E = 1,2,3,...,N_e \quad \text{and} \quad H,L,K = 1,2,3,...,N$$

or in short form, with the Jacobian matrix J_{LK} and the right hand side vector r_L:

$$J_{LK}^{(i)} \, \Delta h_K^{(i)} = r_L^{(i)} \qquad\qquad \text{with} \quad L,K = 1,2,3,...,N_k \qquad (17)$$

The piezometric head is determined by equation 18. If the absolute value of $\Delta h_K^{(i)}$ is lower than the previously defined termination criteria, the solution is acchieved.

$$h_K^{(i+1)} = h_K^{(i)} + \Delta h_K^{(i)} \qquad\qquad \text{with} \quad K = 1,2,3,...,N_k \qquad (18)$$

APPLICATION

The following comparision with a vertical saturated - unsaturated model (Herrling and Leismann, 1984) demonstrates the efficiency of the developed procedure. In the saturated - unsaturated computation, the negative pressure $-p/(\rho g)$, also called as soil moisture tension ψ, describes the dependent value. The tension ψ in return depends on the water content θ and the soil properties, and determines the unsaturated hydraulic conductivity, too. In the numerical computation, the parameterized form of description of these dependencies is used (van Genuchten, 1980). Therefore, the saturated water content θ_s, and residual water content θ_r, as well as the saturated hydraulic conductivity k, and the form parameters α, n, and m are requested. S_e marks the effective soil water saturation.

$$\theta = \theta_r + (\theta_s - \theta_r) \, S_e \qquad\qquad (19)$$

$$k_u = k \, S_e^{1/2} \left[1 - \left(1 - S_e^{1/m}\right)^m \right]^2 \qquad \text{with} \quad S_e = \frac{1}{\left(1 + (\alpha \, \psi)^n\right)^m} \qquad (20)$$

For the comparision, the physical parameters $\theta_s = 0.25$, $\theta_r = 0.05$ and $k = 10^{-3}$ m/s and the form parameters $\alpha = 5.0$, n = 2.5, and m = 0.6 were choosen, which characterizes a soil with a high water retention capacity. Figure 4 shows the water retention curve (θ over ψ) and the relative conductivity curve (k_u/k over ψ). The horizontal computation was carried out with the mean value of saturated hydraulic conductivity $k = 10^{-3}$ m/s, the porosity $n = 0.2$, and the capillary height $r = 0.115$ m (conform to $\psi(0.9\theta_s)$).

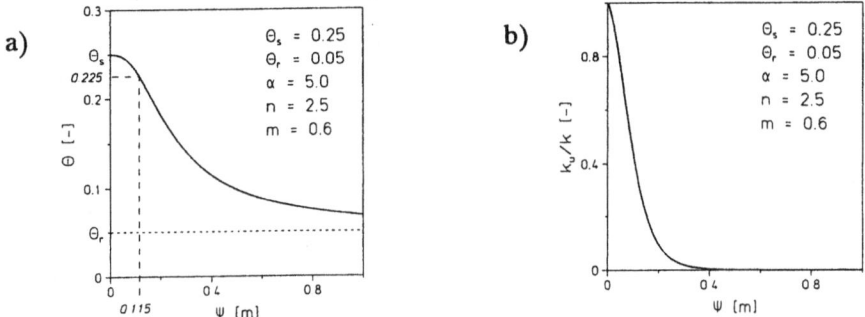

Fig. 4:   a) Water retention curve: water content θ over soil water tension ψ, and
            b) relative hydraulic conductivity k_u/k over soil water tension ψ.

The simulated flow problem represents a homogeneous and isotropic aquifer sequence of 100 m in length and 1.5 m in height, with a linear descending base (I=1%) as shown in figure 5. The vertical model has a discretisation width of $\Delta z = 0.1$ m in vertical and $\Delta x = 1.0$ m in the horizontal direction. The base and the top of the model are defined by Neumann B.C. ($\bar{q}=0$). The horizontal model describes a rectangular aquifer domain with triangular finite elements, using a discretisation width of $\Delta x = 1.0$ m. At the flow parallel borderlines no flow

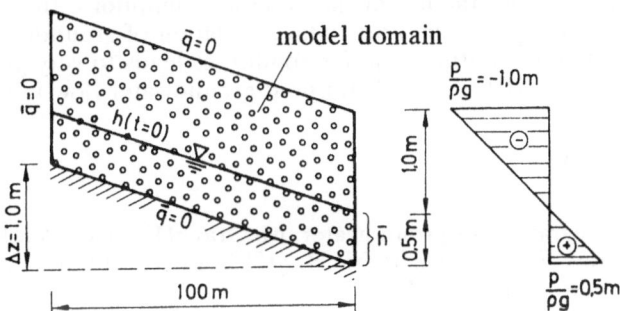

Fig. 5:   Model of the saturated - unsaturated testaquifer with boundary and
            initial conditions.

exchange ($\bar{q}=0$) exists. In both models the outflow is defined to be hydrostatic, using the Dirichlet B.C. $h(x = 100$ m,t$) = 0.5$ m over the total simulation time of 50 d. The initial piezometric head in the whole model is $h(x,t =0) = 0.5$ m, which yields to unsaturated conditions in the left model domain. At the inflow front the discharge $\bar{q}(x = 0, t) = 5$ l/(skm) is prescribed. Due to the sharp wetting front, the vertical computation needs at the begin of the simulation a time step of 0.001 d. The dashed line in figure 6 displays the results of the horizontal simulation and the continuous line represents the vertical simulation.

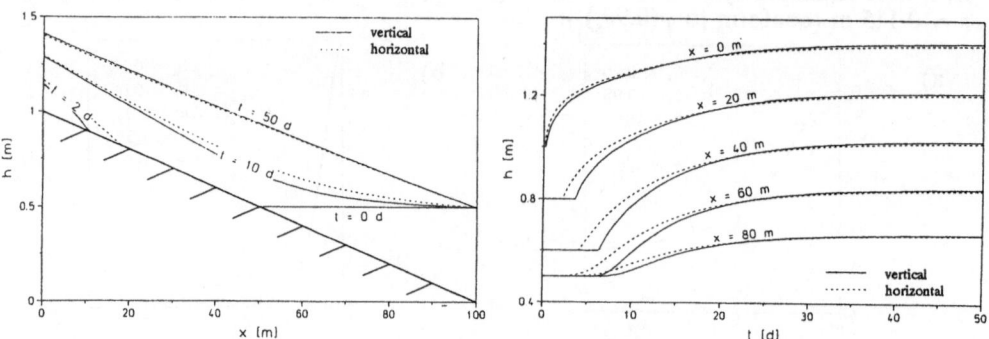

<u>Fig. 6:</u>   a) Free groundwater table in a longitudinal section through a vertical
         respective horizontal model for different time steps, and b) transient
         curves of the groundwater table in different cross sections.

Figure 6 demonstrates an excellent agreement between both model concepts,
even for the used soil with a high water retention capacity. The higher the portion
of sand and gravel in the aquifer, which can be assumed for valley aquifers, the
better the obtained accordance. Furthermore the results show, that the
distribution of water pressure within the different cross sections is rather
hydrostatic, which justifies the use of the Dupuit assumptions. In conclusion, the
presented model concept for the horizontal simulation of groundwater flow for
partially unsaturated or confined aquifer conditions leads to very good results,
and dispose an useful tool to model transient groundwater flow in shallow valley
aquifers.

LITERATURE

Bear, J. (1979) Hydraulics of groundwater, McGraw-Hill, New York, USA.
Gill, P. E., Murray, W. and Wright, M. H. (1981) Praktical Optimization,
Academic Press, New York, USA.
Herrling, B. (1982) "Finite element computations of horizontal groundwater flow
with moving boundaries", in K.P. Holz et al. (eds), Finite Elements in Water
Resources, Springer Verlag, Berlin, 10.25 - 10.39.
Herrling, B. and Leismann, H.M. (1984) "Finite element computation of
unsaturated and saturated groundwater flow in stratified aquifers" in J.P. Laible
et al. (eds), Finite Elements in Water Resources, Springer Verlag, Berlin.
Odenwald, B. (1994) Parameteridentifizierung bei numerischen
Grundwasserströmungsmodellen, VDI-Verlag, Düsseldorf, in press.
Van Genuchten, M. Th. (1980) "A closed-form equation for predicting the
hydraulic conductivity of unsaturated soils", Soil Siences Society of America
Journal, Vol. 44, 892 -898.

QUASI-NEWTON METHODS FOR RICHARDS' EQUATION

MARIO PUTTI[†] and CLAUDIO PANICONI[‡]

[†] *Dept. of Mathematical Methods for Applied Sciences, University of Padua, Italy*

[‡] *CRS4, Via Nazario Sauro 10, 09123 Cagliari, Italy*

Numerical procedures to solve the nonlinear equation governing flow in variably saturated porous media commonly involve Newton or Picard iteration. The former scheme is stable and quadratically convergent in a local sense, but costly and algebraically complex. The latter scheme is simple and cheap, but slower converging and not as robust. To preserve the advantages of the Newton scheme while overcoming some of its drawbacks, we consider the family of quasi-Newton schemes. These methods are based on the use of easily calculated updates to approximations of the Newton Jacobian or its inverse. Two quasi-Newton methods based on sparsity-preserving inverse updates are implemented and applied to solve the one-dimensional Richards equation.

INTRODUCTION

We express Richards' equation, describing fluid flow in partially saturated porous media, in the form

$$\eta(\psi)\frac{\partial \psi}{\partial t} = \nabla \cdot (K_s K_r(\psi) \nabla(\psi + z)) \tag{1}$$

where ψ is pressure head, t is time, z is the vertical coordinate (positive upward), η is the general storage term, and the hydraulic conductivity tensor is expressed as a product of the conductivity at saturation, K_s, and the relative conductivity, $K_r(\psi)$. Equation (1) is highly nonlinear due to pressure head dependencies in the storage and conductivity terms, and is generally solved numerically by iteration. The most popular iterative methods for solving (1) are the Picard and Newton schemes [*Paniconi et al., 1991*]. The Picard scheme applied to Richards' equation yields a very straightforward linearization which preserves symmetry of the discretized system, is easy to implement, and is computationally inexpensive. However, the method may fail or converge very slowly for certain types of problems. The Newton method has better local stability and convergence properties (it is quadratically convergent), but it is more expensive per iteration (requiring evaluation and assembly of the nonsymmetric Jacobian matrix), and its global properties are not well understood (the Newton method often requires a very good initial solution estimate for convergence).

Picard and Newton iteration applied to equation (1) is discussed in *Paniconi and Putti [1993]*. In the present paper we describe the implementation of quasi-Newton

A. Peters et al. (eds.), *Computational Methods in Water Resources X*, 99–106.

methods for solving Richards' equation, and assess the performance of these methods for the case of one-dimensional flow. The basic idea behind quasi-Newton methods is to replace the Jacobian evaluation step of the Newton scheme with a less costly approximation to either the Jacobian or its inverse. The methods are expressed in the form of easily calculated updates to the initial Jacobian or inverse Jacobian approximation. An effective quasi-Newton method will be less expensive than Newton iteration, and it will have good theoretical properties (such as superlinear convergence) that guarantee, at least locally, rapid convergence. Moreover, judicious choice of the initial Jacobian approximation can overcome the Newton method's sensitivity to initial solution estimates. Finally, in finite element applications, it is also important that the updates do not destroy the sparsity pattern of the system matrices. Quasi-Newton methods are discussed in more detail in *Dennis and Moré [1977]*, *Matthies and Strang [1979]*, *Fletcher [1980]*, *Geradin et al. [1981]*, and *Papadrakakis [1993]*.

In the following section we will describe recursion formulae to calculate sparsity-preserving inverse updates for the Broyden and BFGS quasi-Newton methods. To reduce storage and CPU requirements, we will use limited memory implementations for the two quasi-Newton methods. Three different limited memory strategies will be presented.

IMPLEMENTATION

The Newton method for the solution of the discretized form of equation (1) can be written as

$$J(\psi_m^{k+1})s_m = -g(\psi_m^{k+1}) \tag{2}$$

where ψ is the vector of nodal pressure head values, k and m denote the time stepping and iteration indices, respectively, $s_m \equiv \psi_{m+1}^{k+1} - \psi_m^{k+1}$, g_m is the residual vector given by

$$g_m = g(\psi_m^{k+1}) \equiv A(\psi_m^{k+\lambda})\psi_m^{k+\lambda} + F(\psi_m^{k+\lambda})\frac{\psi_m^{k+1} - \psi_m^k}{\Delta t^{k+1}} + b(\psi_m^{k+\lambda}) - q(t^{k+\lambda}),$$

and

$$\begin{aligned}
J_{ij} &= \lambda A_{ij} + \frac{1}{\Delta t^{k+1}} F_{ij} + \sum_s \frac{\partial A_{is}}{\partial \psi_j^{k+1}} \psi_s^{k+\lambda} \\
&\quad + \frac{1}{\Delta t^{k+1}} \sum_s \frac{\partial F_{is}}{\partial \psi_j^{k+1}} (\psi_s^{k+1} - \psi_s^k) + \frac{\partial b_i}{\partial \psi_j^{k+1}} \tag{3}
\end{aligned}$$

is the ij-th component of the Jacobian matrix $J(\psi^{k+1})$. In the previous equations we have denoted with A the stiffness matrix, F the storage or mass matrix, b the vector containing the gravitational gradient component of equation (1), and q the vector containing the specified Darcy flux boundary conditions. The constant λ is the weighting factor used in the finite difference time discretization of (1), so that $\psi^{k+\lambda} = \lambda \psi^{k+1} + (1 - \lambda)\psi^k$. The nodal pressure head vector ψ_{m+1}^{k+1} is calculated at the new iteration as $\psi_{m+1}^{k+1} = \psi_m^{k+1} + s_m$. If we neglect the derivative terms in equation (3), we obtain the Picard scheme [*Paniconi and Putti, 1993*].

The quasi-Newton family is defined by substituting in the Newton equation (2) an approximation K to the Jacobian J, yielding $K(\psi_m^{k+1})s_m = -g(\psi_m^{k+1})$. Rather than calculate the Jacobian at each iteration, as is done in the Newton method, matrix K is updated using recursion formulae that require fewer operations than a full evaluation of J. The update formulae can be derived as follows. Let $y_m = g_{m+1} - g_m$ be the residual difference. Then a Taylor series expansion of the residual vector g gives

$$g_{m+1} = g(\psi_m^{k+1} + s_m) = g_m + J_m s_m + \cdots$$

that is, $y_m = J_m s_m + \ldots$. The approximation K to the Jacobian matrix J is chosen so that it satisfies the previous equation exactly up the the first order terms. We obtain therefore the equation $K_{m+1}s_m = y_m$, which is known as the quasi-Newton or secant condition. This expression enables the direct calculation of K_{m+1} or K_{m+1}^{-1}. In the limit as $m \to \infty$, the matrix K_{m+1} thus defined tends to J. In implementations of the quasi-Newton method, updates for K_{m+1}^{-1} are often used instead of K_{m+1} updates, as these inverse updates avoid the need to solve a linear system at each iteration.

Several update formulae have been proposed in the literature. We will describe only the Broyden and BFGS updates, which are considered to be amongst the most efficient quasi-Newton algorithms. The expression for the Broyden inverse update is

$$K_m^{-1} = K_{m-1}^{-1} + \frac{\left(s_m - K_{m-1}^{-1}y_m\right)s_m^T K_{m-1}^{-1}}{s_m^T K_{m-1}^{-1}y_m} \tag{4}$$

while the BFGS inverse update can be written as

$$K_m^{-1} = \left(I - \omega_m s_m y_m^T\right) K_{m-1}^{-1} \left(I - \omega_m y_m s_m^T\right) + \omega_m s_m s_m^T \tag{5}$$

where $\omega_m = (s_m^T y_m)^{-1}$. In our implementation of the Broyden and BFGS algorithms, we use the exact Jacobian J_0 for the initial update K_0. The inverse is never calculated explicitly. Instead, the LU factorization of K is used, and sparsity-preserving expressions that calculate directly the solution vector s_{m+1} are developed.

For the Broyden update, we define the vectors ρ_m and γ_m as

$$\rho_m = s_m - \prod_{j=m-1}^{1} \left(I + \rho_j \gamma_j^T\right) K_0^{-1} y_m$$

$$\gamma_m = s_m / \left[s_m^T \prod_{j=m-1}^{1} \left(I + \rho_j \gamma_j^T\right) K_0^{-1} y_m \right]$$

and obtain

$$s_{m+1} = -\prod_{j=m}^{1} \left(I + \rho_j \gamma_j^T\right) K_0^{-1} g_m$$

Note that the implementation of this update requires only one matrix-vector product plus $2(m-1)$ vector scalar products for the calculation of ρ_m and γ_m, and one matrix-vector product plus $2m$ vector scalar products and sums for the update. This gives a total of two matrix-vector products and $4m - 2$ vector scalar products and sums.

The expression used in the implementation of the BFGS quasi-Newton method is

$$
\begin{aligned}
s_{m+1} \;=\; & -\prod_{j=m}^{1} \left(I - \omega_j s_j y_j^T\right) K_0^{-1} \prod_{j=1}^{m} \left(I - \omega_j y_j s_j^T\right) g_m \\
& -\sum_{i=1}^{m-1} \left[\prod_{j=m}^{i+1} \left(I - \omega_j s_j y_j^T\right) \omega_i s_i s_i^T \prod_{j=i+1}^{m} \left(I - \omega_j y_j s_j^T\right) \right] g_m \\
& -\omega_m s_m s_m^{\,T} g_m
\end{aligned}
$$

The implementation of this update requires only one matrix-vector product and $m(m+5)$ vector scalar products and sums.

The main difficulty in the implementation of these update formulae is that the vectors s_j and y_j (for BFGS) and ρ_j and γ_j (for Broyden) have to be saved at each iteration. Thus storage requirements become excessive when a large number of iterations is required for convergence. This problem may be circumvented by the use of so-called limited memory quasi-Newton (LMQN) implementations, which are based on different restart strategies after a given number of iterations M has been completed.

We employed three distinct LMQN strategies. In implementation 1 all the previous saved vectors are discarded and the iteration restarts using (s_M, y_M) or (ρ_M, γ_M), and K_0 as the initial Jacobian approximation. Implementation 2 also uses K_0 as the initial Jacobian approximation, but always saves M vectors, with the oldest (s_j, y_j) or (ρ_j, γ_j) vectors replaced by the most recently calculated ones. Implementation (3) is a full restart of the scheme obtained by discarding all the previous vectors and recalculating the Jacobian matrix, that is, $K_{M+1} = J_{M+1}$. A template describing the general procedure for the BFGS and Broyden quasi-Newton methods is given in Figure 1.

Start with ψ_0^{k+1} and K_0; $p = 0$

for i=0,1,...

    1. $p = p + 1$; if $p \geq M$ do a **restart**

    2. $g_i = g(\psi_i^{k+1})$

    3. evaluate $K_i^{-1} = f(K_{i-1}^{-1})$ based on (4) (Broyden) or (5) (BFGS)

    4. $s_{i+1} = -K_i^{-1} g_i$

    5. $\psi_{i+1}^{k+1} = \psi_i^{k+1} + s_i$

    6. if $\left\| \psi_{i+1}^{k+1} - \psi_i^{k+1} \right\| \leq \epsilon$ stop;

end for;

Figure 1: Quasi-Newton algorithm

Table 1. Test Case 1: Number of Iterations to Convergence for the Broyden and BFGS Methods

Iteration Levels Saved (M)	Broyden Method LMQN Implementation			BFGS Method LMQN Implementation		
	1	2	3	1	2	3
4	34	36	10	44	39	25
20	30	30	22	30	34	26
50	29	29	29	30	30	30

NUMERICAL RESULTS

Two numerical examples were used to test the various implementations of the Broyden and BFGS methods, and to compare their convergence behavior against the Picard and Newton schemes. The test cases, whose solutions are shown in Figure 2, involve steady state flow in a one-dimensional soil column. In both test cases the column was discretized into 101 nodes, the pressure head at the base $z = 0$ was kept fixed at $\psi = 0$, a Darcy flux boundary condition of 0.0001 m/h was imposed at the top, and the saturated conductivity K_s was set to 0.01 m/h. For test case 1 we used the exponential relationship $K_r(\psi) = e^{\alpha \psi}$ with $\alpha = 0.1$ m^{-1}, while for test case 2 we used $K_r(\psi) = (1 + \beta)^{-5m/2} [(1 + \beta)^m - \beta^m]^2$ with $\beta \equiv (\psi/\psi_s)^n$, $\psi_s = -3.0$ m, $n = 5.0$, and $m = 0.8$.

The results for the first example are summarized in Table 1, and the convergence profiles for some of the test runs are plotted in Figure 3a. The Picard method failed to converge for this test case, with the errors oscillating between 100 and 1800 m from one iteration to the next. The Newton scheme converged in 7 iterations. Of the various limited memory implementations, strategy 3 with 4 iteration levels saved gave the best results, converging in 10 iterations for the Broyden method and in 25 for the BFGS method. This result is not surprising given the good performance of the Newton scheme for this test problem, since a low-level strategy 3 scheme (with exact initial Jacobian, as in our implementation) is the one which most closely approximates the Newton method. The slowest to converge was the 4-level BFGS-1 method, which required 44 iterations.

The results for the second example are summarized in Table 2, and the convergence profiles for some of the test runs are plotted in Figure 3b. For this test case both the Picard and Newton methods produced oscillations and failed to converge. Several of the quasi-Newton methods also failed or converged very slowly, with particularly poor performance observed for limited memory implementation 2. The best results were obtained with strategy 3 with 20 or 50 iteration levels saved, which converged in 31–71 iterations. The fastest of these was the 20-level Broyden-3 method.

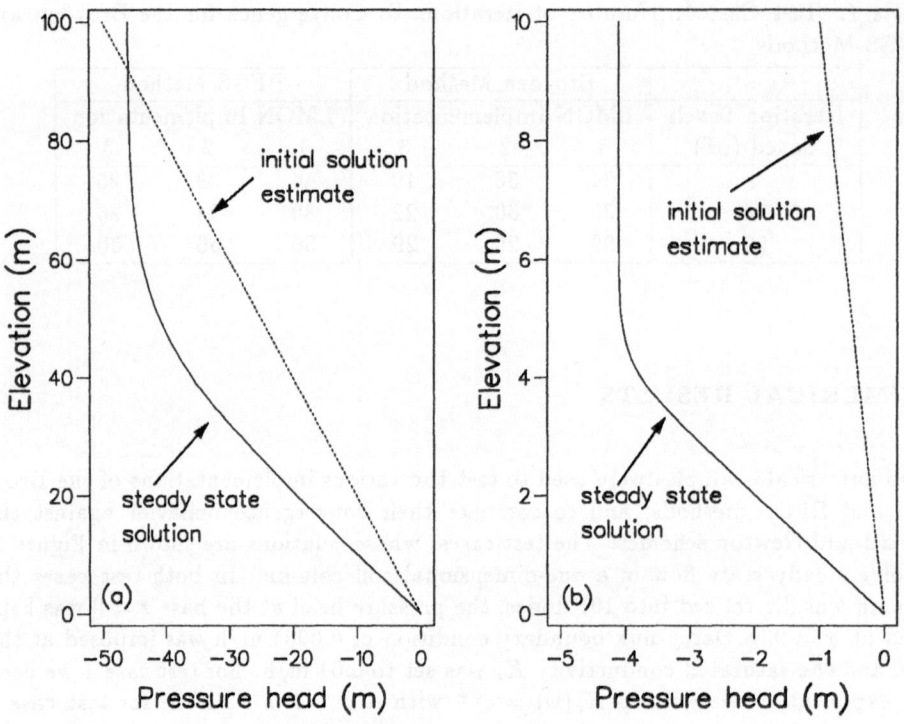

Figure 2. Initial solution estimates and steady state solutions for test cases 1 (a) and 2 (b).

Table 2. Test Case 2: Number of Iterations to Convergence for the Broyden and BFGS Methods

Iteration Levels Saved (M)	Broyden Method LMQN Implementation			BFGS Method LMQN Implementation		
	1	2	3	1	2	3
4	failed	497	failed	221	failed	failed
20	256	527	31	393	failed	56
50	263	968	57	252	failed	71

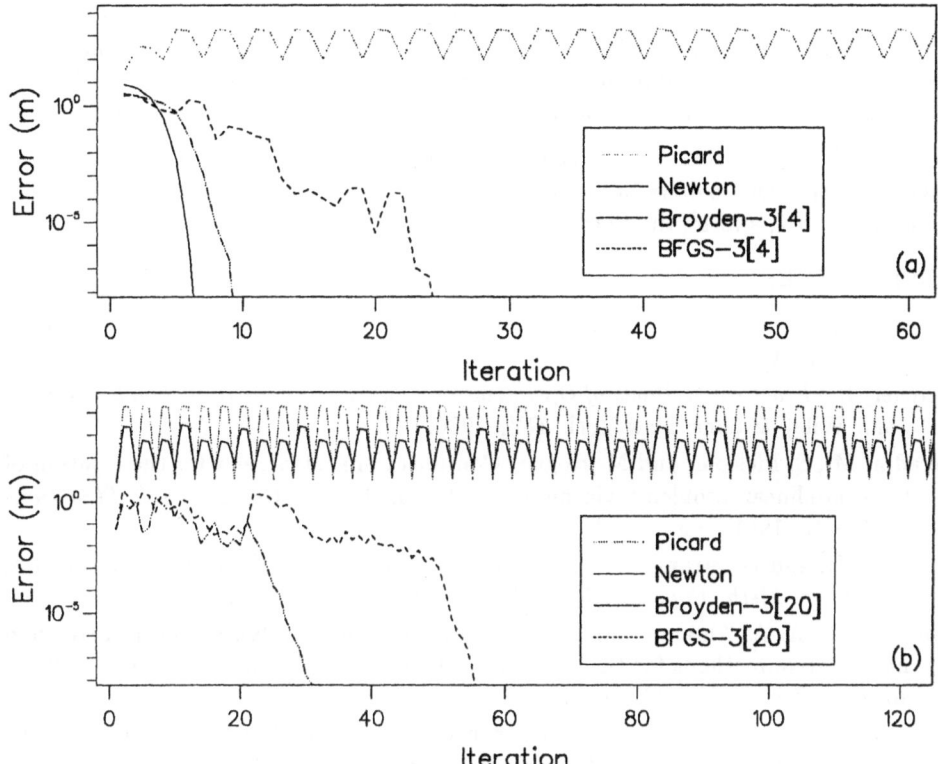

Figure 3. Convergence profiles for test cases 1 (a) and 2 (b). The notation "Broyden-3[4]" refers to limited memory implementation 3 of the Broyden method, with 4 iteration levels saved.

CONCLUSIONS

We have described and tested three limited memory implementations of sparsity-preserving inverse updates for the Broyden and BFGS quasi-Newton methods applied to Richards' equation. Based on convergence behavior, the methods performed well against the Picard and Newton schemes. An assessment of the computational efficiency of these quasi-Newton methods will be made for larger, multi-dimensional flow problems. It will also be interesting to examine: other choices for the initial Jacobian approximation; update formulae based on the Jacobian instead of the inverse Jacobian; and line search procedures to improve the solution after each iteration.

Acknowledgments. This work has been supported by the Italian CNR (Gruppo Nazionale per la Difesa dalle Catastrofi Idrogeologiche, linea di Ricerca n. 4), by Fondi Ministeriali 40%, and by the Sardinia Regional Authorities.

REFERENCES

Dennis, J. E. and J. J. Moré, Quasi-Newton methods, motivation and theory, *SIAM Review* 19(1), 46–89, 1977.

Fletcher, R., *Practical Methods of Optimization, Vol. 1: Unconstrained Optimization.* John Wiley & Sons, New York, NY, 1980.

Geradin, M., S. Idelsohn and M. Hogge, Computational strategies for the solution of large nonlinear problems via quasi-Newton methods, *Computers & Structures* 13, 73–81, 1981.

Matthies, H. and G. Strang, The solution of nonlinear finite element equations, *Int. J. Numer. Meth. Eng.* 14, 1613–1626, 1979.

Paniconi, C., A. A. Aldama and E. F. Wood, Numerical evaluation of iterative and noniterative methods for the solution of the nonlinear Richards equation, *Water Resour. Res.* 27(6), 1147–1163, 1991.

Paniconi, C. and M. Putti, A comparison of Picard and Newton iteration in the numerical solution of multi-dimensional variably saturated flow problems, *Water Resour. Res.* submitted, 1993.

Papadrakakis, M., Solving large-scale nonlinear problems in solid and structural mechanics. In: Papadrakakis, M. (ed.) *Solving Large-Scale Problems in Mechanics* 183–223, John Wiley & Sons, New York, NY, 1993.

A DECOUPLED SOLUTION STRATEGY FOR QUASI-3D FINITE ELEMENT MODEL OF FLOW IN MULTIAQUIFER SYSTEMS

PIETRO TEATINI[†] and GIUSEPPE GAMBOLATI[†]
[†] *Dept. of Mathematical Methods for Applied Sciences, University of Padua, Italy*

When groundwater flow takes place in aquifer-aquitard systems characterized by highly hetereogeneous hydrogeological parameters with nonlinear behaviour, quasi-3D models of flow must be solved by a fully numerical approach. The numerical implementation with finite difference or finite element methods involves systems of large dimension whose solution requires much CPU time and computer storage. A new solution strategy of the resulting algebraic equations is suggested to overcome the problem and is analyzed with linear porus media. The global system is decoupled into a number of smaller subsystems. The aquifers and the aquitards are solved separately, and the final coupled solution is obtained with an iterative procedure. This procedure is naturally suggested by the special sparsity pattern of the coefficient matrix and can be shown to be equivalent to a block SOR (successive over-relaxation) strategy. The SOR blocks correspond to the aquifer and aquitard equations. The convergence property of the new scheme is analyzed, the optimum over-relaxation factor is computed, and the asymptotic convergence rate is theoretically analyzed in relation to the hydrogeological parameters of the multiaquifer system.

INTRODUCTION

The hydrologic response of multiaquifer systems to groundwater pumping is highly dependent on the difference between the physical parameter values of sandy and clayey layers. In most cases a high contrast in the hydraulic conductivity exists and water flow can be assumed essentially horizontal in the aquifers and vertical in the aquitards. Quasi three dimensional models, initially presented by *Bredehoeft and Pinder [1970]*, take computational advantage of this specific hydrogeological setting, which discards the vertical hydraulic gradient in the aquifers and the horizontal hydraulic gradient in the aquitards. Neglecting point and distributed sources, the governing flow equation in the i-th aquifer confined between aquitards j and $j + 1$ can be written, as:

$$\frac{\partial}{\partial x}\left(T_{xi}\frac{\partial h_i}{\partial x}\right) + \frac{\partial}{\partial y}\left(T_{yi}\frac{\partial h_i}{\partial y}\right) = S_i\frac{\partial h_i}{\partial t} + q_j - q_{j+1} \qquad (1)$$

where T_{xi}, T_{yi}, S_i indicate the components of the anisotropic trasmissivity and the elastic storage coefficient, respectively, h_i is the hydraulic potential head, t is

107

A. Peters et al. (eds.), *Computational Methods in Water Resources X*, 107–114.

time, and q_j, q_{j+1} represent leakage from the overlying and underlying aquitards. In aquitard j, flow is governed by the 1D vertical diffusion equation:

$$K_{zj}\frac{\partial^2 h_j}{\partial z^2} = S_{sj}\frac{\partial h_j}{\partial t} \tag{2}$$

where K_{zj} is the vertical permeability, S_{sj} is the specific elastic storage and h_j is the hydraulic potential head.

When eqs. (1) and (2) are both numerically integrated, systems of large dimension are involved. Special numerical techniques to efficiently solve these large problems have been consequently devised. *Chorley and Frind* [*1978*] treat the resulting equations with an iterative technique, "solving aquifers and aquitards alternately at each time step. A favorable rate of convergence was obtained by staggering aquifer and aquitard solution by half a time step". The "adaptive explicit-implicit" technique implemented by *Neuman at al.* [*1982*] takes advantage of the large ratio between the hydraulic aquifer-aquitard diffusivities, so that for small time step values Δt, most of the element equations connected to the nodes of the aquitards can be solved explicitly with low computational cost. The remaining equations are solved implicitly with a direct scheme.

A new strategy to solve the discrete system arising from a quasi-3D flow model is described. If the equations are linear, the procedure coincides with a block SOR iterative method, in which each aquifer block is solved with the MCG (modified conjugate gradient) method and each aquitard block is solved with the direct Thomas alghoritm for a three-diagonal matrix.

The convergence property of the new scheme is analyzed in relation to the physical properties of the multiaquifer system. Two numerical codes are implemented. In the first code (labeled "coupled"), the resulting equations are solved simultaneously with the MCG method, and in the second one (labeled "decoupled"), the new strategy is used. The two codes are applied to the same linear test problems and compared in terms of computational efficiency. Some preliminary results are presented and discussed.

NUMERICAL MODEL

Consider a multiaquifer system consisting of M aquifers separated by $M-1$ aquitards. Equations (1) and (2) are solved numerically with a finite element model and coupled together by consideration of mass conservation.

Applying the finite element method, each aquifer is discretized into the same triangular finite element grid of N nodes, and each vertical column within an aquitard, extending from node i in the lower aquifer to the corresponding node i in the upper aquifer, is represented by k 1D elements ($k + 1$ nodes overall, $k - 1$ inside the aquitard plus 2 at the interfaces). Applying the Galerkin procedure to eq. (1), using linear basis functions and considering the leakage terms as known quantities, yield the following finite element equation for each individual aquifer in matrix form:

$$[H_1]\{h\} + [P_1]\left\{\frac{\partial h}{\partial t}\right\} + \{Q\} = 0 \tag{3}$$

where $[H_1]$ and $[P_1]$ are the N-dimensional aquifer stiffness and capacity matrices, and $\{Q\}$ is a N-dimensional vector accounting for flow at the aquifer-overlying aquitard and aquifer-underlying aquitard interfaces. For the i-th aquifer confined between aquitards j and $j + 1$, the r-th term of $\{Q\}$ is given by:

$$Q_r = \frac{q_{j,r}}{3} \sum_e \Delta_e + \frac{-q_{j+1,r}}{3} \sum_e \Delta_e \qquad (4)$$

where $q_{j,r}$ and $q_{j+1,r}$ are specific leakages from aquitards j and $j + 1$ at node r, Δ_e represents the area of the triangle e and the summation is taken over all elements e sharing node r.

Since the vertical elements, into which the aquitards are discretized, are not coupled, aquitard flow equation (2) is numerically integrated independently over each vertical element column. Analyzing the r-th vertical column within the aquitard j, the solution of the aquitard flow equation is obtained by prescribing the flow continuity condition at the aquitard-aquifer interfaces. Applying the Galerkin procedure and using linear basis functions leads to the matrix equation:

$$[H_2]\{h\} + [P_2]\left\{\frac{\partial h}{\partial t}\right\} + \{q'_{j,r}\} = 0 \qquad (5)$$

where $[H_2]$ and $[P_2]$ are $(k + 1)$-dimensional global stiffness and capacity matrices and $\{q'_{j,r}\}$ is a $(k + 1)$-dimensional specific leakage vector representing outflow from the aquitard column per unit horizontal area. It is $q'_{j,r} = 0$ for all column nodes except for those at the top and bottom of the column, belonging to both the aquifer and the aquitard. If we consider the node at the bottom of the column (shared by aquitard j and aquifer i), because of the flow continuity condition $q'_{j,r} = q_{j,r}$, where $q_{j,r}$ is the leakage term used in eq. (4). The term $q_{j,r}$ appears in both eq. (3) and eq. (5), and couples the aquifer and aquitard flow together.

Writing eq. (3) for the M aquifers and eq. (5) for the N columns of the $(M - 1)$ aquitards, and applying the Crank-Nicolson scheme for the time integration, yields a system of linear equations [Neuman et al., 1982]:

$$[H]\{h\}_{t+\Delta t} = \{b\} \qquad (6)$$

where $[H]$ is a symmetric positive defined matrix.

BLOCK GAUSS-SEIDEL AND SOR ITERATIVE SOLUTION SCHEMES

Consider the sample multiaquifer system of Figure 1(a). Each aquifer is discretized into 6 triangular elements (7 nodes) and each aquitard column into 4 linear elements (3 nodes inside the aquitard). With the nodal numbering indicated in Figure 1(a), the sparsity pattern of the global matrix $[H]$ (eq. (6)) is shown in Figure 1(b). We can observe the following:

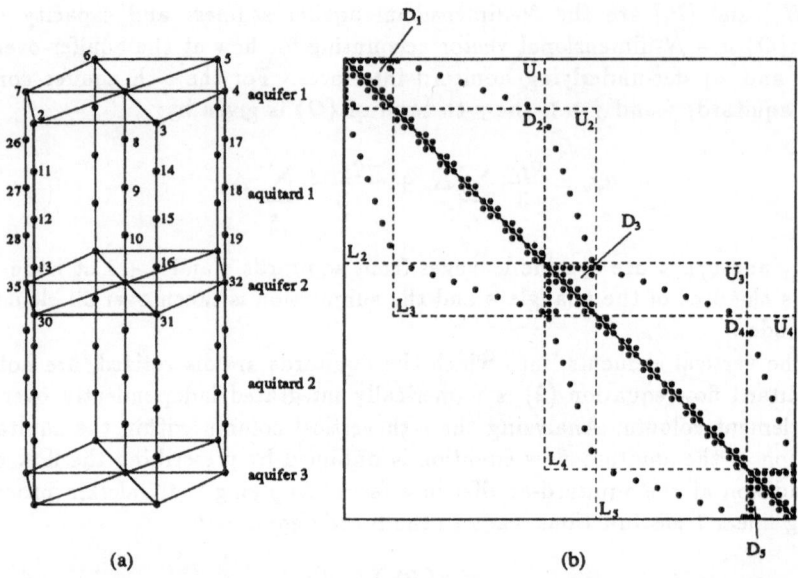

(a)   (b)

Figure 1. Discretization of a sample multiaquifer system (a), and the related matrix sparsity pattern (b).

- the non-zero structure of the aquifer matrices ($[D_1]$, $[D_3]$ and $[D_5]$ in Figure 1(b)) and that of the aquitard matrices ($[D_2]$ and $[D_4]$) are easily located within the global matrix;
- each aquifer matrix is sparse and each matrix corresponding to an aquitard nodal column is tridiagonal;
- the number of "coupling terms" is relatively small and their position ($[U_1] \div [U_4]$ and $[L_2] \div [L_5]$ in Figure 1(b)) is out of the square submatrices ($[D_1] \div [D_4]$) associated with of each layer.

These specific sparsity properties of $[H]$ naturally suggests the iterative decoupled solution strategy. Each layer, beginning from the upper one, is solved separately, decoupling the global system into a number of smaller subsystems. If we solve at iteration level k the subsystem of the i-th unit, $[L_i]$, $[D_i]$ and $[U_i]$ are the matrix blocks that must be considered (Figure 1(b)). Decoupling is accomplished with the elimination of the coupling terms from the subsystem matrix: the terms of $[L_i]$ are multiplied by the corresponding nodal head values known at the current iteration k because they belong to the $(i-1)$-th layer (the $(i-1)$-th layer has already been solved at the current iteration). The terms of $[U_i]$ are multiplied by the corresponding head values at the previous iteration level $k-1$. At the initial iteration, the previous time step values are used. Iterations are continued until the Euclidean norm $|r_k|$ of the residual of the global system:

$$\{r\}^{(k)} = \{b\} - [H]\{h\}^{(k)}_{t+\Delta t}$$

is less than a specified tolerance TOL. A faster rate of convergence is achieved starting the cycle from the pumped aquifer. If aquifers 1 and 3 of Figure 1(a) are pumped, the most convenient solution sequence is:

aquifer 1 - aquitard 1 - aquifer 3 - aquitard 2 - aquifer 2

The aquifer subsystems are solved by the MCG method, which is best suited to large sparse positive definite set of equations [*Gambolati and Perdon, 1984*]. The direct Thomas alghoritm is used to solve the aquitard subsystems; since the vertical elements into which an aquitard is discretized, are not coupled, the alghoritm is separately applied to each sub-block corresponding to each single vertical column.

The iterative solution scheme outlined above can easily be shown to be equivalent to a block Gauss-Seidel iterative method [*Varga, 1962*]. It is well known that the block Gauss-Seidel procedure can be mathematically accelerated and generalized into a block SOR by a relaxation factor ω. If system (6) is rewritten under the form:

$$
\begin{bmatrix}
H_{1,1} & H_{1,2} & \cdots & H_{1,N} \\
H_{2,1} & H_{2,2} & \cdots & H_{2,N} \\
\cdot & \cdot & & \cdot \\
\cdot & \cdot & & \cdot \\
\cdot & \cdot & & \cdot \\
H_{N,1} & H_{N,2} & \cdots & H_{N,N}
\end{bmatrix}
\begin{bmatrix}
X_1 \\ X_2 \\ \cdot \\ \cdot \\ \cdot \\ X_N
\end{bmatrix}
=
\begin{bmatrix}
B_1 \\ B_2 \\ \cdot \\ \cdot \\ \cdot \\ B_N
\end{bmatrix}
$$

where $H_{i,i}$, $1 \leq i \leq N$, are square and non void matrices, the associated block SOR iterative scheme can be defined as [*Varga, 1962*]:

$$
H_{i,i}X_i^{(k+1)} = H_{i,i}X_i^{(k)} + \omega \left\{ -\sum_{j=1}^{i-1} H_{i,j}X_j^{(k+1)} - \sum_{j=i+1}^{N} H_{i,j}X_j^{(k)} + B_i - H_{i,i}X_i^{(k)} \right\}
$$

$$k \geq 0 \qquad (7)$$

If $[H]$ is symmetrical and positive definite, the block SOR scheme (7) is convergent for all $\{X\}^{(0)}$ if and only if $0 < \omega < 2$ [*Varga, 1962*].

PRELIMINARY RESULTS

The sample multiaquifer system used to analyze the convergence properties of the new solution scheme is composed of 6 circular aquifers and 5 intervening aquitards. The aquifers are discretized into 324 triangles and 169 nodes and each vertical aquitard column into 6 (test case 1) or 11 (test case 2) linear elements, with 5 or 10 interior nodes, respectively. The boundary conditions are zero hydraulic head on the boundary nodes, and a unit pumping rate from the central nodes of the shallowest and lowest aquifer. Initial conditions are assumed to be static. The capacity matrix is lumped into a diagonal matrix. Constant hydrogeological parameters are used in each formations.

Some steady-state simulations are run for various values of the dimensionless parameter $G = (K_z b)/T_x$, which is representative of the degree of coupling. Decrease

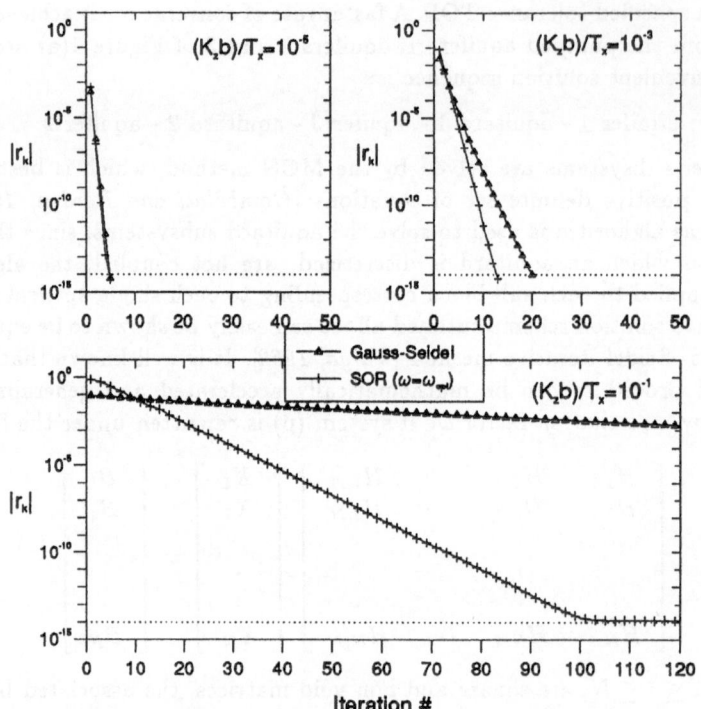

Figure 2. Test case 2: convergence profiles in steady state conditions for three values of the dimensionless parameter $(K_z b)/T_x$. On the vertical axis the Euclidean norm of the residual $|r_k|$ is given.

of G implies a relative decrease of the aquitard permeability, hence flow is more restricted to the pumped layers and coupling is less important.

Figure 2 shows the convergence profiles for the test case 2 of the block Gauss-Seidel and block SOR scheme with the optimum relaxation factor ω_{opt}. Table 1 gives the spectral radius ρ and ω_{opt} versus G of both schemes for the two test cases. Note that ρ is computed as $|r_{k+1}|/|r_k|$ (for sufficiently large k) between two consecutive Euclidean norms of the residual while ω_{opt} is evaluated by the equation:

$$\omega_{opt} = \frac{2}{1 + \sqrt{1 - \rho_{GS}}}$$

where ρ_{GS} is the spectral radius of the block Gauss-Seidel iterative matrix. The numerical computation of ω_{opt} has provided a very close value.

The results of Figure 2 and Table 1 point out that the asymptotic rate of convergence of the block Gauss-Seidel and SOR schemes is related to G. If coupling between the layers increases (the importance of the terms of the $[L_i]$ and $[U_i]$ matri-

		ρ_{GS}	ρ_{SOR}	ω_{opt}
	$(K_z b)/T_x = 10^{-5}$	0.0011	0.0009	1.0003
Test	$(K_z b)/T_x = 10^{-4}$	0.0119	0.0092	1.0030
case 1	$(K_z b)/T_x = 10^{-3}$	0.1598	0.0512	1.0435
	$(K_z b)/T_x = 10^{-2}$	0.6561	0.2967	1.2607
	$(K_z b)/T_x = 10^{-1}$	0.9461	0.6544	1.6232
	$(K_z b)/T_x = 10^{-5}$	0.0018	0.0010	1.0004
Test	$(K_z b)/T_x = 10^{-4}$	0.0207	0.0142	1.0052
case 2	$(K_z b)/T_x = 10^{-3}$	0.2601	0.0827	1.0752
	$(K_z b)/T_x = 10^{-2}$	0.7790	0.3697	1.3605
	$(K_z b)/T_x = 10^{-1}$	0.9698	0.7346	1.7039

Table 1. Maximum eigenvalue of the iteration matrix and ω_{opt} versus $(K_z b)/T_x$.

ces increases compared to that of the $[D_i]$ matrices), the rate of convergence becomes worse. If flow in the various layers is weakly coupled, optimal SOR tends to coincide with Gauss-Seidel ($\omega_{opt} = 1$). By contrast convergence increases with optimal SOR for higher G values in which case ω_{opt} is far from 1.

Figure 3 compares the computational efficiency of the coupled and decoupled solution strategy in terms of CPU time. If coupling is important ($G > 10^{-3}$), the coupled (MCG) strategy is more efficient than the block decoupled strategy. In this condition, a marked difference between the performance of optimal SOR and Gauss-Seidel exists (if $G = 10^{-1}$, $\text{CPU}_{GS}/\text{CPU}_{SOR} > 8$). On the contrary, when flow in the various layers is not much coupled ($G < 10^{-3}$), the efficiency of SOR and Gauss-Seidel approaches that of MCG. If $G < 10^{-4}$ and test case 2 is considered (where a large number of nodes for the aquitard discretization are used), both SOR and Gauss-Seidel schemes perform even better than MCG.

CONCLUSIONS

A decoupled solution strategy for a quasi-3D finite element model of flow in multiaquifer systems is developed. The procedure, which is naturally suggested by the special sparsity pattern of the coefficient matrix, is equivalent to a block Gauss-Seidel iterative method and can be accelerated by the use of a SOR factor. Results from some preliminary tests show that the asymptotic rate of convergence of optimal SOR scheme increases if the contrast between aquifer and aquitard permeability increases thus making coupling less important. When flow is weakly coupled, optimal SOR and Gauss-Seidel schemes tend to coincide. By distinction, SOR shows a superior convergence and a better computational efficiency if the degree of coupling between layers increases. The decoupled solution strategy is generally more time consuming than the coupled (MCG) strategy in linear problems. However, the performance of both Gauss-Seidel and SOR schemes approaches that of the MCG when coupling is weak.

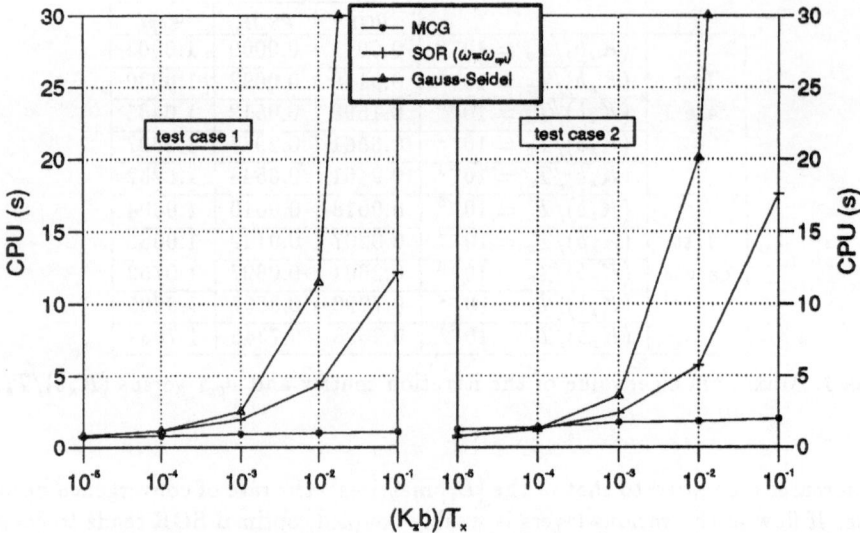

Figure 3. Comparison between the computational efficiency of the coupled (MCG) and decoupled (Gauss-Seidel and SOR) solution strategies with linear test problem.

The new solution strategy will be applied to nonlinear multiaquifer systems, where an iterative solution scheme of the kind developed in the present paper is naturally designed to cope with the non linear behavior of the soil properties.

Acknowledgments. This work was developed in the CNR project "Sistema Lagunare Veneziano", Linea di Ricerca 2-7, U.O. 2.

REFERENCES

Bredehoeft, J. D. and G. F. Pinder, Digital analysis of areal flow in multiaquifer groundwater systems: a quasi three-dimensional model, *Water Resour. Res.* 6, 883–888, 1970.

Chorley, D. W. and E. O. Frind, An iterative quasi three-dimensional finite element model for heterogeneous multiaquifer systems, *Water Resour. Res.* 14(5), 943–952, 1978.

Gambolati, G. and A. M. Perdon, The conjugate gradients in flow and land subsidence modeling, In: Bear, J. and Y. Corapcioglu (eds.) *Fundamentals of Transport Phenomena in Porous Media.* NATO-ASI Series, Applied Sciences 82, Martinus Nijoff B.V., The Hague, pp 953–984, 1984.

Neuman, S. P., C. Preller and T. N. Narasimhan, Adaptive explicit-implicit quasi three-dimensional finite element model of flow and subsidence in multiaquifer systems, *Water Resour. Res.* 18(5), 1551–1561, 1982.

Varga, R. S., *Matrix Iterative Analysis.* Prentice-Hall, Englewood Cliffs, New Jersey, 1962.

COUPLING OF 2D AND 3D GROUND WATER MODELS: A CASE STUDY

M. VOGT, R. HORST and M. GERDING
Lahmeyer International GmbH ·
Lyoner Strasse 22
60528 Frankfurt/Main
Germany

Within the framework of plans for the new ICE high-speed railway line, the German Railways Board investigated the influence of tunnel alternatives which intersected the ground water aquifer at three locations on the right bank of the Rhine in the Cologne-Bonn area. In order to investigate not only the three-dimensional influence of flow disruptions on specific local areas but also on overall ground water management strategies, a numerically-coupled 2D/3D stationary/transient finite element ground water model was developed. This enabled the investigation to be carried out with the accuracy required and within an acceptable budget.

INTRODUCTION

A large number of practical problems in the field of applied ground water modelling arise from the influence of specific local changes in ground water flow or quality on overall ground water management strategies. The problem is often magnified because not only do the effects of local changes have consequences for a larger area but in addition, boundary conditions, such as flooding, in the larger area also have a considerable influence on the local changes which are to be determined.

Such a situation arose during an investigation for the German Railways Board, in which the three-dimensional effects of restrictions in aquifer flow resulting from the construction of three tunnel alternatives occurred. The influence of the structures was to be investigated not only for the near vicinity but also for a wider area and was to be optimized in terms of minimizing these effects. The task was rendered more difficult because a hydrological flood event within the study area, the course of which had an influence on the flow-conditions in the areas near the various tunnel structures, was to be included in the evaluation procedure.

In the following paper, some general points for the solution of such problems are discussed. Finally, the cost-saving solution which was developed for the study, a numerical coupling of a 2D model of the whole area with one or more detailed 3D models of specific areas of interest is presented. The practical procedure is explained in conjunction with the above-mentioned investigations of tunnel alternatives for the Cologne-Rhine/Main ICE railway line.

A. Peters et al. (eds.), Computational Methods in Water Resources X, 115–122.
© 1994 Kluwer Academic Publishers. Printed in the Netherlands.

DISCUSSION OF ALTERNATIVE SOLUTIONS

Fundamentally, a number of alternative solutions are possible for this type of coupled problem. These provide more or less acceptable results, according to the type of problem and precision required (Fig. 1):

- Preparation of a 2D model of the whole area only, with refinement of the discretization in the respective vicinity of ground water changes. An estimate and application of the 3D effects and their influence such as flow disruptions near a structure is made by selecting a proportionally-altered permeability coefficient. The validity of the selected approximation and its application must be investigated in detail, e.g. through vertical plane observations, and the reliability of this procedure demonstrated. No exact data on local conditions can be obtained using this method. The influence over a larger area cannot be determined, as the higher precision required also renders this procedure unsuitable.

- Preparation of a 3D model for the whole area. Limitations in computer capacity, processing speed, preparation costs and ease of handling are likely to make this procedure impractical if extensive or multiple ground water disturbances are to be investigated. If a coarser 3D discretization is used to surmount the above-mentioned hurdles, the reliability and accuracy of the model suffers, at least at a local level, and is no longer acceptable.

- Preparation of a general model (usually 2D; possibly multi-layer or 3D) and one or more detailed 3D models. The interactions between the models can be taken into consideration by the input of relevant boundary conditions and iteration between the models. Practical limitations are encountered at the transition to transient computations.

model type	regional flow regime	local 3D flow components	data aquisition and preparation	handling of model	required calculation time	costs
2D regional (Transmissivity)	+ +	– –	+ +	+ +	+ +	+ +
3D regional	+ +	+	– –	– –	– –	– –
3D Detail	– –	+ +	+ / –	+ / –	+ / –	+ / –
2D regional + 3D Detail	+ +	+ +	+ / –	– –	+ / –	+ / –
2D regional / 3D Detail	+ +	+ +	+ / –	+	+ / –	+ / –

Fig. 1   Comparison of alternative solutions

- Preparation of an overall model (usually 2D) and one or more detailed 3D models with numerical coupling of the general model with each of the detailed 3D models. The theoretical basis and practical applications of this procedure will be explained in the following.

NUMERICAL COUPLING OF 2D AND 3D MODELS

The 2D model is a transmissivity model based on the geometry of triangular elements; the 3D model is based on tetrahedral elements. It includes the implicit determination of the free surface through iterative sectioning of elements with fixed discretization (Vogt, 1986).

A number of equations (equivalent to the number of nodes) is elaborated for each triangular (prism) and tetrahedral element for the coupled model. The combination of these equations into an overall equilibrium system is based on the local mass balance, which must be rigidly adhered to at each node, and the continuity of the respectively approximated ground water potentials. By this means, the node equations for all elements, independent of their dimension and number, can be added to an equilibrium system.

Coupling results in the formation of a equilibrium system, in which the 3D node fraction is appended behind the 2D node fraction. A node index vector for the 2D discretization is elaborated, in which the following are given:
- whether the given node can be treated as "normal", i.e. the node has no contact with 3D nodes,
- whether the node is a contact node to 3D nodes (if so, to which?),
- whether the node lies in the area of the 3D model and consequently requires no further treatment in the equilibrium system (in order to avoid numerical inconsistency, the node is allocated a DIRICHLET boundary condition).

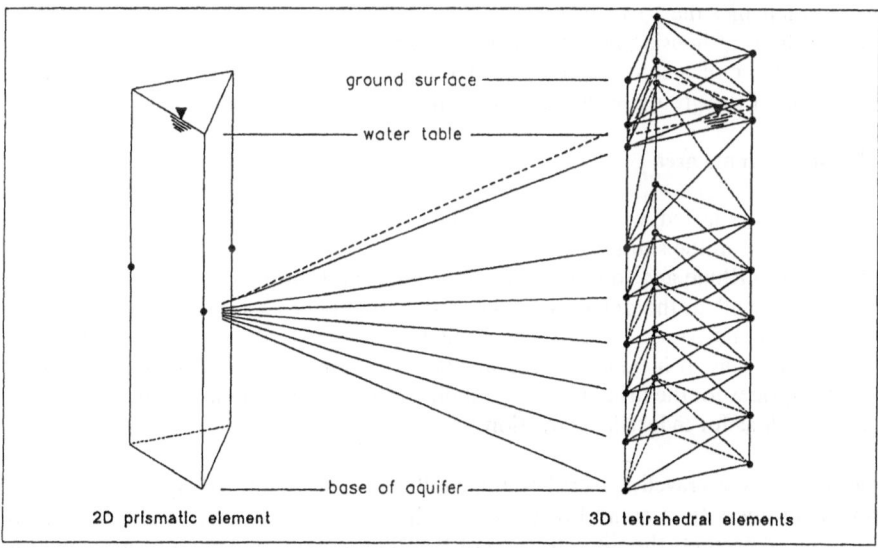

Fig. 2   Scheme of the coupling of the 2D- with the 3D-elements

This index vector, which may easily be defined by an appropriate preprocessor, represents the only difference between the separate functional 2D and 3D models and the coupled model. Only the 2D model nodes found at the coupling nodes of the prism elements must be coupled with the corresponding 3D model nodes of the superimposed tetrahedral elements in the equilibrium system (Fig. 2). At the free surface area, a precise transfer of the elements or node fraction must be undertaken, particularly in the case of a transient moving ground water interchange zone.

The coupled symmetrical equilibrium system can be resolved in this case using a solver which stores only the effective skyline of the given band width at each node. This guarantees that unfavourable band widths occurring at the coupling nodes do not lead to an excessively high memory requirement in the coupled equilibrium system.

CASE STUDY: COUPLED MODEL FOR THE GERMAN RAILWAYS BOARD

Data base

The German Railways Board investigated the effect of three different tunnel alternatives, a Cologne/Bonn airport rail link, a subway in the Troisdorf-Spich city area and a tunnel below the River Sieg near Siegburg/Sankt Augustin on the ground water hydraulics within the framework of plans to develop the ICE high-speed railway line between Cologne and the Rhine/Main area. The depth and alignment of the tunnels with regard to ground water hydraulics were to be optimized, if necessary. Tunnel construction represents consequences for:
- the overall ground water management in the "Kölner Bucht" (the tunnels are driven in some cases through the protected areas for drinking water extraction),
- the ground water flow characteristics as a result of the disruption of the ground water flow which accompanies the River Sieg,
- the foundations of housing near the new structures,
- the sealing layer of a household waste depot because of possible submersion due to modification of the ground water transfer zone.

2D model of the whole area

The area to be investigated, which is presented in Fig. 3, extends from the Bonn-Siegburg/-Sankt Augustin line in the south to the level of Cologne-Deutz in the north and is bordered to the west by the Rhine and to the east by the geological transit from the low flood plains of the "Kölner Bucht" to the devonian rocks of the "Bergisches Land". The area is crossed by the rivers Sieg and Agger and includes a number of gravel pits, some of which are still operated commercially. It includes a large number of water works, emergency water works and protection zones for the drinking water supply to the heavily populated areas of Cologne and Bonn as well as for industrial extraction.

Geologically, a single gravelly ground water aquifer can be identified in this area, which increases in thickness from some few meters in the south-east to more than 20 m in the north-west. Fig. 4 presents the discretization of the 2D ground water model (projected onto the calculated ground water surface) and the exact position of the relevant discretizations of the three detailed 3D models. To the west, the Rhine is defined as a leakage boundary condition. To the east and south NEUMANN's boundary conditions with magnitudes equivalent to the estimated inflows from the devonian rocks ot the "Bergisches Land" are assumed. Only to the north over a narrow band of model boundary potential boundary occurs.

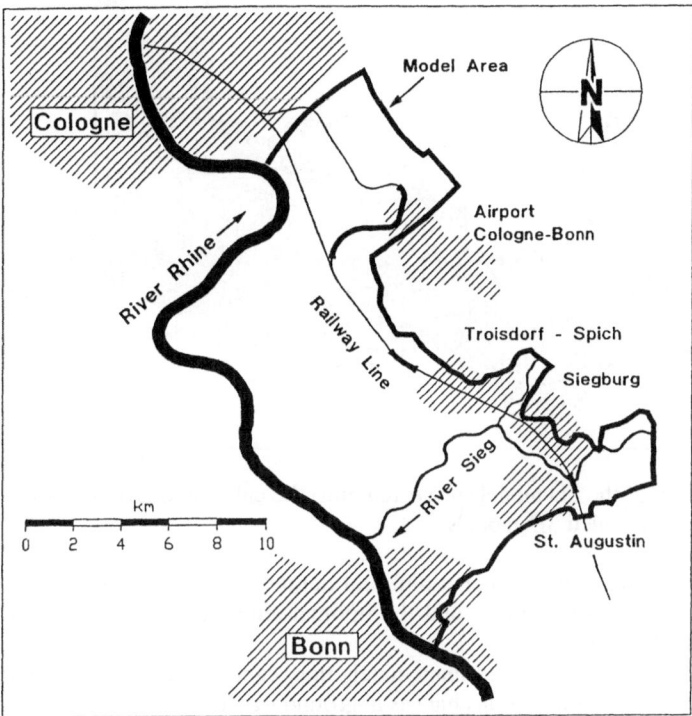

Fig. 3   Area investigated, railway alignment and tunnels

A stationary calibration for mean hydrological conditions was undertaken as well as a verification based on low water conditions in the Rhine combined with a low level of ground water replenishment. Approximately 400 observation wells were available for comparisons. A south-north flow associated with the Rhine was determined from the ground water contours found under calibration conditions. This flow is additionally fed by the devonian rocks to the east (Fig. 4). Permeabilities between 0.002 m/s and 0.02 m/s were determined for the ground water aquifer. Furthermore, a transient calibration for a flood event on the Rhine and the other rivers, combined with higher ground water replenishment over a period of 6 months, was undertaken.

Detailed 3D model of tunnel driven below the Sieg River

In parallel to the calibration of the model for the whole area, the three detailed 3D models for the areas near the tunnels were elaborated. In the following, the 3D model representing the tunnel below the River Sieg and the influence of the tunnel are given in detail. A number of constraints were taken into account by the model:

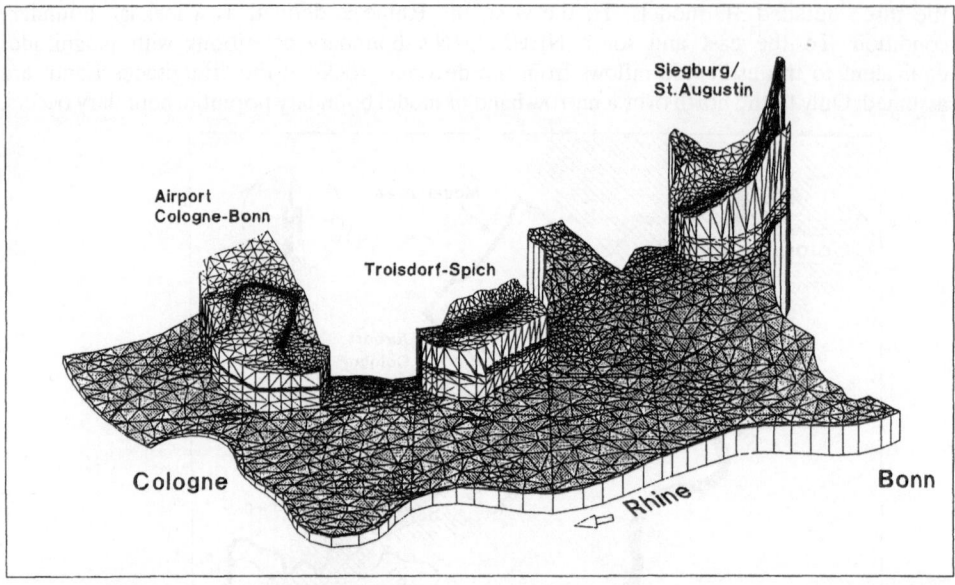

Fig. 4   Discretization of the 2D model (projected onto the calculated ground water surface)
and the three detailed 3D models.

- Six alternatives for driving the tunnel at different depths were taken into account as well
  as the appurtenant ramp structures, and the optimal alternative with regard to ground
  water hydraulics elaborated.
- A household waste depot is located immediately upstream of the tunnel. A submersion
  of the foundation seal resulting from changes in ground water conditions must be avoided.
  Particular attention was paid to this area in the discretization process.
- In certain sections, the tunnel and/or the ramps completely disrupt the ground water flow
  associated with the River Sieg.

A discretization process comprising tetrahedral elements, which are finely graded near the
tunnel in both horizontal and vertical directions, was elaborated for the above-mentioned
reasons. Within the framework of the vertical subdivisions, simulation of all tunnels and ramp
locations was planned beforehand and undertaken by variation of the permeability coefficients
of superimposed layers and thus undertaken on a time-saving basis (Fig. 5). The uppermost
element series is extended up to the ground level over the whole model area. As a rule, this
area remains dry for the entire computation.

As the areas of equal permeability coefficient, determined during the calibration of the
general model, were integrated into the discretization during its elaboration, only fine
calibrations of the 3D model following coupling with the general model were necessary. It was
only necessary to extrapolate the internal and external boundary conditions of the general
model to the detailed model and to distribute factors such as inflows vertically onto a profile
of nodes. The detailed model was only coupled with the general model by nodes on the east

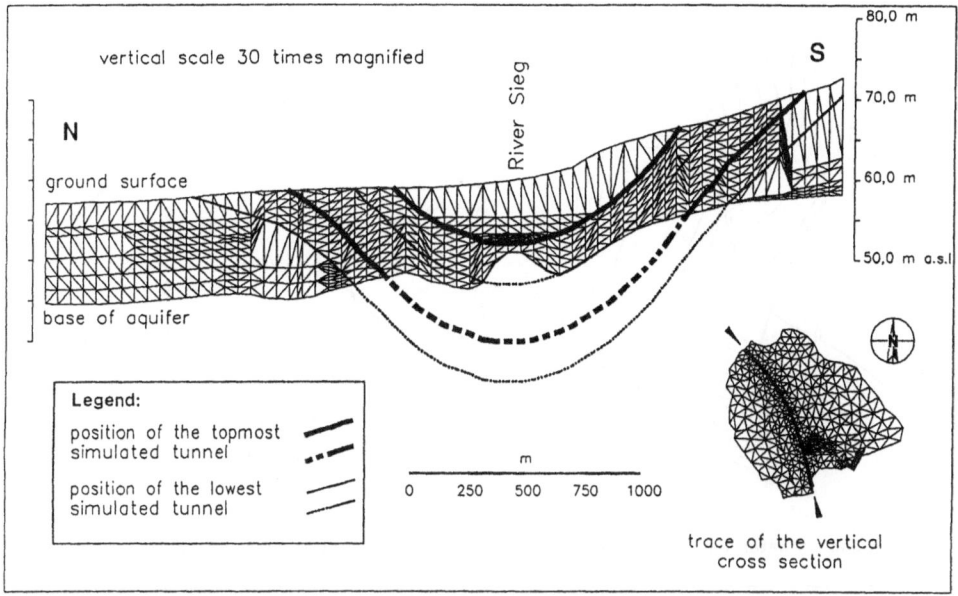

Fig. 5 Vertical section through the discretization parallel to the tunnel alignment
(magnified)

and west. To the north and south, the boundary of the detailed model formed the boundary
to the devonian rocks, where the temporally-variable inflows determined from the general
model could be applied.

Simulations with the coupled model

During the course of computations for stationary events, an initial investigation of six
different depths for the tunnel alternatives and the resultant effects on the immediate and
distant areas was undertaken with the assistance of the coupled model. Contrary to initial
fears arising from the assumption of a large-scale vertical disruption of the ground water flow
parallel to the River Sieg, the analysis of the first results showed only a small rise in water
table upstream of the tunnel of the order of max. 30 cm (Fig. 6). A precise analysis of these
results and comparison with the ground water flow provided the following explanation for
these small changes:

- A complete disruption of ground water flow, and subsequent elevation of the ground
  water table upstream, is only found in the ramp areas.
- Near the Sieg, the tunnel cuts into the tertiary strata, and the ground water can
  consequently flow above it.
- In the south tunnel ramp area, the ground flow is practically parallel to the tunnel
  alignment.
- The north tunnel ramp disrupts the ground water flow vertically across the direction of
  flow, but only over a short distance. To either side, the ground water can flow either
  above or below the tunnel or the ramp.

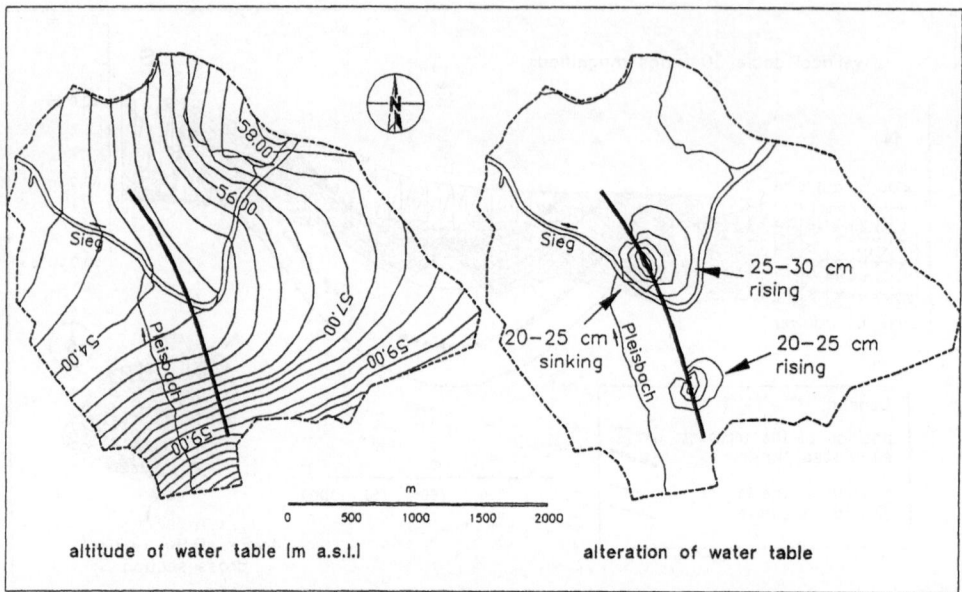

Fig. 6   Original flow situation (left) and differences in water table (right)

With the assistance of this and further transient computations, encouraging results were found in these investigations. Firstly, the most economic alternative for the German Railways Board, the most shallow and shortest alignment, is also the most favourable with regard to ground water flow. Secondly, the waste depot floor seal does not come into contact with the ground water to any significantly greater degree, even in the case of a flood event. Furthermore, by virtue of the small changes in flow characteristics, overall ground water management is not detrimentally affected by tunnel construction.

SUMMARY

By the development of a numerically coupled 2D and 3D ground water model, a cost and time-saving method was elaborated which enabled sufficiently accurate investigations of the effects of three tunnel alternatives on ground water conditions near the tunnels and on overall ground water management strategies in the region to be undertaken within reasonable limits for the client, the German Railways Board.

ACKNOWLEDGEMENT

The authors thank the client, the Deutsche Bundesbahn, Bundesbahndirektion Köln, for the permission to publish the results of the investigation.

REFERENCES

Vogt, Matthias (1986) "Dreidimensionale Sickerströmung mit freier Oberfläche" Diplom-Thesis, 165 p., RWTH Aachen, unpublished.

DEVELOPMENT AND APPLICATION OF ADAPTIVE FE-MODELS FOR THE SIMULATION OF FLOW IN RUBBLE MOUND BREAKWATERS

H. WIBBELER and U. MEISSNER
Institute for Numerical Methods and Informatics in Civil Engineering
University of Darmstadt
Petersenstraße 13, 64287 Darmstadt
Germany

The contribution presents an adaptive finite element model for the analysis of wave-induced flow processes in rubble mound breakwaters. This model simulates the two-dimensional non-linear groundwater flow within the structure.
The approximation errors of the numerical calculation are estimated by an a posteriori analysis of finite element patches. By use of this error estimator mesh refreshments are carried out.
The application of the numerical model is shown by presenting results about several numerical studies of two rubble mound structures.

INTRODUCTION

Rubble mound breakwaters represent the most commonly used type of structure for the protection of coastal areas and harbours against wave action. The traditional approach to study the stability of such structures is the use of physical models. To minimize the costs of such experiments numerical models can be used to carry out parameter studies. For this purpose the numerical model must be calibrated by experimental data from the large-scale model tests.

The reliability of finite element results depends on the magnitude of approximation errors. Therefore an a posteriori analysis of approximation errors is of great importance for practical applications. In the presented finite element model the quantitative and qualitative distribution of approximation errors are calculated by an error estimator. The error distribution can be used for mesh refinement techniques to achieve optimal numerical solutions with a minimum of computation effort.

A. Peters et al. (eds.), Computational Methods in Water Resources X, 123–130.

SIMULATION OF FLOW WITHIN THE BREAKWATER

The wave-induced flow inside the rubble mound breakwater is described by a two-dimensional non-linear groundwater flow. This flow is characterized by the following equation:

$$\phi\gamma\frac{\partial S}{\partial p}\frac{\partial h}{\partial t} - k_r\,k_n\,div\,(\;K\;grad\;h\;) - q = 0 \; . \tag{1}$$

The non-linearity of the flow is described by the permeability parameter k_n which can be calculated from the equation

$$k_n = \frac{2}{1 + \sqrt{1 + 4b/a\,|K\;grad\;h|}} \; . \tag{2}$$

The parameters of equations (1) and (2) are:

a, b	FORCHHEIMER-constants
γ	specific weight
h	piezometric head
K	permeability tensor
k_n	non-linear coefficient of permeability
k_r	relative permeability
p	pressure
ϕ	porosity
q	discharge
S	saturation
t	time

An important part of the numerical model is a reliable algorithm for the evaluation of the position of the phreatic surface [3]. The calculations must be carried out by an iterative procedure in each time step. For this procedure the cross-section of the breakwater is devided into two parts: the wet part beneath and the dry part above the phreatic surface. The two parts are separated by a transition area. During the iteration, the permeability of the structure above the phreatic surface is set to a very small value, to behave as almost impermeable.

In each iteration step the element and saturation matrices are calculated by using the position of the phreatic surface from the previous step.

The matrices of elements which are intersected by the phreatic surface have to be calculated by a special integration technique [4]. The iteration continues until the position of the phreatic surface remains constant.

SIMULATION OF WAVE MOTION

The boundary conditions of the finite element model are governed by the wave motion on the seaward slope of the breakwater. At each boundary node the wave height is prescribed by a function $h_{(t)}$, where $h_{(t)}$ denotes the piezometric head. The wave motion can be simulated by several methods. In the present version of the numerical model, following methods are available:

1. Data from measurements of the wave field;

2. Analytical solutions of wave theories:
   linear wave theory,
   non-linear wave theory (Stokes);

3. One-dimensional linear numerical wave model.

If the wave loads are calculated by a wave theory, a dynamic pressure can be evaluated and used to modify the boundary conditions.

SIMULATION OF TWO-PHASE FLOW

Air entrainment, essentially resulting from wave breaking, can decrease the hydraulic conductivity of a porous structure significantly. A numerical model for rubble mound breakwaters must be able to describe such a two-phase flow. In the presented finite element model the air entrainment ist simulated by using modified FORCHHEIMER-constants. The FORCHHEIMER-constants are reduced by the air-fraction α as shown by HANNOURA/McCORQUODALE [1].
The factor α must be defined as a function $\alpha(t)$ for all elements of the finite element mesh. This can be achieved by defining polygons with functions $\alpha(t)$ within the cross-section of the breakwater. For this purpose a procedure is used to recognize if the actual element is located inside or outside of the polygon. The maximum of air entrainment occurs during wave run-up in the area shown in Fig. 1.

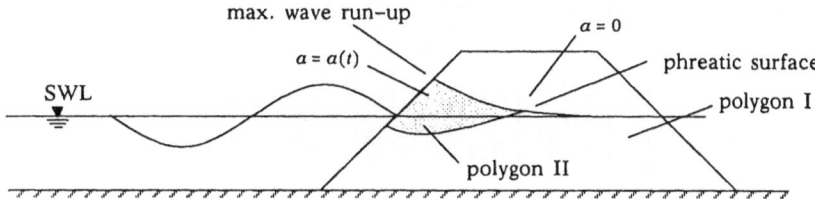

Fig. 1: Location of air entrainment in a breakwater

RELIABILITY OF THE NUMERICAL MODEL

The numerical errors of the finite element solution are calculated by use of an error estimator. The theory of the error estimation has been published in an earlier paper [2].

This contribution presents an extension of the published procedure and its application to triangle meshes.

In the finite element model the piezometric head h is first approximated by polynomial functions. After the initial calculation a higher order approximation is constructed by local refinements of the original mesh (h-version). This technique is applied to all individual patches which are cut out of the finite element mesh as illustrated in Fig. 2.

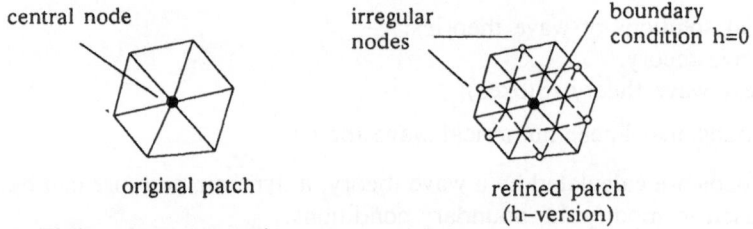

Fig. 2: Finite element patch

The adaptive algorithm estimating the approximation errors is carried out by the following steps:

1. Global finite element analysis of the original mesh,
2. Assembling of local element patches around the central node,
3. Calculation of error loads from all elements of the patch,
4. Local finite element analysis of the patch,
5. Calculation of improved velocities within the individual patches,
6. Computation of the error estimator from mean velocities.

During this procedure the steps 2 to 6 are carried out for all nodes of the finite element mesh. At the end of the algorithm an error estimator distribution is available for all nodes. These data are transferred to the nodes of a rectangular grid as shown in Fig. 3.

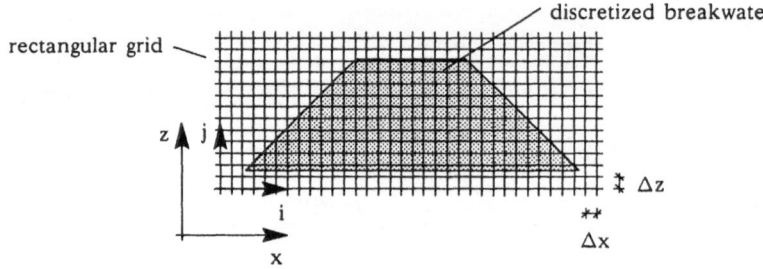

Fig. 3: Rectangular grid overlying the cross-section of a breakwater

By use of this grid the mesh generator can construct a new adaptive mesh. The values of Δx and Δz depend on the element size of the original finite element mesh. They should be as small as possible so that no informations about the error estimator distribution is lost. The calculations are continued with step 1 of the algorithm by use of the adaptive mesh.

The adaptive algorithm has to be carried out, until the numerical error is smaller than a given tolerance in the energy norm and it can be activated during a wave period at each time step or at pre–selected time steps.

APPLICATION OF THE NUMERICAL MODEL

The application of the numerical model is demonstrated by tests on two porous structures.

Example 1

In the first example the results of the adaptive algorithm are displayed and discussed. Subject of the numerical calculation is a rectangular homogeneous structure. The discretization of this structure with 768 elements and 425 nodes is shown in Fig. 4. The boundary conditions of the finite element calculations are given by the presented sine wave with a period of Tw=4.0 s.

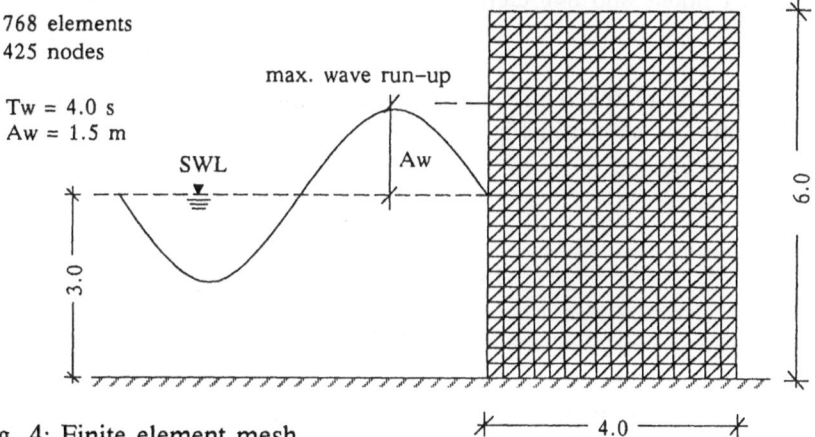

Fig. 4: Finite element mesh

Using a time step of Δt=0.1 s the numerical analysis is carried out from t=0.0 to 1.0 s. In each time step the error estimator is calculated for all nodes of the mesh by activating the adaptive algorithm. Each error estimator is compared to that of the previous step and stored if it is greater. The result of this procedure is an error distribution as illustrated in Fig. 5. By use of these data the mesh generator creates the shown adaptive finite element mesh. In this way an optimized mesh is constructed for a complete wave period.

If a numerical simulation is carried out for more than one wave period, the numerical effort might be reduced by calculating the error estimator only over the first wave period. This means, that the finite element mesh is optimized initially and remains constant over the following wave periods.

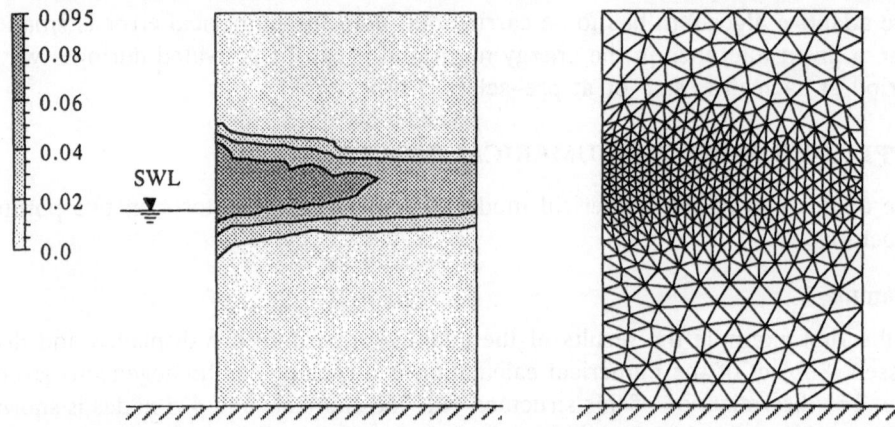

Fig. 5: Error distribution and adaptive finite element mesh
with 451 nodes and 840 elements

Example 2

A second application of the numerical model is demonstrated for a trapezoidal shaped breakwater, as shown in Fig. 6.

Lw=18.96 m

Aw=1.5 m

Tw=4.0 s

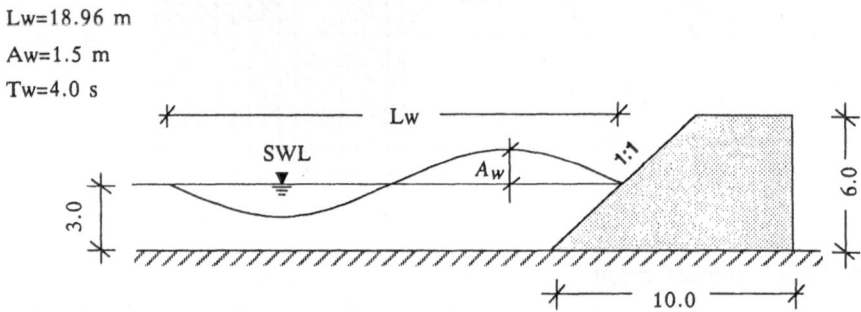

Fig. 6: Cross-section of the breakwater, example 2

For the discretization of this structure, 897 triangular elements with 485 nodes were used (see Fig. 7). Some typical results from the numerical analysis of this breakwater are presented in Figures 8–11.

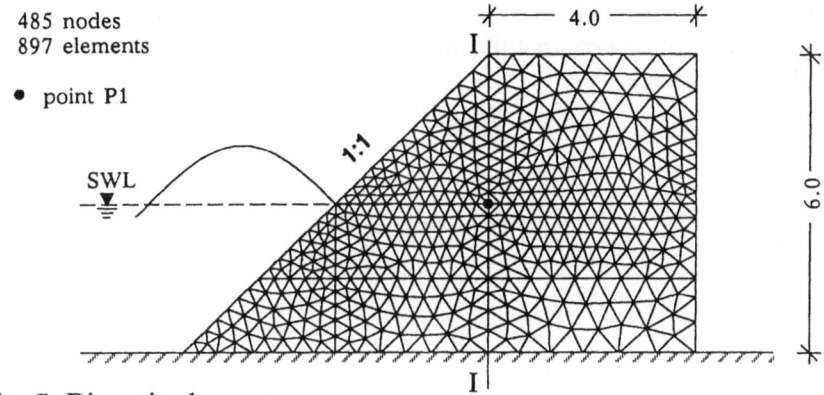

Fig. 7: Discretized structure

In Fig. 8 the phreatic surface and the vectors of velocity during wave run-down are plotted. These results provide a qualified understanding of the flow processes within the breakwater. The time variations of pore pressure p and the velocity v during two wave periods are displayed in Figures 9 and 10.

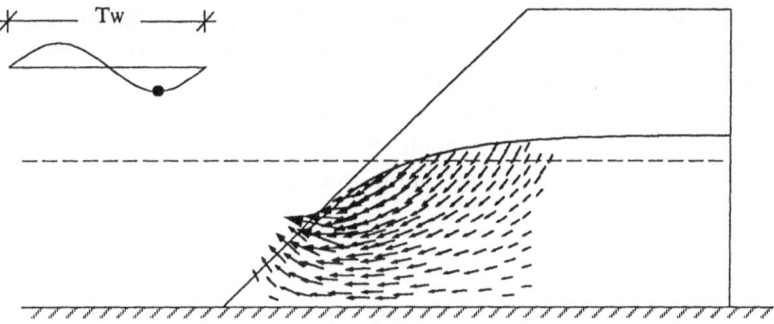

Fig. 8: Velocity vectors during wave run-down

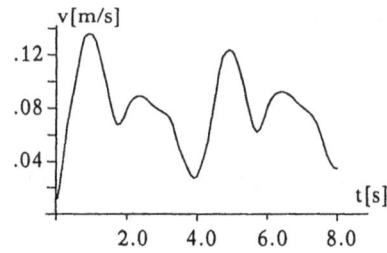

Fig. 9: Velocity v at point P1

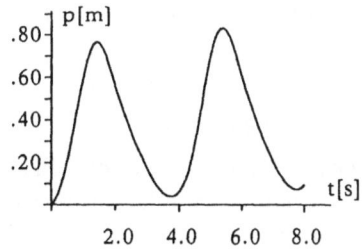

Fig. 10: Pore pressure at point P1

Fig. 11 illustrates the distribution of velocity v_x in section I–I during wave run–up and run–down. It can bee seen, that the maximum of v_x occurs during wave run–up near the phreatic surface.

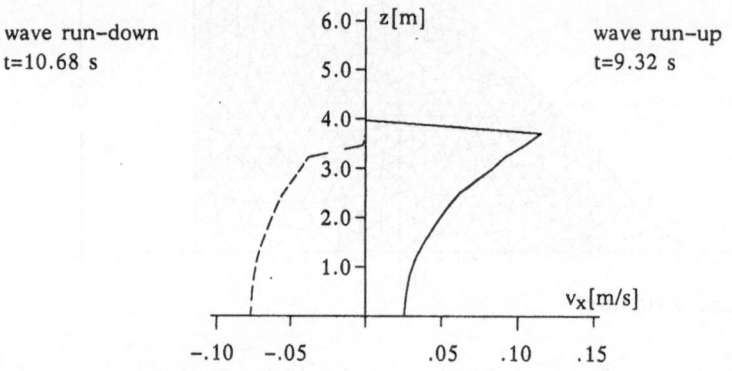

Fig. 11: Velocity v_x in section I–I

Further detailed results can be gained from the transient numerical model. The evaluation of such results are of great practical value for an optimization of this type of coastal structures.

The presented FE–model is very flexible in handling boundary conditions and structures with complicated geometries and different porous layers. This flexibility in conjunction with the mesh generator and the graphical post–processor makes the model very suitable for extensive parameter studies; e.g. variations of hydraulic conductivity, air entrainment or slope angle. Such numerical simulations are helpful to study the flow processes within rubble mound breakwaters with reliability and efficiency.

REFERENCES

1. **Hannoura A.A., McCorquodale J.A.** (1978) Air–Water Flow in Coarse Granular Media, Journal of the Hydraulics Division, Vol. 104, No. HY7.

2. **Meißner U., Wibbeler H.** (1990) A Posteriori Errors of Finite Element Models in Groundwater and Seepage Flow, Computational Methods in Subsurface Hydrology, Comp. Mech. Publications / Springer–Verlag, pp. 83–101.

3. **Wibbeler H., Oumeraci H.** (1992) Finite Element Simulation of Wave–induced Internal Flow in Rubble Mound Structures, 23rd International Conference on Coastal Engineering, Venice.

4. **Wibbeler H., Meißner U.** (1992) FEM–Model for Seepage Flow through Rubble Mound Breakwaters, IX International Conference on Computational Methods in Water Resources, Denver.

A THREE-DIMENSIONAL FINITE-ELEMENT MODEL OF TRANSIENT FREE SURFACE FLOW IN AQUIFERS

GOUR-TSYH YEH (*) and JING-RU CHENG (*)
JACOB A. BENSABAT (**)
(*) The Pennsylvania State University
University Park, PA 16802
USA
(**) Environmental & Water Resources Engineering LtD
6 Wedgewood Street
Haifa, 34633, Israel
ISRAEL

Modeling and calibration of three-dimensional complex, heterogeneous and anisotropic aquifers often require the consideration of a moving phreatic surface because the error induced by not considering it may be higher than standard calibration criteria. Many numerical models are available to deal with three-dimensional flow in phreatic aquifers. Most of these models, for example the widely used MODFLOW, employed ad hoc approaches to treat the change of free surface. In these ad hoc approaches, the net fluxes in the most upper cells are used to estimate the phreatic surface. On the other hand, there are many two-dimensional models to deal with the free surface under steady-state conditions using boundary element methods. Very few "true" transient, moving free surface models have been developed. In this paper, we present the development and verification of a three-dimensional moving boundary value model to treat the transient free surface in complex aquifers. The seepage face is also determined along with the tracking of the free surface. The implicit implementation of the moving boundary conditions enables the use of a large time-step size. Both steady state and transient problems are used to verify the model. For all examples, global mass balance is maintained to ensure the accuracy of simulations.

INTRODUCTION

Moving water table is caused by recharge and injection to and pumping from saturated zones in an unconfined aquifer. The shape, elevation, and position of the water table depend on the geometry and the rate of recharge and/or pumping, the geological structure, and the ambient groundwater flow in the aquifer. The interaction between groundwater flow field and the moving free surface is needed to predict the concentration distribution of chemicals added by the recharge, injection, and pumping. Models and methods for predicting the phreatic surface and flow pattern due to recharge and injection to and pumping from aquifers include: complete treatment of saturated-unsaturated flow (Yeh, 1987); Dupuit-Forchheimer assumption (Hantush, 1967); linearized potential flow (Dagan, 1967); boundary integral equation (Liggett and Liu, 1983); complex variables (Schmitz and Edenhofer, 1988), finite differences (McDonald and Harbaugh); and finite element methods (Huyakorn, 1983). While the saturated-unsaturated flow approaches provide the most rigorous treatment of modeling

A. Peters et al. (eds.), Computational Methods in Water Resources X, 131–138.
© 1994 Kluwer Academic Publishers. Printed in the Netherlands.

moving free surfaces, in many engineering applications, the data of unsaturated characteristics are not available. Furthermore, because of highly nonlinearity in unsaturated flow, convergency problems often occur, especially under dry conditions. As a result, a vast majority of groundwater flow models with moving phreatic surfaces deal with saturated zones only. While there are many three-dimensional flow models for phreatic aquifers, there are many more two-dimensional ones. Most of the three-dimensional models, for example, the widely used MODFLOW (McDonald and Harbaugh, 1985) employed ad hoc approaches to treat the change of free surface. In these ad hoc approaches, the net fluxes in the most upper cells are used to estimate the phreatic surface. The majority of two-dimensional models use either the Dupuit-Forchheimer approximations to result in vertically integrated flow equations (Yeh and Huff, 1983) or the assumption of negligible flow in the unsaturated zones to result in free surface flow (De Wiest, 1962; Dicker, 1969;Tsay, et al, 1993). Very few "true" transient, three-dimensional, moving free surface models has been developed. In this paper, we present the development and verification of a three-dimensional moving boundary model to treat the transient free surface in complex aquifers. The seepage face is also determined along with the tracking of free surfaces. The free-surface boundary condition is rigorously implemented in the finite element approximation of groundwater flow equations. Both transient and steady state problems are used to verify the model. For all examples, the mass balance is maintained to ensure the accuracy of simulations.

MODEL FORMULATION

Derivation of flow equations along with the specification of initial and boundary conditions can be found elsewhere (Huyakorn, 1983; Yeh, 1987). The governing equation, derived by substituting Darcy's law for specific discharge into the mass balance equation, is given by

$$F\frac{\partial h}{\partial t} = \nabla \cdot [\mathbf{K} \cdot (\nabla h + \nabla z)] + q \tag{1}$$

where h is the pressure head, t is time, $\mathbf{K}$ is the saturated hydraulic conductivity tensor, z is the potential head, q is the source and/or sink, and F is the storativity given by

$$F = \alpha' + n_e \beta' \tag{2}$$

In Eq. (2) α' is the modified compressibility of the media, n_e is the effective porosity, and β' is the modified compressibility of water. To complete the formulation of the problem, Eq. (1) must be supplemented with initial and boundary conditions. The initial condition can be stated as

$$h = h_i(x,y,z) \quad \text{in} \quad R, \tag{3}$$

where R is the region of interest and h_i is the prescribed initial condition, which can be obtained by either field measurements or solving the steady state version of Eq. (1).

Many types of boundary conditions can be specified for the simulation of groundwater flow problems with moving free surfaces. The first type is the Dirichlet boundary conditions in which the head is prescribed on the boundary as

$$h = h_d(x_b,y_b,z_b,t) \quad \text{on} \quad B_d \tag{4}$$

The second type is the Neumann boundary conditions in which the gradient of the pressure head is specified as

$$-\mathbf{n} \cdot \mathbf{K} \cdot \nabla h = q_n(x_b, y_b, z_b, t) \quad \text{on } B_n, \tag{5}$$

The third type is the Cauchy boundary condition in which the normal flux to boundary is prescribed as given by

$$-\mathbf{n} \cdot (\mathbf{K} \cdot \nabla h + \mathbf{K} \cdot \nabla z) = q_c(x_b, y_b, z_b, t) \quad \text{on } B_c, \tag{6}$$

In Eqs. (4) through (6), (x_b, y_b, z_b) is the spatial coordinate on the boundary; $\mathbf{n}$ is an outward unit vector normal to the boundary; h_d, q_n, and q_c are the prescribed Dirichlet functional value, Neumann flux, and Cauchy flux, respectively; B_d, B_n, and B_c are the Dirichlet, Neumann, and Cauchy boundary, respectively.

The fourth type of boundary conditions is the variable moving boundary conditions. The boundary can be either a ground surface when the moving boundary reaches the ground or a phreatic surface when the moving boundary stays within the media. When the moving boundary reaches the ground surface, a Dirichlet boundary condition with zero pressure head is imposed, i.e.

$$h = 0 \quad \text{on } B_v \tag{7}$$

When the moving boundary remains in the media, the following free phreatic conditions are imposed,

$$S_y \frac{\partial \zeta}{\partial t} + \left(u_s \frac{\partial \zeta}{\partial x} + v_s \frac{\partial \zeta}{\partial y} - w_s \right) = -q_v \quad \text{on } B_v \big|_{z = \zeta(x,y,t)} \tag{8}$$

$$h = 0 \quad \text{on } z = \zeta(x,y,t) \tag{9}$$

In Eqs. (7) through (9), B_v is the variable moving boundary, it can be either the ground surface or the phreatic free surface; q_v is the outward flux rate across the moving surface given by $z = \zeta(x,y,t)$; S_y is the specific yield; u_s, v_s, and w_s are x-, y-, and z-component of the velocity of the free surface; and ζ is the elevation of the free phreatic surface from the datum. It should be noted that the term in the parenthesis in Eq. (8) is equal to the inward flux normal to the free surface, i.e.,

$$\mathbf{n} \cdot \mathbf{K} \cdot \nabla (h + z) \big|_{z = \zeta(x,y,t)} = \left(u_s \frac{\partial \zeta}{\partial x} + v_s \frac{\partial \zeta}{\partial y} - w_s \right) \tag{10}$$

Equations (1) through (10) form a closed system to describe the pressure head distribution and the moving surface as function of space and time. Analytical solutions for this system can be obtained only for very specific and simplified problems. In general, numerical methods must be applied to approximate the system to yield a matrix equation for the discrete spatial points at discrete time. Either finite difference or finite element methods can be used to solve the system. We elect to use the finite element method because of its ease to treat the variable moving boundary conditions.

IMPLEMENTATION OF FREE PHREATIC BOUNDARY CONDITIONS

Boundary conditions on the free phreatic surface can be implemented rigorously with the finite element methods. Application of the Galerkin finite element procedure to Eq. (1) involves the following integration over the variable moving boundary,

$$B_i = \int_{B_v} N_i \mathbf{n} \cdot \mathbf{K} \cdot \nabla(h + z) \, dB \tag{11}$$

where N_i is the base function of the i-th node. Combing Eqs. (8) and (1) and using the fact that $\frac{\partial \zeta}{\partial t}$ equals to $\frac{\partial h}{\partial t}$, we obtain

$$B_i = -\int_{B_v} N_i \left(S_y \frac{\partial h}{\partial t} \right) dB - \int_{B_v} N_i q_v \, dB \tag{12}$$

The first term on the right-hand side of Eq. (12) contributes to the coefficient matrix and the second term to the load vector of the matrix equation for h.

The complication for the variable moving boundary conditions is that the free surface is not known a priori. The boundary conditions on the free surface must satisfy both Eqs. (8) and (9). Thus, the position of the free surface is usually determined by iteration methods. The algorithm to determine the free surface is as follows.

(i)   Assume the initial position of the free surface at the k-th iteration is known and is designated by ZETA(k).
(ii)   Assemble the finite element equations and solve them for h.
(iii)   For all nodes on the free surface, check if ABS (h) < TOL? If it is, the free surface is correctly located and the simulation continue for the next time step. If it is not, go to Step (iv).
(iv)   Correct the free surface position by ZETA(k+1) = ZETA(k) + OMEGA*h, where OMEGA is a positive number between 0 and 1. Update the iteration counter k and go to Step (1).

The above four steps complete the iteration procedure for the determination of the moving free surface of transient flow problems. The model described here is call 3DFEWA: A **3**-Dimensional **F**inite **E**lement Model of **W**ater Flow through **A**quifers with Variable Free Surfaces.

EXAMPLES

To very 3DFEWA, two examples dealing with unconfined aquifer problems are presented. These examples represent transient and steady-state problems, respectively.

Problem No. 1 - Transient Problem: This example is selected to represent the simulation of a two-dimensional drainage problem in a unconfined aquifer with 3DFEWA. The region of interest is bounded on the left and right by parallel drains fully penetrating the medium, on the bottom by an impervious aquifuge, and on the top by an air-soil interface (Fig. 1). The distance between the two drains is 20 m apart (Fig. 1). The medium is assumed to have a saturated hydraulic conductivity of 0.01 m/d, a porosity and specific yield of 0.25. Because of the symmetry, the region for numerical simulation will be taken as 0 < x < 10 m and 0 < z < 10 m, and 10 m wide along the y-direction will be assumed. The boundary conditions are given as: no flux is imposed on the left (x = 0), front (y = 0), back (y = 10), and bottom (z = 0) sides of the region; pressure head is assumed to vary from zero at the water surface (z = 2.5) to 2.5 m at the bottom (z = 0) on the right side (x = 10); the remaining right side is considered as a potential seepage face; and variable moving boundaries

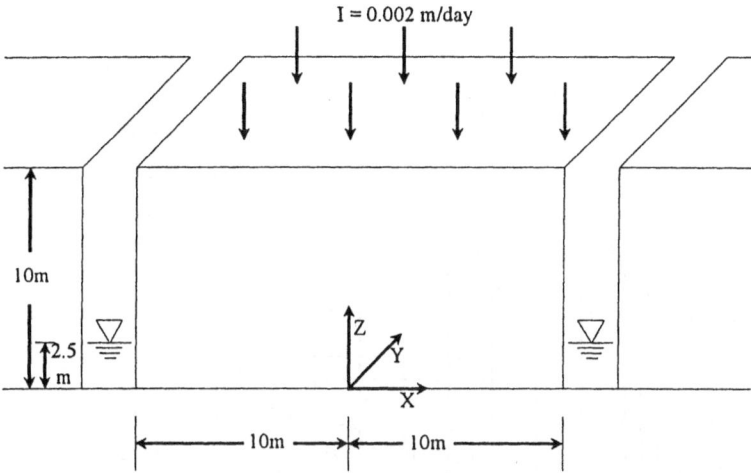

Fig. 4.1 Problem definition and sketch for example 1

specifying the free surface boundary conditions. Fluxes across the variable moving boundary from the outside world are assumed equal to 0.002 m/d. The right side above the water surface are specified as a seepage face which is represented as Dirichlet boundary condition with negative profile type. A transient state solution will be sought. A pre-initial condition is set, h = 10-z. The initial condition is obtained by simulating steady state problem without any infiltration.

The region of interest is discretized with 10 x 1 x 10 x 4 = 400 triangular prism elements with element size = 0.5 cm^2 (= 0.5x1x5) x 1 cm, resulting in 11 x 2 x 11 + 11 x 10 = 352 node points (Fig. 2). For 3DFEWA simulation, each of the three vertical planes will be considered a subregion. Thus, the total of the three subregions, each with 121, 121, and 110 node points, respectively, is used for the subregional block iteration simulation.

The pressure head tolerance is $2 \cdot 10^{-4}$ m for cycle iteration to locate water table and is 2×10^{-3} m for block iteration. The relaxation factors for both the nonlinear iteration and block iteration are set equal to 0.5. The results of free surface location simulated with 3DFEWA are shown in Figure 3. Computer output indicate that this simulation is quite reasonable as the mass balance errors ranges from less than 3% for the first time step to less than 7% for the 500-th time step. These errors are considered acceptable.

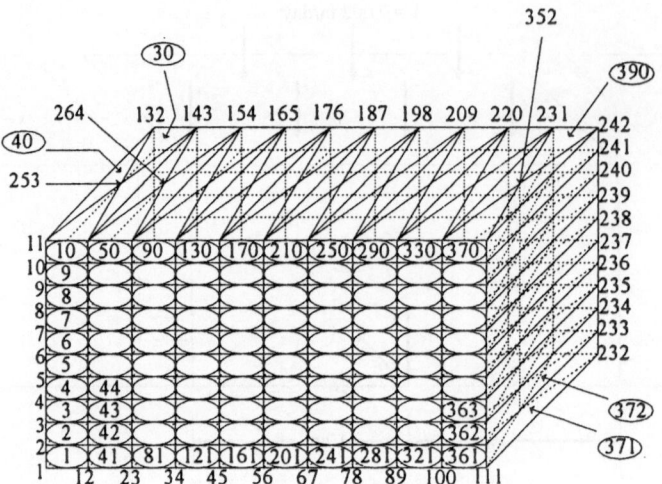

Fig. 2 Finite element discretization for example 1

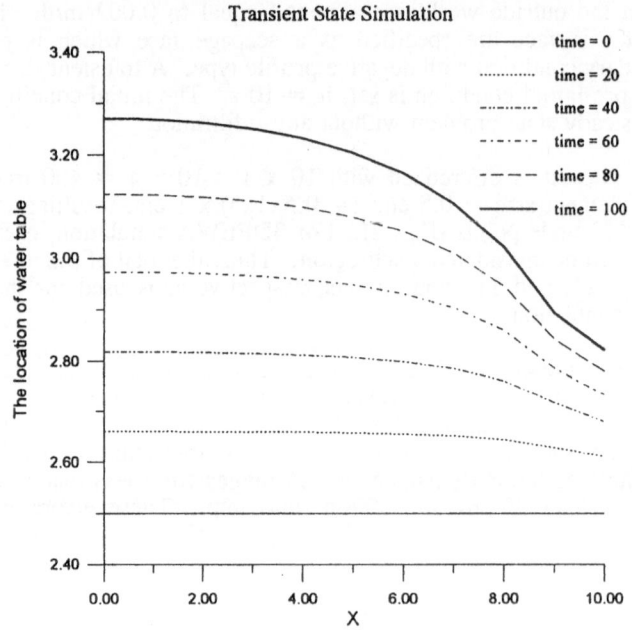

Fig. 3 The results of Problem No. 1

Problem No. 2 - Steady-State Problem: This example is selected to represent the simulation of a steady-state problem with 3DFEWA. All the problem definition and parameter specification are the same as the description in Problem No. 1. The only difference is that steady-state simulations are desired. The results from Case 1 with the rainfall rate 2×10^{-3} and Case 2 with the rainfall rate 4×10^{-3} are compared and shown (Fig. 4). It is seen that the free surface is correctly located when the recharge rate is 0. When the recharge rate is 0, the free surface is horizontal, as should be, and its exit point is at $z = 2.5$ because the specified pressure head at $z = 2.0$ is 0.5. When the recharge rate is not zero, the simulation is partially verified as the mass balance errors are smaller than 10% for both cases.

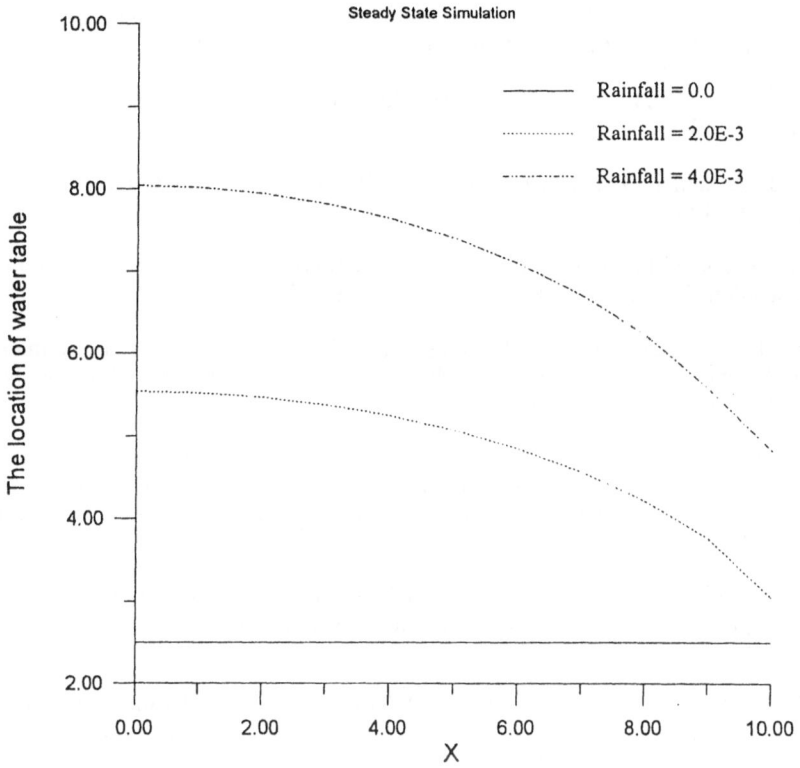

Fig. 4 The results of Case 1 and Case 2 for Problem No. 2.

ACKNOWLEDGEMENT

This research is supported in part by the Office of Health and Environmental Research, U. S. Department of Energy under Contract No. DE-FG02-91ER61197 with the Pennsylvania State University.

REFERENCES

Dagan, G. (1967) "Linearized solutions of free-surface groundwater flow with uniform recharge." J. Geophy. Res. Vol 72, No. 4, 1183-1193.

De Wiest, R. J. M. (1962) "Free surface flow in homogeneous porous mediums." Trans. Ameri. Soc. Civil Engrs., Vol. 127, 1045-1084.

Dicker, D. (1969) "Transient free-surface flow in porous media." Flow through Porous Media, ed. R. J. M. De Wiest. Academic Press, New York, pp. 293-330.

Hantush, M. S. (1967) "Growth and decay of groundwater-mounds in response to uniform percolation." Water Resour. Res. Vol. 3, No. 1, 227-234.

Huyakorn, P. (1983) "Computational Methods in Subsurface Flow." Academic Press, New York.

Liggett, J. A. and P. L-F Liu (1983) "The Boundary Integral Equation Method for Porous Media Flow." Allen & Union Inc., 255 pp.

Schmitz, G. and J. Edenhofer (1988) "Semi-analytical solution for the groundwater mound problem." Adv. Water Resources Vol. 11, 21-24.

McDonald M. G. and A. W. Harbaugh (1985) "A Modular Three-Dimensional Finite-Difference Groundwater Flow Model." U. S. Geogological Survey, National Center, Reston, Virginia. 528 pp.

Tsay, T. T., J. Hoopes, D. Olson, and S. Rashad (1993) "Modeling ground water mounding in a heterogeneous unconfined aquifer." Proc. Hydraulic Engineering'93, Volume 1, ed. H. W. Shen, S. T. Su, and F. Wen, American Soc. of Civil Engineers, 345 East 47th Street, New York, N. Y. 10017-2398. pp. 198-203.

Yeh, G. T. and D. D. Huff (1983) "FEWA: A Finite Element Model of Water Flow through Aquifers." ORNL-5976. Oak Ridge National Laboratory, Oak Ridge, TN 37830. 216 pp.

Yeh, G. T. (1987) "3DFEMWATER: A Three-Dimensional Finite Element Model of WATER Flow through Saturated-Unsaturated Media." ORNL-6386. Oak Ridge National Laboratory, Oak Ridge, TN 37830. 310 pp.

NEW ITERATIVE SCHEMES FOR MIXED FINITE–ELEMENT MODELS OF GROUNDWATER UNSTEADY FLOWS

WENLIANG   ZHAO and CHUNHONG   XIE

Department of Mathematics,Naning University,
Nanjing, 210008, P.R. China.

ABSTRACT

In this paper we proposed a new iterative technique for solving disrete approximations to equations of unsteady groundwater flow by mixed finite element method. With this technique, the computational work is algebraic computation, it doesn't need to solve a linear systems, calculation speed is high and the computer storage requirement are greatly reduceed. In addition, we present some numerical results for this algorithm and compare with analtical solution shows the relative error of Darcy velocities is 3%, it is better than Galarkin finite element[3] (Yeh. 1981).

1. Introduction We consider methods for solving discrete the equations governing groundwater flow. If the flow is unsteady and two–dimensional, Darcy's law and mass balance take the following forms:

$$u^x = -T\frac{\partial H}{\partial x}, \qquad in \quad \Omega \qquad (1.1)$$

$$T \geqslant T_0 > 0$$

$$u^y = -T\frac{\partial H}{\partial y}, \qquad in \quad \Omega \qquad (1.2)$$

$$S\frac{\partial H}{\partial t} + \mathrm{div}\,U = f \qquad in \quad \Omega. \qquad (1.3)$$

Here $U = (u^x, u^y)$, H and f represent the Darcy velocity, water head and source term, respectively. For simplicity, we take the spatial domain to be a square, scaled so that $\Omega = (0,1) \times (0,1)$. T is the transmissibility, s is coefficitent of storage.

The boundary conditions are $H = H_b$ along $\partial\Omega$ $\qquad (1.4)$
where

$$H_b = \begin{cases} g_1(y), & x = 0, & 0 \leqslant y \leqslant 1 \\ g_2(y), & x = 1, & 0 \leqslant y \leqslant 1 \\ g_3(x), & y = 0, & 0 \leqslant x \leqslant 1 \\ g_4(x), & y = 1, & 0 \leqslant x \leqslant 1. \end{cases}$$

A. Peters et al. (eds.), Computational Methods in Water Resources X, 139–145.

The initial condition is given as

$$H = H_0 \qquad\qquad \text{when} \quad t = 0, \quad (x,y) \in \Omega \qquad\qquad (1.5)$$

2.   A mixed finite–element method.

We begin with a brief review of the mixed finit–element method. Let $H(\text{div},\Omega) = \{ V \in L^2(\Omega) \times L^2(\Omega) : \text{div} V \in L^2(\Omega) \}$. Multiplying (1.1), (1.2) by $T^{-1} v^x, T^{-1} v^y$, respectively and parts yields

$$\int_\Omega \frac{u^x v^x}{T} dxdy = -\int_{\partial\Omega} v^x H \cos(n,x) ds + \int_\Omega H \frac{\partial v^x}{\partial x} dxdy. \qquad (2.1)$$

$$\int_\Omega \frac{u^y v^y}{T} dxdy = -\int_{\partial\Omega} v^y H \cos(n,y) ds + \int_\Omega H \frac{\partial v^y}{\partial y} dxdy. \qquad (2.2)$$

$$\forall \; V \equiv (v^x, v^y) \in H(\text{div},\Omega)$$

Multiplying (1.3) by q and integrate yields

$$\int_\Omega (s\frac{\partial H}{\partial t} + \text{div} U - f) q \, dxdy = 0 \qquad\qquad \forall \; q \in L^2(\Omega) \qquad (2.3)$$

Find a pair $(U,H) \in H(\text{div},\Omega) \times L^2(\Omega)$ such that (2.1), (2.2)(2.3) hold.

To discretize the systems (2.1), (2.2), (2.3), Let $\Delta_x = \{0 = x_0 < x_1 < \cdots < x_m = 1\}$ be a set of points on the x–axis and $\Delta_y = \{0 = y_0 < y_1 < \cdots < y_n = 1\}$ a set of points on the y–axis. Let $\Delta_h = \Delta_x \times \Delta_y$.

$$h = \max\{x_i - x_{i-1}, y_j - y_{j-1}\} \qquad (2.4)$$

We assume throughout the paper that Δ_x and Δ_y are quasi–uniform in the sense that $x_i - x_{i-1} \geqslant \alpha h$ and $y_j - y_{j-1} \geqslant \alpha h$ for some fixed $\alpha \in (0,1)$. With Δ_h we associate a finite–element subspace $Q_h \times W_h$ of $H(\text{div},\Omega) \times L^2(\Omega)$. The "velocity space" is $Q_h = Q_h^x \times Q_n^y$. In particular, we use the lowest–order Raviart–Thomas spaces in which Q_h^x contains functions that are piecewise linear and continuous on Δ_x and piecewise constant on Δ_y. Similarly, Q_h^y contains functions that are piecewise linear and contiuous on Δ_y and piecewise constant on Δ_h. The "head water space" W_h consists of functions that are piecewise constant on Δ_h.

Given these approximation spaces, the corresponding mixed finite–element method for solving (2.1), (2.2), (2.3) is as follows:

Find a pair $(U_h, H_h) \in Q_h \times W_h$ such that

$$\int_\Omega \frac{u_h^x v_h^x}{T} dxdy = -\int_{\partial\Omega} v_h^x H_h \cos(n,x) ds + \int_\Omega H_h \frac{\partial v_h^x}{\partial x} dxdy \qquad (2.5)$$

$$\int_\Omega \frac{u_h^y v_h^y}{T} dxdy = -\int_{\partial\Omega} v_h^y H_h \cos(n,y)ds + \int_\Omega H \frac{\partial v_h^y}{\partial y} dxdy \tag{2.6}$$

$$\forall V_h \equiv (v_h^x, v_h^y) \in H(\text{div},\Omega)$$

$$\int_\Omega (s\frac{\partial H_h}{\partial t} + \text{div}U_h - f)q_h dxdy = 0, \qquad \forall q_h \in W_h \tag{2.7}$$

Let

$$W_{i,j} = \begin{cases} 1 & in \quad (x,y)\in[x_{i-1},x_i]\times[y_{j-1},y_j]. \\ 0 & (x,y)\bar{\in}[x_{i-1},x_i]\times[y_{j-1},y_j]. \end{cases} \tag{2.8}$$

$$Q_{o,j}^x = \begin{cases} \dfrac{-x+x_1}{x_1-x_0}, & (x,y)\in[x_0,x_1]\times[y_{j-1},y_j]. \\ 0, & (x,y)\bar{\in}[x_0,x_1]\times[y_{j-1},y_j]. \end{cases} \tag{2.9}$$

$$Q_{i,j}^x = \begin{cases} \dfrac{x-x_{i-1}}{x_i-x_{i-1}}, & (x,y)\in[x_{i-1},x_i]\times[y_{j-1},y_j]. \\ \dfrac{-x+x_{i+1}}{x_{i+1}-x_i}, & (x,y)\in[x_i,x_{i+1}]\times[y_{j-1},y_j]. \end{cases} \tag{2.10}$$

$$Q_{m,j}^x = \begin{cases} \dfrac{x-x_{m-1}}{x_m-x_{m-1}}, & (x,y)\in[x_{m-1},x_m]\times[y_{j-1},y_j] \\ 0, & (x,y)\bar{\in}[x_{m-1},x_m]\times[y_{j-1},y_j] \end{cases} \tag{2.11}$$

$$Q_{i,0}^y = \begin{cases} \dfrac{-y+y_1}{y_1-y_0}, & (x,y)\in[x_{i-1},x_i]\times[y_0,y_1]. \\ 0, & (x,y)\bar{\in}[x_{i-1},x_i]\times[y_0,y_1] \end{cases} \tag{2.12}$$

$$Q_{i,j}^y = \begin{cases} \dfrac{y-y_{j-1}}{y_j-y_{j-1}}, & (x,y)\in[x_{i-1},x_i]\times[y_{j-1},y_j]. \tag{2.13} \\[2mm] \dfrac{y_{j+1}-y}{y_{j+1}-y_j}, & (x,y)\in[x_{i-1},x_i]\times[y_j,y_{j+1}] \tag{2.14} \\[2mm] 0, & (x,y)\bar{\in}[x_{i-1},x_i]\times[y_{j-1},y_j] \tag{2.15} \end{cases}$$

$$Q_{i,n}^y = \begin{cases} \dfrac{y-y_{n+1}}{y_n-y_{n-1}}, & (x,y)\in[x_{i-1},x_i]\times[y_{n-1},y_n] \\ 0, & (x,y)\bar{\in}[x_{i-1},x_i]\times[y_{n-1},y_n] \end{cases} \tag{2.16}$$

Here, $W_{i,j}, Q_{i,j}^x, Q_{i,j}^y$ Signify element in the standard nodal bases for W_h, Q_h^x and Q_h^y.

Suppose U_h, H_h have the expansions

$$U_h = (u_h^x, u_h^y) = (\sum_{i=0}^m \sum_{j=1}^n U_{i,j}^x Q_{i,j}^x, \sum_{i=1}^m \sum_{j=0}^n U_{i,j}^y Q_{i,j}^y) \tag{2.17}$$

$$H_h = \sum_{i=1}^m \sum_{j=1}^n H_{i,j} W_{i,j} \qquad (2.18)$$

$T = T_{i,j} = \text{Constant} \quad in \ [x_{i-1}, x_i] \times [y_{j-1}, y_j], \quad g_1 = g_1^j, \quad in \ [y_{j-1}, y_j],$

$g_2 = g_2^j \quad in \ [y_{j-1}, y_j], g_3 = g_3^i \quad in \ [x_{i-1}, x_i], \quad g_4 = g_4^i \quad in \ [x_{i-1}, x_i]. \qquad (2.18)$

Let $v_h^x = Q_{i,j}^x$, in (2.5), $i = 0,1,2,\cdots,m$, we obtain by calculating these integrals,

$$\frac{1}{3}\frac{1}{T_{1,j}}(x_1 - x_0)U_{0,j}^x + \frac{1}{6T_{1,j}}(x_1 - x_0)U_{1,j}^x + H_{1,j} = -g_1^j \qquad (2.19)$$

$$\frac{1}{6}\frac{1}{T_{i,j}}(x_i - x_{i-1})U_{i-1,j}^x + \frac{1}{3}[\frac{1}{T_{i,j}}(x_i - x_{i-1}) + \frac{1}{T_{i+1,j}}(x_{i+1} - x_i)]U_{i,j}^x + \frac{1}{6}\frac{1}{T_{i,j}}(x_i$$

$$- x_{i-1})U_{i+1,j}^x - H_{i,j} + H_{i+1,j} = 0 \qquad (2.20)$$

$$\frac{1}{3T_{m,j}}(x_m - x_{m-1})U_{m-1,j}^x + \frac{1}{3T_{m,j}}(x_m - x_{m-1})U_{m,j}^x - H_{m,j} = g_2^j \qquad (2.21)$$

Similarly, we obtain ($v_h^y = Q_{i,j}^y$ in (2.6))

$$\frac{1}{3T_{i,j}}(y_1 - y_0)U_{i,0}^y + \frac{1}{6T_{i,j}}(y_1 - y_0)U_{i,j}^y + H_{i,j} = -g_3^i \qquad (2.22)$$

$$\frac{1}{6}\frac{1}{T_{i,j}}(y_j - y_{j-1})U_{i,j-1}^y + \frac{1}{3}[\frac{1}{T_{i,j}}(y_j - y_{j-1}) + \frac{1}{T_{i,j+1}}(y_{j+1} - y_j)]U_{i,j}^y +$$

$$\frac{1}{6}\frac{1}{T_{i,j+1}}(y_{j+1} - y_j)U_{i,j+1}^y - H_{i,j+1} = 0 \qquad (2.23)$$

$$\frac{1}{6}\frac{1}{T_{i,m}}(y_n - y_{n-1})U_{i,n-1}^y + \frac{1}{3}\frac{1}{T_{i,n}}(y_n - y_{n-1})U_{i,n}^y - H_{i,n} = g_4^i \qquad (2.24)$$

let $W = W_{i,j}$, in (2.7) than (2.7) become

$$\int_{x_{i-1}}^{x_i} \int_{y_{j-1}}^{y_j} (s\frac{\partial H}{\partial t}\frac{\partial U^x}{\partial x} + \frac{\partial U^y}{\partial y} - f)dxdy = 0. \qquad (2.25)$$

We use a forward difference approximation to the time derivative

$$\frac{\partial H}{\partial t} = \frac{H(,t_n) - H(,t_{n-1})}{\Delta t}, \qquad \Delta t = t_n - t_{n-1}$$

Integrate (2.95), We obtain

$$H_{i,j}^n = H_{i,j}^{n-1} - \frac{\Delta t}{s(x_i - x_{i-1})}(U_{i,j}^x - U_{i-1,j}^x) - \frac{\Delta t}{s(y_j - y_{j-1})}(U_{i,j}^y - U_{i,j-1}^y) + \frac{\Delta t}{s}F_{i,j}^n \qquad (2.26)$$

where $F_{i,j}^n = \frac{1}{(x_i - x_{i-1})(y_j - y_{j-1})}\int_{x_{i-1}}^{x_i} \int_{y_{j-1}}^{y_j} f(x,y,t_n)dxdy.$

Let $(U^{(\cdot)}, H^{(\cdot)})^T$ by initial guesses for the value of $(U,H)^T$, then then $(k+1)$th iterate for $(U,H)^T$ is the solution of

$$(U_{0,j}^x)^{(k+1)} = -\frac{1}{2}(U_{1,j}^x)^{(k)} - \frac{3T_{1,j}}{x_1 - x_0}H_{1,j}^{(k)} - g_1^i\frac{3T_{1,j}}{x_1 - x_0} \tag{2.27}$$

$$(U_{i,j}^x)^{(k+1)} = -\frac{\overset{1}{\alpha}_{i,j}}{\beta_{i,j}^1}(U_{i-1,j}^x)^{(k)} - \frac{\alpha_{i+1,j}^1}{\beta_{i,j}^1}(u_{i+1,j}^x)^{(k)} + \frac{1}{\beta_{i,j}^1}(H_{i,j})^{(k)} - \frac{1}{\beta_{i,j}^1}H_{i,j}^{(k)} \tag{2.28}$$

$$(U_{m,j}^x)^{(k+1)} = -\frac{1}{2}(U_{m-1,j}^x)^{(k)} + \frac{3T_{m,j}}{x_m - x_{m-1}}(H_{m,j})^{(k)} + g_2^i\frac{3T_{m,j}}{x_m - x_{m-1}} \tag{2.29}$$

$$(U_{i,0}^y)^{(k+1)} = -\frac{1}{2}(u_{i,1}^y)^{(k)} - \frac{3T_{i,1}}{y_1 - y_0}(H_{i,1})^{(k)} - g_3^i\frac{3T_{i,1}}{y_1 - y_0} \tag{2.30}$$

$$(U_{i,j}^y)^{(k+1)} = -\frac{\overset{2}{\alpha}_{i,j}}{\beta_{i,j}^2}(U_{i,j-1}^y)^{(k)} - \frac{\alpha_{i,j+1}^2}{\beta_{i,j}^2}(u_{i,j+1}^y)^{(k)} + \frac{1}{\beta_{i,j}^2}(H_{i,j})^{(k)} - \frac{1}{\beta_{i,j}^2}H_{i,j+1}^{(k)}. \tag{2.31}$$

$$(U_{i,n}^y)^{(k+1)} = -\frac{1}{2}(U_{i,n-1}^y)^{(k)} + \frac{3T_{i,n}}{y_n - y_{n-1}}H_{i,n}^{(k)} + g_4^i\frac{3T_{i,n}}{y_n - y_{n-1}} \tag{2.32}$$

$$H_{i,j}^{(k+1)} = -r_i^1((U_{i,j}^x)^{(k+1)} - (U_{i-1,j}^x)^{(k+1)}) - r_j^2((U_{i,j}^y)^{(k+1)} - (U_{i,j-1}^y)^{(k+1)}) + H_{i,j}^{n-1}$$
$$+ F_{i,j}\frac{\Delta t}{s} \tag{2.33}$$

where $\alpha_{i,j}^1, \beta_{i,j}^1, \cdots\cdots r_i^1, r_j^2$ is position constant, respectively.

Compute a new approximation $(U^{(k+1)}, H^{(k+1)})^T$ using the following steps:

(i)   compute $U^{(k+1)}$ by (2.27)–(2.32).

(ii)   compute $H^{(k+1)}$ by (2.33)

In each iteration, the computational work (i)or(ii) is algebraic computation, it doesn't need to solve a linear systems. Therefore it's colculation spead is high and the computer storage requirement are greatly reduced.

We have been shown that the iteration method is convergence for appropriate small time step.

4.   Example.

We consider the following groundwater unsteady flows problem:

$$\begin{cases} \dfrac{\partial H}{\partial t} - (\dfrac{\partial^2 H}{\partial x^2} + \dfrac{\partial^2 H}{\partial y^2}) = e^{-t}\sin x\sin y \\ H(0,y,t) = H(\pi,y,t) = 0 \\ H(x,0,t) = H(x,\pi,t) = 0 \\ H(x,y,0) = \sin x\sin y \qquad 0 \leqslant x \leqslant \pi, \qquad 0 \leqslant y \leqslant \pi \end{cases}$$

This problem has an exact analytical solution

$$H(x,y,t) = e^{-t}\sin x \sin y$$

According to Darey's law, the analytical velocity is as follows

$$u^x = e^{-t}\cos x \sin y, \qquad u^y = e^{-t}\sin x \sin y.$$

Analytical solutions are valuable for checking the accuracy of everplicated numerieal methods. The domain was discretized into squares, $\Delta x = \Delta y = \frac{\pi}{10}$, $\Delta t = 0.002$. 16 observation nodal point $(a_1, a_2, \cdots, a_{16})$ were selected to compare Darcy velocities $U_{i,j}^x$ (cf. Table1). 16 observation nodal point $(b_1, b_2, \cdots, b_{16})$ to compare Darcy' velocities $U_{i,j}^y$ (cf. Table2). Form table1, 2, it is not difficult to discover that we present algorithm and compare with analytial solution shows the absolute error of Darcy velocites is $0.7 \sim 3\%$, the relative error of Darcy velocities is 3%, it is better than Galarkin finite element [3] (Yeh. 1981)

U_{ij}^x–TABLE 1

point	Analytical	mixd–finite	absolute–err	relative–err
a1	−3.600133E−01	−3.710776E−01	1.106435E−02	3.073317%
a2	−1.375128E−01	−1.417371E−01	4.224241E−03	3.071889%
a3	1.375129E−01	1.417379E−01	4.225045E−03	3.072473%
a4	3.600133E−01	3.710760E−01	1.106268E−02	3.072854%
a5	−7.065659E−01	−7.282816E−01	2.171570E−02	3.073415%
a6	−2.698841E−01	−2.781745E−01	8.290350E−03	3.071819%
a7	2.698842E−01	2.781769E−01	8.292735E−03	3.072701%
a8	7.065659E−01	7.282779E−01	2.171195E−02	3.072883%
a9	−7.832342E−01	−8.073058E−01	2.407157E−02	3.073356%
a10	−2.991688E−01	−3.083586E−01	9.189725E−03	3.071752%
a11	2.991689E−01	3.083610E−01	9.192079E−03	3.072538%
a12	7.832343E−01	8.073018E−01	2.406752E−02	3.07283%
a13	−5.607338E−01	−5.779670E−01	1.723319E−02	3.073329%
a14	−2.141812E−01	−2.207606E−01	6.579310E−03	3.071842%
a15	2.141813E−01	2.207622E−01	6.580889E−03	3.072579%
a16	5.607339E−01	5.779642E−01	1.723039E−02	3.072829%

$U_{i,j}^{y}$ −TABLE 2

point	analytical	mixed−finite	absolute−err	relative−err
b1	−5.607339E−01	−5.779672E−01	1.72331E−02	3.073350%
b2	−7.832342E−01	−8.073059E−01	2.407169E−02	3.073371%
b3	−7.065659E−01	−7.282813E−01	2.171540E−02	3.073373%
b4	−3.600132E−01	−3.710775E−01	1.106429E−02	3.073302%
b5	−2.141812E−01	−2.207604E−01	6.579131E−03	3.071759%
b6	−2.991689E−01	−3.083585E−01	6,579131E−03	3.071759%
b6	−2.991689E−01	−3.083585E−01	9.189665E−03	3.071732%
b7	−2.698841E−01	−2.781743E−01	8.290231E−03	3.071775%
b8	−1.375128E−01	−1.417368E−01	4.223987E−03	3.071705%
b9	2.141813E−01	2.207622E−01	6.580859E−03	3.072565%
b10	2.991689E−01	3.083615E−01	9.192586E−03	3.072708%
b11	2.98842E−01	2.781765E−01	8.292317E−03	3.072547%
b12	1.375128E−01	1,417379E−01	4.225105E−03	3.072517%
b13	5.067339E−01	5.779645E−01	1.723063E−02	3.072871%
b14	7.832343E−01	8.073025E−01	2.406818E−02	3.072922%
b15	7.065659E−01	7.282774E−01	2.171159E−02	3.072833%

References

(1) P. A. Raviart and J. M. Thomas. A mixed finite element method for 2nd order elliptic problems, in Mathematical Aspects of Finite Element Methods, Lecture Notes in Mathematics 606, I. Galhgani and E. Magenes; eds.

Springer−Verlag, Berlin, 1977. PP292−315.

(2) M. B. Allen, R. E. Ewing and P. Lu. Well−conditioned iterative schemes for mixed finite−element models of porous−media flows. SIAM. J. SCI. STAT. comput vol. 13, N0. 3.PP794−814, May 1992

(3) Yeh. G. T. on the computation of Darcian Velocity and mass balance in the finite element modeling of groundwater flow, Water Resour. Res 17(5), 1529−1534 1981.

2. SUBSURFACE TRANSPORT

LOGICALLY RECTANGULAR MIXED METHODS FOR GROUNDWATER FLOW AND TRANSPORT ON GENERAL GEOMETRY

TODD ARBOGAST, MARY F. WHEELER, and IVAN YOTOV

Department of Computational and Applied Mathematics
Rice University
Houston, Texas 77251–1892
U.S.A.

ABSTRACT

We consider an extended mixed finite element formulation for groundwater flow and transport problems with either a tensor hydraulic conductivity or a tensor dispersion. While the aquifer domain can be geometrically general, in our formulation the computational domain is rectangular. The approximating spaces for the mixed method are defined on a smooth curvilinear grid, obtained by a global mapping of the rectangular, computational grid. The original problem is mapped to the computational domain, giving a similar problem with a modified tensor coefficient. Special quadrature rules are introduced to transform the mixed method into a simple cell-centered finite difference method with a 9 point stencil in 2-D and 19 point stencil in 3-D. The resulting scheme is locally mass conservative. In the case of flow, linear Galerkin procedures give first order accurate velocities, while our method is second order accurate. Both computational and theoretical results are given.

1. INTRODUCTION

We develop a numerical scheme for groundwater flow and transport problems with tensor coefficients on a geometrically general domain Ω in $\mathbf{R}^d$ ($d = 2$ or 3). In the flow problem, we solve for the pressure p and the velocity $\mathbf{u}$ satisfying

$$(1.1) \qquad \nabla \cdot \mathbf{u} = q, \quad \mathbf{u} = -K\nabla p,$$

where K is the hydraulic conductivity tensor. In the transport problem, we solve for the concentration c such that

$$(1.2) \qquad \phi\frac{\partial c}{\partial t} + \nabla \cdot \big(\mathbf{u}c - D(\mathbf{u})\nabla c\big) = q_c(c),$$

where ϕ is the porosity and D is the dispersion tensor with components

$$(1.3) \qquad D_{i,j}(\mathbf{u}) = \phi d_m \delta_{i,j} + |\mathbf{u}|\left\{ \frac{u_i u_j}{|\mathbf{u}|^2}(d_l - d_t) + d_t \delta_{i,j} \right\},$$

149

A. Peters et al. (eds.), Computational Methods in Water Resources X, 149–156.
© 1994 *Kluwer Academic Publishers. Printed in the Netherlands.*

where d_m is the molecular diffusivity and d_l, and d_t are the longitudinal and transverse dispersivity, respectively. For simplicity of the presentation, we consider homogeneous Neumann boundary conditions.

Our numerical scheme is based on mixed finite element methods, because they conserve mass locally and give a good approximation of the flux variable. However, mixed methods can be difficult to implement directly.

By using special quadrature rules for evaluating the integrals, Russell and Wheeler [6] showed that the standard cell-centered finite difference method was equivalent to the lowest order, RT_0 mixed method [5]. Thus, the RT_0 mixed method can be easily implemented as a five or seven point finite difference method. Weiser and Wheeler [8] obtained superconvergence results for this scheme; that is, if h is the maximum grid spacing, both the pressure and velocity are approximated to order h^2. A Galerkin method will approximate the velocity only to order h.

Those two works were limited to the case of a diagonal tensor and a rectangular grid. To solve our flow and transport problems with tensor coefficients on fairly geometrically general domains, we develop a new scheme to overcome these limitations. We will not sacrifice the ease of implementation, the accuracy, or the local mass conservation property of the approximation. In the notation of the flow problem, we present in the next section some of the necessary background of the expanded mixed finite element method that is the basis of our scheme [4]. We derive our finite difference procedure [3] from it in Section 3. We summarize the convergence results [4] in Section 4. Transport is discussed briefly in Section 5, and computational results are given in Section 6.

Our main requirement is that there be a smooth mapping F of a rectangular, computational domain $\hat{\Omega}$ onto the aquifer domain Ω. Given a rectangular grid $\hat{\mathcal{T}}_h$ on $\hat{\Omega}$, F defines a smooth, logically rectangular, curvilinear grid $\mathcal{T}_h$ on Ω (see Fig. 1). The Jacobian matrix of F is $DF = (\partial F_i / \partial x_j)$, and the Jacobian of the mapping is $J = |\det(DF)|$. (There are grid generation codes available for creating F and its Jacobian matrix.)

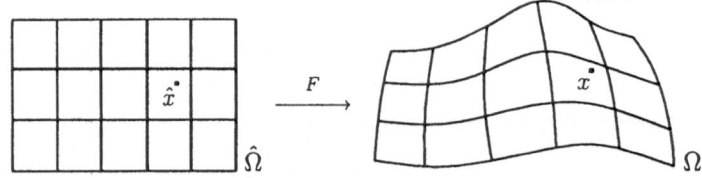

Fig. 1. The computational domain $\hat{\Omega}$ and the physical domain Ω.

2. THE EXPANDED MIXED FINITE ELEMENT METHOD ON GENERAL GEOMETRY

Following [3] and [4], we introduce an unknown $\tilde{\mathbf{u}}$ such that

$$(2.1) \qquad M\tilde{\mathbf{u}} = -\nabla p, \quad \mathbf{u} = KM\tilde{\mathbf{u}},$$

where $M = J(DF^{-1})^T DF^{-1}$. Note that M is a symmetric, positive definite matrix. It is introduced to simplify the computations significantly, after mapping to the rectangular grid on $\hat{\Omega}$.

To define the RT_0 mixed space on the curvilinear grid, we need first the standard definition of this space on rectangles [5]. Let $\hat{V}_h$ and $\hat{W}_h$ be the velocity and the pressure space, respectively. On any rectangular element $\hat{E} \in \hat{\mathcal{T}}_h$,

$$\hat{V}_h(\hat{E}) = \left\{ (\alpha_1 x_1 + \beta_1, \alpha_2 x_2 + \beta_2, \alpha_3 x_3 + \beta_3)^T : \alpha_i, \beta_i \in \mathbf{R} \right\},$$
$$\hat{W}_h(\hat{E}) = \left\{ \alpha : \alpha \in \mathbf{R} \right\}$$

(where the last component in $\hat{V}_h(\hat{E})$ should be deleted if $d = 2$). Then,

$$\hat{V}_h = \left\{ \hat{\mathbf{v}} = (v_1, v_2, v_3) : \hat{\mathbf{v}}|_{\hat{E}} \in \hat{V}_h(\hat{E}) \text{ for all } \hat{E} \in \hat{\mathcal{T}}_h, \text{ and each } v_i \text{ is} \right.$$
$$\left. \text{continuous in the } i\text{th coordinate direction} \right\},$$
$$\hat{W}_h = \left\{ \hat{w} : \hat{w}|_{\hat{E}} \in \hat{W}_h(\hat{E}) \text{ for all } \hat{E} \in \hat{\mathcal{T}}_h \right\};$$

thus, if $\hat{\nu}$ denotes the normal direction, $\hat{\mathbf{v}} \cdot \hat{\nu}$ is well defined on the boundaries of each element $\hat{E}$, and $\hat{w}$ is piecewise discontinuous. We use the standard nodal basis, where for $\hat{V}_h$, the nodes are at the midpoints of the edges or faces of the elements, and for $\hat{W}_h$, the nodes are at the midpoints of the elements (cell-centers).

Let V_h and W_h be the RT_0 spaces on $\mathcal{T}_h$, defined as follows [7]. For each $\hat{\mathbf{v}} \in \hat{V}_h$ and $\hat{w} \in \hat{W}_h$, we define $\mathbf{v} \in V_h$ and $w \in W_h$ at $F(\hat{x}) = x \in \Omega$ by

$$(2.2a) \qquad \mathbf{v}(x) = \frac{1}{J(\hat{x})} DF(\hat{x})\hat{\mathbf{v}}(\hat{x}),$$

$$(2.2b) \qquad w(x) = \hat{w}(\hat{x}).$$

The velocity space is defined by the Piola transformation; it preserves the normal component of the velocity across the element boundaries, and is therefore locally mass conservative. The key property is that $\nabla \cdot \mathbf{v} = \frac{1}{J}\hat{\nabla} \cdot \hat{\mathbf{v}}$.

We have the following mixed formulation for approximating the flow equation (1.1). Find $\mathbf{u}_h \in V_h$, $\tilde{\mathbf{u}}_h \in V_h$, and $p_h \in W_h$ such that

$$(2.3a) \qquad \int_E \nabla \cdot \mathbf{u}_h \, dx = \int_E q \, dx, \qquad\qquad E \in \mathcal{T}_h,$$

$$(2.3b) \qquad \int_\Omega M\tilde{\mathbf{u}}_h \cdot \mathbf{v} \, dx = \int_\Omega p_h \nabla \cdot \mathbf{v} \, dx, \qquad \mathbf{v} \in V_h,$$

$$(2.3c) \qquad \int_\Omega M\mathbf{u}_h \cdot \mathbf{v} \, dx = \int_\Omega MKM\tilde{\mathbf{u}}_h \cdot \mathbf{v} \, dx, \quad \mathbf{v} \in V_h.$$

The existence and uniqueness of a solution is shown in [4].

We now transform (2.3) to the rectangular, computational domain by the map F. The Piola transform (2.2a), (2.2b), and the definition of M (2.1) imply that

$$(2.4a) \qquad \int_{\hat{E}} \hat{\nabla} \cdot \hat{\mathbf{u}}_h \, d\hat{x} = \int_E q \, dx = \int_{\hat{E}} \hat{q} J \, d\hat{x}, \qquad \hat{E} \in \hat{T}_h,$$

$$(2.4b) \qquad \int_{\hat{\Omega}} \hat{\mathbf{u}}_h \cdot \hat{\mathbf{v}} \, d\hat{x} = \int_{\hat{\Omega}} \hat{p}_h \hat{\nabla} \cdot \hat{\mathbf{v}} \, d\hat{x}, \qquad \hat{\mathbf{v}} \in \hat{V}_h,$$

$$(2.4c) \qquad \int_{\hat{\Omega}} \hat{\mathbf{u}}_h \cdot \hat{\mathbf{v}} \, d\hat{x} = \int_{\hat{\Omega}} J DF^{-1} K (DF^{-1})^T \tilde{\hat{\mathbf{u}}}_h \cdot \hat{\mathbf{v}} \, d\hat{x}, \quad \hat{\mathbf{v}} \in \hat{V}_h,$$

where $\hat{q}(\hat{x}) = q(F(\hat{x}))$. Note that (2.4) is similar to the original problem (2.3) with $M = I$ and the modified tensor coefficient

$$(2.5) \qquad\qquad \mathcal{K} = J DF^{-1} K (DF^{-1})^T.$$

All computations are performed on the rectangular grid of $\hat{\Omega}$; we recover the true pressure and velocity on Ω using (2.2).

3. THE CELL-CENTERED FINITE DIFFERENCE METHOD

To simplify the finite element method (2.4), we use special quadrature rules for approximating the integrals. The two divergence integrals can be computed exactly, since the divergence of any $\hat{\mathbf{v}} \in \hat{V}_h$ is piece-wise constant. The trapezoidal rule is used for evaluating the three integrals involving a vector-vector product. This enables us to express $\hat{\mathbf{u}}_h$ and $\tilde{\hat{\mathbf{u}}}_h$ in terms of $\hat{p}_h$, and therefore obtain a single equation for the pressure. Herein, we describe this stencil for $d = 2$; a straightforward generalization gives the stencil for $d = 3$.

We need some relatively standard cell-centered finite difference notation. Denote the grid points by

$$(\hat{x}_{i+1/2}, \hat{y}_{j+1/2}), \quad i = 0, ..., N_1, \; j = 0, ..., N_2,$$

and define, for $i = 1, ..., N_1$ and $j = 1, ..., N_2$,

$$\hat{x}_i = \tfrac{1}{2}(\hat{x}_{i+1/2} + \hat{x}_{i-1/2}), \qquad \hat{y}_j = \tfrac{1}{2}(\hat{y}_{j+1/2} + \hat{y}_{j-1/2}),$$
$$\hat{h}_i^{\hat{x}} = \hat{x}_{i+1/2} - \hat{x}_{i-1/2}, \qquad \hat{h}_j^{\hat{y}} = \hat{y}_{j+1/2} - \hat{y}_{j-1/2}.$$

We write $\mathbf{v} = (v^{\hat{x}}, v^{\hat{y}})$ for $\mathbf{v} \in \mathbf{R}^2$, and for any function $g(\hat{x}, \hat{y})$, let g_{ij} denote $g(\hat{x}_i, \hat{y}_j)$, let $g_{i+1/2,j}$ denote $g(\hat{x}_{i+1/2}, \hat{y}_j)$, etc.

If $\hat{\mathbf{v}}$ in (2.4b) is the basis function at node $(i + 1/2, j)$ or at node $(i, j + 1/2)$, then

$$(3.1) \qquad \tilde{\hat{u}}_{h,i+1/2,j}^{\hat{x}} = -\frac{\hat{p}_{h,i+1,j} - \hat{p}_{h,i,j}}{\frac{1}{2}(\hat{h}_i^{\hat{x}} + \hat{h}_{i+1}^{\hat{x}})}, \qquad \tilde{\hat{u}}_{h,i,j+1/2}^{\hat{y}} = -\frac{\hat{p}_{h,i,j+1} - \hat{p}_{h,i,j}}{\frac{1}{2}(\hat{h}_j^{\hat{y}} + \hat{h}_{j+1}^{\hat{y}})},$$

which is a finite difference approximation of $\hat{\tilde{\mathbf{u}}} = -\hat{\nabla}\hat{p}$. The same choice of $\hat{\mathbf{v}}$ in (2.4c) gives

$$(3.2) \quad \hat{u}^{\hat{x}}_{h,i+1/2,j} = \frac{1}{2}\left[(\mathcal{K}_{11})_{i+1/2,j-1/2} + (\mathcal{K}_{11})_{i+1/2,j+1/2}\right]\hat{\tilde{u}}^{\hat{x}}_{h,i+1/2,j}$$

$$+ \frac{1}{2(\hat{h}^{\hat{x}}_i + \hat{h}^{\hat{x}}_{i+1})}\Bigg\{$$

$$\left[(\mathcal{K}_{12})_{i+1/2,j-1/2}\hat{\tilde{u}}^{\hat{y}}_{h,i+1,j-1/2} + (\mathcal{K}_{12})_{i+1/2,j+1/2}\hat{\tilde{u}}^{\hat{y}}_{h,i+1,j+1/2}\right]\hat{h}^{\hat{x}}_{i+1}$$

$$+ \left[(\mathcal{K}_{12})_{i+1/2,j-1/2}\hat{\tilde{u}}^{\hat{y}}_{h,i,j-1/2} + (\mathcal{K}_{12})_{i+1/2,j+1/2}\hat{\tilde{u}}^{\hat{y}}_{h,i,j+1/2}\right]\hat{h}^{\hat{x}}_i\Bigg\},$$

with a similar expression for $\hat{u}^{\hat{y}}_{h,i,j+1/2}$; this is a finite difference approximation of $\hat{\mathbf{u}} = \mathcal{K}\hat{\tilde{\mathbf{u}}}$. Finally, for $E = E_{ij}$ in (2.4a), we have

$$(3.3) \quad \left\{\frac{\hat{u}^{\hat{x}}_{h,i+1/2,j} - \hat{u}^{\hat{x}}_{h,i-1/2,j}}{\hat{h}^{\hat{x}}_i} + \frac{\hat{u}^{\hat{y}}_{h,i,j+1/2} - \hat{u}^{\hat{y}}_{h,i,j-1/2}}{\hat{h}^{\hat{y}}_j}\right\}\hat{h}^{\hat{x}}_i\hat{h}^{\hat{y}}_j$$

$$= \int_{\hat{E}_{ij}} \hat{q}J\,d\hat{x} = \int_{E_{ij}} q\,dx.$$

The combination of (3.1), (3.2), and (3.3) gives our finite difference stencil for the pressure, approximating the elliptic equation $-\hat{\nabla} \cdot \mathcal{K}\hat{\nabla}\hat{p} = \hat{q}J$. This in turn is an approximation of the original problem (1.1). The stencil is 9 points in two dimensions and 19 points in three dimensions.

4. A SUMMARY OF THE CONVERGENCE RESULTS

Let $\|\cdot\|$ denote the L^2-norm; that is, for a scalar or vector function φ,

$$\|\varphi\| = \sqrt{\int_\Omega |\varphi(x)|^2\,dx}.$$

Let $\||\cdot\||_M$ and $\||\cdot\||_T$ denote the L^2-norm approximated, respectively, by the midpoint and trapezoidal quadrature rules over our mesh on Ω defined either directly or as induced by F from the computational domain. The proof of the following theorem is given in [4].

Theorem. *There exists a constant C depending on the smoothness of F and the solution, but independent of the maximum grid spacing h, such that*

$$\|\mathbf{u} - \mathbf{u}_h\| + \|\bar{\mathbf{u}} - \bar{\mathbf{u}}_h\| \leq Ch, \qquad \||\mathbf{u} - \mathbf{u}_h\||_T + \||\bar{\mathbf{u}} - \bar{\mathbf{u}}_h\||_T \leq Ch^2,$$

$$\|p - p_h\| \leq Ch, \qquad \||p - p_h\||_M \leq Ch^2,$$

$$\|\nabla \cdot (\mathbf{u} - \mathbf{u}_h)\| \leq Ch.$$

This theorem implies optimal order convergence in the L^2-norms, and, moreover, superconvergence in L^2 for the computed pressure at the cell-centers, and for the computed velocity at the grid points. Furthermore, the normal component of the flux at the midpoints of the edges or faces is also superconvergent (see [4]).

5. TRANSPORT BY THE CHARACTERISTICS-MIXED METHOD

The characteristics-mixed method of Arbogast and Wheeler [1], [2] is a method for approximating the transport problem (1.2)–(1.3). It treats the advection terms $\phi(\partial c/\partial t) + \mathbf{u} \cdot \nabla c$ in (1.2) as a characteristic derivative; that is, essentially (1.2) is viewed as

$$\phi \frac{\partial c}{\partial \tau} - \nabla \cdot D(\mathbf{u})\nabla c = q_c(c) - qc,$$

where τ is the characteristic direction. The characteristics are the solution to

$$\frac{d\breve{x}}{dt} = \frac{\mathbf{u}(\breve{x},t)}{\phi(\breve{x})}$$

on the physical domain Ω, or after the change of variables,

$$(5.1) \qquad \frac{d\hat{\breve{x}}}{dt} = \frac{\hat{\mathbf{u}}(\hat{\breve{x}},t)}{J(\hat{\breve{x}})\phi(F(\hat{\breve{x}}))}$$

on the computational domain $\hat{\Omega}$; thus, only a factor of J enters into the computation of the characteristics.

The dispersion term, $-\nabla \cdot D(\mathbf{u})\nabla c$, is then treated by a mixed method, as in the case for flow. The dispersion tensor D is a function of the true velocity. It can be expressed on the computational grid by the analogue of (2.5) in terms of $\mathbf{u}$, or in terms of $\hat{\mathbf{u}}$ as

$$(5.2) \qquad \mathcal{D}_{i,j}(\hat{\mathbf{u}}) = J\phi d_m \delta_{i,j} + |DF\hat{\mathbf{u}}| \left\{ \frac{\hat{u}_i \hat{u}_j}{|DF\hat{\mathbf{u}}|^2}(d_l - d_t) + d_t \delta_{i,j} \right\},$$

where $\hat{\phi}(\hat{x}) = \phi(F(\hat{x}))$. There is a small change in the way the speed times J, $|DF\hat{\mathbf{u}}|$, needs to be calculated, and J multiplies $\hat{\phi}$. The other, nondispersion terms are treated on the computational grid after multiplying by the Jacobian factor J.

6. SOME COMPUTATIONAL RESULTS

Tests show that the predicted rates of convergence are obtained by the method, including the superconvergence [3], [4]. The only requirement is that the map F be reasonably smooth; otherwise, there is a degeneration in the convergence rate.

We now present some results of a code that uses the finite difference scheme presented in this paper to solve the coupled flow and transport equations (1.1)–(1.3). The characteristics-mixed method [1], [2] is used for transport. The code is designed to run on a massively parallel, distributed memory computer.

We present first a 2-D example on a circular domain with a single well at the center injecting a tracer solute. Porosity and permeability are uniform; thus, the true solution is radially symmetric. We set the dispersion to zero in this example. Fig. 2A–B show concentration fronts at six equally spaced times. Fig. 2A shows the solution as it appears on the square computational domain. Fig. 2B shows the solution after mapping back to the true domain according to (2.2b). As one can see, the solution on the computational domain is distorted just enough to give concentric circles on the true domain. The radii of these circles increase at the expected rate as measured by the volume swept. Fig. 2C shows the curvilinear grid on the true domain that results from mapping the uniform computational grid.

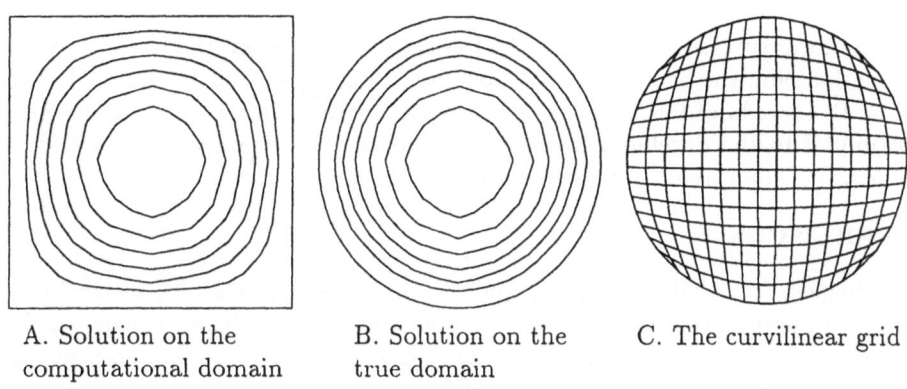

A. Solution on the computational domain

B. Solution on the true domain

C. The curvilinear grid

Fig. 2. A single injection well at the center of a circular domain.
The solution shows concentration fronts at six equally spaced times.

In Fig. 3, we demonstrate tracer injection at the WAG–6 (Waste Area Grouping #6) site of the Oak Ridge National Laboratory (ORNL). Porosity and permeability are again uniform. A uniform pressure drop occurs from the left face (Haw Ridge) to the right face (White Oak Lake and Copper ridge); there is no-flow across the other faces. Tracer injection occurs at the left face. We show the surface topography and the concentration front at four equally spaced times. The peak of Haw ridge is at the left corner; thus, fluid flow is slower there.

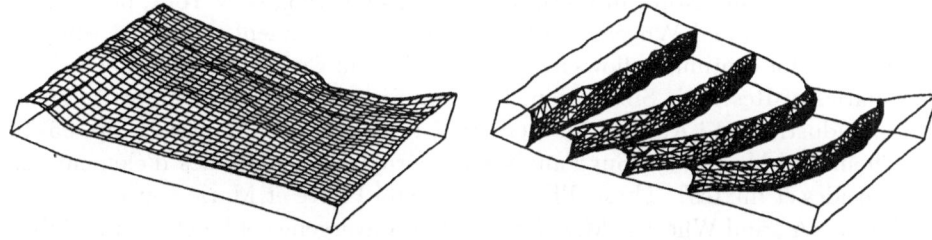

Fig. 3. A demonstration of tracer injection at ORNL WAG–6.

7. CONCLUSIONS

We presented a cell-centered finite difference mixed method that is locally mass conservative and highly accurate, especially for the velocity. General geometry can be handled by a mapping of a rectangular, computational domain to the physical domain. The net result of this mapping is a simple transformation of the tensor conductivity or dispersivity, and possibly the multiplication of certain other coefficients by the Jacobian factor. The scheme is easily implemented, since the data structures need only reflect the rectangular, computational grid.

ACKNOWLEDGEMENTS

This work was supported in part by the Department of Energy, the National Science Foundation, and the State of Texas Governor's Energy Office. L. Toran, L. West, and E. D'Azevedo of ORNL provided the WAG-6 topography data. Many people contributed to the development of the computer code, most notably, beside the authors, A. Chilakapati, L. Cowsar, and D. Moore.

REFERENCES

[1] Arbogast, T., Chilakapati, A., and Wheeler, M.F. (1992) "A characteristic-mixed method for contaminant transport and miscible displacement", in Russell, Ewing, Brebbia, Gray, and Pindar (eds.), Computational Methods in Water Resources IX, Vol. 1: Numerical Methods in Water Resources, Computational Mechanics Publications, Southampton, U.K., pp. 77-84.

[2] Arbogast, T., and Wheeler, M.F. (in press) "A characteristics-mixed finite element method for advection dominated transport problems", SIAM J. Numerical Analysis.

[3] Arbogast, T., Wheeler, M.F., and Yotov, I. (in preparation) "Mixed finite elements for elliptic problems with tensor coefficients as finite differences".

[4] Arbogast, T., Wheeler, M.F., and Yotov, I. (in preparation) "Mixed finite element methods on general geometry".

[5] Raviart, R.A., and Thomas, J.M. (1977) "A mixed finite element method for 2nd order elliptic problems", in Mathematical Aspects of the Finite Element Method, Lecture Notes in Math. 606, Springer-Verlag, New York, pp. 292–315.

[6] Russell, T.F., and Wheeler, M.F. (1983) "Finite element and finite difference methods for continuous flows in porous media", in Ewing, R.E. (ed.), The Mathematics of Reservoir Simulation, Frontiers in Applied Mathematics 1, Society for Industrial and Applied Mathematics, Philadelphia, Chapter II, pp. 35–106.

[7] Thomas, J.M. (1977) "Sur l'analyse numérique des méthodes d'éléments finis hybrides et mixtes", Thèse d'Etat, Université Pierre et Marie Curie.

[8] Weiser, A., and Wheeler, M.F. (1988) "On convergence of block-centered finite-differences for elliptic problems", SIAM J. Numerical Analysis 25, pp. 351–375.

A FINITE VOLUME METHOD FOR THE TRANSPORT OF RADIONUCLIDES IN POROUS MEDIA.

*A. BALAGUER, **C. CONDE, *J.A. LÓPEZ, *V. MARTÍNEZ
*Universitat Jaume I, Departament de Matemàtiques,
Campus Penyeta Roja, 12071 Castellón. Spain
** UPM, ETSIM, Departamento de Matemática Aplicada y Métodos informáticos
C./ Rios Rosas,21 28003 Madrid. Spain

One of the most probable and dangerous technical problems in an underground radioactive waste disposal is the dissolution and transport of radioactive substances by the water around the site. The solution of the partial differential equations system that governs the transport model depends on the radionuclide rates release to water. Therefore, we have to take into account two models in our study: a source term model giving the radioactive substances rate dissolved into water and a transport model considering conservative and non-conservative physico-chemical mechanisms. A numerical method designed for solving both models in a 1-D saturated porous media is presented in this paper. The numerical scheme of the transport model is based on finite volume method.

INTRODUCTION.

The radiactive wastes, generated in several activities of our present society, show a activity that is reflected on the electromagnetic radiations emission, which do them very dangerous for the humanity. This radioactive activity can last until several millions of years. Hence it is necessary that the radiaoactive wastes are isolated and your underground disposal seems the most feasible and the safest protection device. Nevertheless there is a moment at which the amoured packages that isolate the matrix structure incorporating the radionuclides are broken and the water gets in touch with this matrix.

We consider a radioactive waste consisting of "N" radionuclides dispersed through a solid matrix. We assume that the waste matrix and its contained radionuclides start

A. Peters et al. (eds.), Computational Methods in Water Resources X, 157–164.

to dissolve and diffuse into water at time $"t = t_c"$ and each nuclide leaches from the waste matrix with a fractional rate which may be in general time dependent and different for all nuclides. Moreover it also depends on the waste matrix geometric form. We are going to assume that the waste matrix is spherical and this form keeps during all leaching process.

The radionuclides dissolution in not always complete, so that an intermediate phase or storage arises. The radioactive substances will be removed from the intermediate phase by secondary dissolution in water flow which finally will carry them to the geosphere.

The model of this process will be denoted by *source term*. Such model allows to know at every time value the radioisotope amount which can be carried away by water. We assume that dissolved substances into water are carried in a one-dimensional spatial domain $[0, L]$. The equations that govern the transport model will be solved subject to initial and boundary conditions defined by source term model and their physical parameters can be space and time dependent. For solving such model a numerical scheme based on the finite volume method, efficient for convection dominate cases will be derived and discussed. Time integration is performed by a θ-scheme.

NUMERICAL MODELLING OF SOURCE TERM.

The temporal evolution of radionuclides stored in underground repositories from the waste packages to the geosphere can be divided in three phases. The first one arises until $"t_c"$ time at which water penetrates into waste containers and starts to leach the waste matrix. We assume a waste matrix consisting of different radionuclides making up a radioactive chain. The amount of each radionuclide $u_1(t), \ldots, u_N(t)$ will be evalue (in time) checking the next ordinary equations system:

$$\frac{du_i(t)}{dt} = \lambda_{i-1}\, u_{i-1}(t) \ - \ \lambda_i\, u_i(t) \quad i = 1, \ldots, N \ \ t \in [0, t_c]$$

with the initial condition: $u_1(0) = u_1^0, \ldots, u_n(0) = u_n^0$
where λ_i is the radioactive decay constant ($\lambda_0 = 0$) and u_i^0 is the initial concentration of each radionuclide.

The analytical solution of this system is know and it is given by the next equality:

$$(u_1(t), \ldots, u_n(t))^t = M\ (\exp(-\lambda_1 \cdot t), \ldots, \exp(-\lambda_N \cdot t))^t \quad t \in [0, t_c]$$

where M is a $N \times N$ lower triangular matrix whose elements are defined by:

$$M_{1,1} = u_1^0;\ M_{i,j} = \frac{\lambda_{i-1} \cdot M_{i-1,j}}{\lambda_i - \lambda_j} \quad \begin{matrix} i = 2, \ldots, N \\ j = 1, \ldots, i-1 \end{matrix}\ ;\ M_{i,i} = u_i^0 - \sum_{j=1}^{i-1} M_{i,j}\ i = 2, \ldots, N$$

The second phase starts at "t_c" time and your length depends on a lixiviation parameter "L_0", which expresses the waste matrix amount that is leached by water at one time and surface unity. Letting "p" the waste matrix density, as we assume that the waste matrix is spherical whose radius is equal to "R_0" then:

$$t_L = t_c + \frac{p \cdot R_0}{L_0}$$

is the waste matrix life time.

There is not transport of radiactive mass in this phase. There will just a process of corrosion of the waste packages and dissolution of radionuclides contained which pass to water or to intermediate storage. As we suppose that the amount of each radionuclide "$c_l(t)$" which stays into packages is dispersed through the waste matrix then your disappearance along time by leaching effect will be given by:

$$g_i(t) = \frac{3 \cdot c_i(t)}{t_L - t} \quad t_c \le t \le t_L$$

(we assume a spherical form in the waste matrix during all leaching process.)

Thus the differential system that govern the amount of each radioisotope which stays in the waste matrix, to take into account both decay radiactive and leaching process (we assume $\lambda_0 = 0$):

$$\frac{dc_i(t)}{dt} = \lambda_{i-1} \, c_{i-1}(t) \; - \; \lambda_i \, c_i(t) \; - \; g_i(t) \quad i = 1, \ldots, N \;\; t \in [t_c, t_L]$$

with the initial condition: $c_1(t_c) = u_1(t_c) = u_1^{t_c}, \ldots, c_n(t_c) = u_n(t_c) = u_n^{t_c}$
Your analytical solution is also know and it is yielded by:

$$(c_1(t), \ldots, c_n(t))^t = \left(\frac{t_L - t}{t_L - t_c} \right)^3 D \; (\exp(-\lambda_1 \cdot t), \ldots, \exp(-\lambda_N \cdot t))^t \;\; t \in [t_c, t_L]$$

where D is a $N \times N$ lower triangular matrix defined by:

$$D_{1,1} = u_1^{t_c}; \;\; D_{i,j} = \frac{\lambda_{i-1} \cdot D_{i-1,j}}{\lambda_i - \lambda_j} \;\; \begin{array}{l} i = 2, \ldots, N \\ j = 1, \ldots, i-1 \end{array} \; ; \; D_{i,i} = u_i^{t_c} - \sum_{j=1}^{i-1} D_{i,j} \;\; i = 2, \ldots, N$$

Now we have to compute the radionuclides amount that is dissolved and carry along water. The quantity of each radioisotope that water may be able to carry, at every moment, is always smaller or equal than: $C_{s,i} \, Q(t)$, where $C_{s,i}$ is the solubility limit of isotope "i" into water and $Q(t)$ is the water flow into waste packages. Hence it can to exist some moments at which the amount of radioactive isotope that is leached by water is greater than admissible. Then, spare amount will be precipitated around the site so that it arise the called intermediate deposit or intermediate phase.

We will assume all quantity $g_i(t)$ leached from the waste matrix leads in the intermediate phase and we denote by $s_i(t)$ the amount of an radionuclide "i" that is dissolved into water from intermediate phase. Thus the concentration "w_i" of nuclide "i" in the intermediate phase is yielded by:

$$\frac{dw_i(t)}{dt} = \lambda_{i-1}\, w_{i-1}(t) + g_i(t) - \lambda_i\, w_i(t) - s_i(t) \quad i = 1,\ldots,N \ \ t \geq t_c$$

with the initial condition: $w_1(t_c) = 0, \ldots, w_n(t_c) = 0$
where:

$$s_i(t) = \begin{cases} C_{s,i}\, Q(t) & if \ \ w_i(t) > 0 \\ \lambda_{i-1}\, w_{i-1}(t) + g_i(t) & if \ \ w_i(t) \leq 0 \end{cases} \tag{1}$$

For the approximation to the solution of this equations we have applied a numerical method based on the predictor-corrector Adams method (see Burden & Faires [4]) designed for stiff equations. We consider a time discretization of length Δt.

NUMERICAL MODELLING OF TRANSPORT MECHANISM.

The evolution of a substance through a porous medium suffers two different sorts of physico-chemical mechanisms: conservative and non-conservative. Here the phenomenon we are going to considerer are, among the conservative ones, advection, molecular diffusive and mechanical dispersion and among the non-conservative ones, adsorption, dissolution, radioactive decay and filiation. Considering all these phenomenons in one spatial dimension, the partial derivative equations system that governs them is: (see Balaguer [1]):

$$\frac{\partial(a_l(x,t)\, z_l)}{\partial t} + \frac{\partial(v(x,t)\, z_l)}{\partial x} - \frac{\partial}{\partial x}\left(k(x,t)\, \frac{\partial z_l}{\partial x}\right) + \lambda_l\, a_l\, z_l - \lambda_{l-1}\, a_{l-1}\, z_{l-1} = f_l$$
$$\forall l = 1,\ldots,N \quad \forall x \in]0,L[\quad \forall t \in]t_c,T[\tag{2}$$

where $z_l(x,t)$ is the concentration of nuclide "l", $v(x,t)$ is the pore velocity, $k(x,t)$ is the dispersion coefficient and $a_l(x,t)$ is the retention factor (v, k, a_l, f_l known).

On the above partial differential equations system we have supposed that the transport process starts at $x = 0$. We are going to study the change in the concentration of each radionuclide along a spatial segment $[0, L]$ and a temporal interval $[t_c, T]$ by the transport mechanisms. The point $x = L$ has been chosen so that there is not diffusive and dispersive transport through it. Thus we rely on a Neumann boundary condition at $x = L$ that may be expressed in this terms:

$$\frac{\partial z_l}{\partial x}(L,t) = 0 \ \ \forall t \in [t_c,T] \ \ \forall l = 1,\ldots,N$$

On the other hand the source term model gives the concentration value for each radioactive isotope at $x = 0$ so that the value of $z_l(0,t)$ will be equal to $s_l(t)$ $\forall l$

defined in (1). But in addition we needs a initial condition by means of which we know the concentration value for each radionuclide at $t = t_c$. We will assume that:

$$z_l(x, t_c) = 0 \quad \forall x \in]0, L[\quad \forall l = 1, \dots, N$$

The developed code solves (2) by the finite volume method which is a discretization method for conservation laws. In order to perform a space finite volume discretization of equations (2) a mesh of the domain $[0, L]$ is introduced such that $[0, L] = \cup_{i=0}^{NX} E_i$, where $E_i = [x_{i-\frac{1}{2}}, x_{i+\frac{1}{2}}]$ $i = 1, \dots, NX - 1$, $E_0 = [0, x_{\frac{1}{2}}]$, $E_{NX} = [x_{NX-\frac{1}{2}}, L]$ are the called discretization cell or control volume. In this work we assume $H = x_{i+\frac{1}{2}} - x_{i-\frac{1}{2}}$, $x_{\frac{1}{2}} = \frac{H}{2}$, $L - x_{NX-\frac{1}{2}} = \frac{H}{2}$ but it can consider a non-uniform grid.

The basic principle of the classical finite volume method is to integrate equations (2) over each cell E_i which yields the conservation law under its non-local form for the volume $[x_{i-\frac{1}{2}}, x_{i+\frac{1}{2}}]$. Letting $t^0 = t^c < t^1 = t_c + \triangle t < \dots < t^{NT} = t_c + NT \cdot \triangle t = T$ $\theta \in [0, 1]$ and considering a θ-scheme for the discretization in time we obtain the following equations (we define $x_{NX+\frac{1}{2}} = L$):

$$\int_{x_{i-\frac{1}{2}}}^{x_{i+\frac{1}{2}}} a_l^{n+1}(x) \, z_l^{n+1}(x) \, dx \;+\; \theta \, \triangle t \, \Pi_{l,i}^{n+1}$$

$$= \int_{x_{i-\frac{1}{2}}}^{x_{i+\frac{1}{2}}} a_l^n(x) \, z_l^n(x) \, dx \;-\; (1 - \theta) \, \triangle t \, \Pi_{l,i}^n \qquad (3)$$

$$l = 1, \dots, N \qquad i = 1, \dots, NX \qquad n = 0, \dots, NT - 1$$

where:

$$\Pi_{l,i}^n = -\int_{x_{i-\frac{1}{2}}}^{x_{i+\frac{1}{2}}} f_l(x, t^n) \, dx \;+\; v^n(x_{i+\frac{1}{2}}) \, z_l^n(x_{i+\frac{1}{2}}) \;-\; v^n(x_{i-\frac{1}{2}}) \, z_l^n(x_{i-\frac{1}{2}})$$

$$-\; k^n(x_{i+\frac{1}{2}}) \, \frac{\partial z_l^n(x_{i+\frac{1}{2}})}{\partial x} \;+\; k^n(x_{i-\frac{1}{2}}) \, \frac{\partial z_l^n(x_{i-\frac{1}{2}})}{\partial x}$$

$$+\; \lambda_l \int_{x_{i-\frac{1}{2}}}^{x_{i+\frac{1}{2}}} a_l^n(x) \, z_l^n(x) \, dx \;-\; \lambda_{l-1} \int_{x_{i-\frac{1}{2}}}^{x_{i+\frac{1}{2}}} a_{l-1}^n(x) \, z_{l-1}^n(x) \, dx \qquad (4)$$

and the overindex "n" indicates the function values at time "t^n".

The Neumann boundary condition at $x = L$ will allow us to know the value of $\frac{\partial z_l}{\partial x}(L, t) \, \forall t \in [0, T]$. On the other hand, the integration over E_0 is replaced by next equations:

$$z_i^n(x_0) = s_i(t^n)$$

($s_i(t^n)$ defined in (1)).

Integrals of (3) and (4) will be approximated by a trapezium rule and for spatial derivatives of concentration we will assume that concentration varies linearly in space between adjacent nodes. We take a node into each cell that will be denoted by $"x_i"$. In this work we will consider that $x_i = \frac{x_{i+\frac{1}{2}} + x_{i-\frac{1}{2}}}{2}$ $\forall i = 1, \ldots, NX - 1$, $x_0 = 0$, $x_{NX} = L$. Therefore the concentration in the extremes of cells will be interpolated linearly based on values of $z_l(x_i)$ $\forall i = 0, \ldots, NX$. Thus we obtain a linear system:

$$A^{n+1} Z^{n+1} = f^n + B^n Z^n \tag{5}$$
$$n = 0, \ldots, NT - 1$$

where:

$$Z^n = (Z_1^n, Z_2^n, \ldots, Z_l^n, \ldots, Z_{NE}^n)^t$$
$$Z_l^n = (z_l^n(x_0), z_l^n(x_1), \ldots, z_l^n(x_i), \ldots, z_l^n(x_{NX}))^t$$

Because $\frac{\partial z_l}{\partial x}(L, 0) = 0$ and $z_l(x, 0) = 0$ $\forall x \in]0, L[$ we take $z_l(L, 0) = 0$ $\forall l$. Thus as $z_l(0, 0) = s_l(t_c)$ then Z^0 is known and hence solving (5) we may know the solution for every one of radionuclides of the radioactive chain at each node and at each time t^{n+1}.

This numerical scheme has been verified by comparison with numerical results of code COLUMN2 (see Balaguer et al. [2]) based on a combination of characteristics and finite difference methods (see Bo et al. [3]). Tests results demostrate that the behaviour of both schemes is almost the same in terms of accuracy and efficiency for transport problems that are dominated by advection. Nevertheless our finite volume scheme may be coupled easier with the source term model and it can considerer a variable velocity.

The above finite volume scheme is always consistent in the finite difference sense but it is only stable (in the sense $\| (A^{n+1})^{-1} \cdot B^n \| \leq 1$) for $\theta \geq 0.5$. We will take $\theta = 0.5$ because a value of θ higher than 0.5 introduces numerical diffusion (see Balaguer [1]). On the other hand values $\triangle t_n > H$ give spurious numerical oscillations. Therefore we take $\triangle t = H$.

4. EXAMPLE.

We have solved a problem that consists in one radioactive chain with 5 radioactive species embedded in a stabilising matrix:

isotope	λ	initial concentration (mol./l.)	C_s (mol/l.)
Cm-245	8.15E-5	3.4E-3	1.0E-4
Am-241	1.60E-3	2.7E-1	1.0E-4
Np-237	3.24E-7	3.6E+0	1.0E-8
U-233	4.30E-6	1.0E-3	2.5E-7
Th-229	9.40E-5	2.0E-6	1.6E-6

We assume:

$$L_0 = 3.65 \text{ E-4 Kg.}/(m^2 \cdot \text{ year}) \quad p = 2700 \text{ Kg. } / m^3 \quad R_0 = 0.021 \text{ m.}$$
$$Q(t) = 4.2 \ l./\text{year} \quad t_c = 1000 \text{ years} \quad a_l = 1.0 \ \forall l$$
$$v = 0.01 \text{ m./year} \quad k = 0.03 \ m^2/\text{year}$$

Transport of these elements in the geosphere has been simulated. Concentration versus space graphicae are shown in Figure 1 for several times.

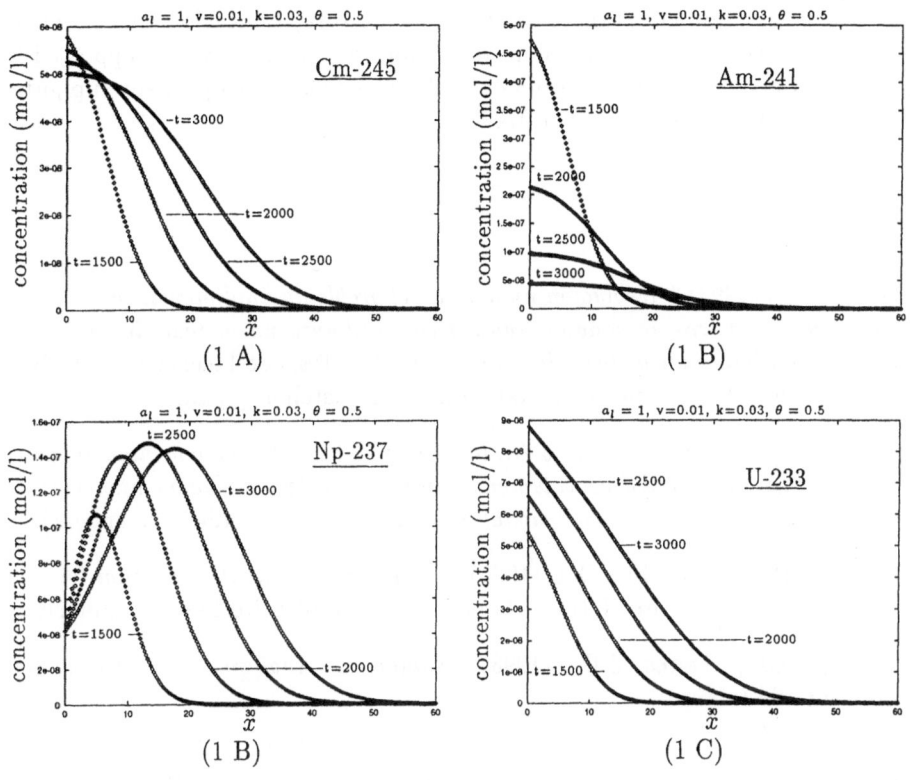

Figure 1: $H = \frac{1}{4}, \triangle t = \frac{1}{4}$

CONCLUSIONS.

It has simulated the radioactive substances flight process in an underground waste disposal and your transport by underground water running in an 1-D saturated porous media. A finite volume method allows the simulation of transport process and it can be easily coupled with a source term model that gives the radioactive substances release rate from the repository. To verify the developed code we have solved a problem composed by 5 radionuclides and we have obtained satisfactory umerical and physical results. A problem physically more realist with several radioactive chains it can find in Hartley [7].

ACKNOWLEDGEMENTS

Angel Balaguer acknowledges the Consellería de Cultura, Educació i Ciencia of the Generalitat Valenciana for a fellowship. This work has been partially supported by the Fundació Caixa Castelló. The work of Carlos Conde has been partially supported by ENRESA under contract number 0700442.

References

[1] Balaguer A. (1993) *"Resolución numérica del problema unidimensional de transporte en la geosfera de residuos radiactivos mediante un método de volúmenes finitos y mediante un método de Random Walk"*. Tesis de Licenciatura. Universitat Jaume I, Departament de Matemàtiques, Castellón (Spain).

[2] Balaguer A., Conde C., Lopez J.A., Martinez V. (1993) *"Implementación de un método C.N. trapezoidal de volúmenes finitos para la simulación numérica del proceso de migración unidimensional de una cadena radiactiva"* (to appear).

[3] Bo P., Carlsen L., Nielsen O.J. (1985) *"Column2, a computer program for simulating migration"*. Riso National Laboratory, DK-4000 Roskilde (DanmarK).

[4] Burden R.L., Faires J.D. (1993) *Numerical Analysis*. PWS-Kent cop. Boston.(USA)

[5] Conde C., Elorza F.J., Ferragut L., Ruiz A., Alonso J., Heras F., Mustieles F.J., Fidalgo A., Lagunas M., Alvarez I., Kindelan U., De Lope L.F., Prieto J., Santos A., De Miguel M.J. (1989): *Revision sobre los modelos numéricos relacionados con el almacenamiento de residuos radiactivos*. Publicación técnica Núm. 01/91, ED. ENRESA. Madrid (Spain).

[6] Gallouet T. (1992) *"An introduction to finite volume methods."* (Support de cours) Ed. CEA. EDF, INRIA. Clamart (France).

[7] Hartley, R.W. (1985) *"Release of radionuclides to the geosphere from a repository for high-level waste. Matematical modelling, results."* Polydinamics Ltd, Zurich.

TWO-DIMENSIONAL EULERIAN LAGRANGIAN LOCALIZED ADJOINT METHOD FOR THE SOLUTION OF THE CONTAMINANT TRANSPORT EQUATION IN THE SATURATED AND UNSATURATED ZONES.

P. BINNING[1] and M.A. CELIA[2]

[1] Department of Civil Engineering and Surveying, The University of Newcastle, University Drive, Callaghan, N.S.W., 2308, Australia. E-mail: philip@firebird.newcastle.edu.au
[2] Water Resources Program, Department of Civil Engineering, Princeton University, Princeton, NJ 08544, U.S.A. E-mail celia@karst.princeton.edu

An Eulerian Lagrangian Localized Adjoint Method (ELLAM) has been developed to solve the contaminant transport equation in two dimensions. The ELLAM allows the treatment of the advective and diffusive parts of the equation in a Lagrangian and Eulerian framework respectively. It provides a consistent method for developing a mass conservative numerical scheme and for treating the boundary conditions.

The ELLAM uses finite volume test functions with local space-time support. The support is a subdomain found by requiring that the test function satisfy the adjoint of the governing equation locally in the subdomain. The advection term is incorporated in the ELLAM through the definition of the subdomain boundaries, which follow the hyperbolic characteristics of the governing equation. The diffusion term appears as a series of boundary integrals around each subdomain. When used with the conservative form of the transport equation, the method conserves contaminant mass both locally in each subdomain and globally.

The method was tested for contaminant transport in saturated groundwater systems and compared with the Galerkin finite element method. Test problems included both a constant flow field where analytical solutions are available and more complex problems. The ELLAM was found to overcome the time step limitations of the finite element method while perfectly conserving contaminant mass.

The authors are currently extending the method to the case of multiphase flow and transport in the unsaturated zone. The transport equations are being solved in the air and water phases with equilibrium partitioning between the phases. In this case, special consideration must be given to the differences between the hyperbolic characteristics of the multiphase flow and contaminant transport equations.

Introduction

The major difficulty in numerically solving the governing equation for contaminant transport has been to devise effective methods for the treatment of the advective or hyperbolic part of the equation. There have been two major classes of methods used. The first are Eulerian methods, for example the finite difference or finite element method which are typically restricted to fine space-time grids in order to guarantee good solutions. The sec-

A. Peters et al. (eds.), Computational Methods in Water Resources X, 165–172.
© 1994 Kluwer Academic Publishers. Printed in the Netherlands.

ond class of numerical methods are Lagrangian Methods. These methods exploit the hyperbolic characteristics and gain the advantage of being able to use less restricted temporal grids, while still obtaining accurate solutions. Examples of Lagrangian methods are the method of characteristics or the Eulerian Lagrangian Localized Adjoint Method (ELLAM).

The Eulerian Lagrangian Localized Adjoint Method (ELLAM) was devised in the late 1980's (Celia et. al. 1990) and resolves the issues of the treatment of boundary conditions and of conservation of mass, both difficult in standard Lagrangian techniques such as the Method of Characteristics. The ELLAM is a formalism through which all other Lagrangian methods can be described (Celia 1994). The method has been applied by various authors to single phase contaminant transport in one dimension (see Binning and Celia 1994 for a review of these). Recently a finite volume ELLAM has been developed which uses piecewise continuous test functions in space, instead of the piecewise linear basis functions used by previous authors (Celia et. al. 1994, Celia and Ferrand 1993, Healy and Russell 1993). The finite volume approach leads to simplified integrals in the ELLAM formulation and has the desirable property of ensuring local contaminant mass balance. This development is viewed as being particularly favourable for the solution of higher dimensional problems and was used by Binning (1994) and Binning and Celia (1994) to solve the two-dimensional multiphase contaminant transport problem.

Binning and Celia (1994) considered contaminant transport in unsaturated two-phase (air-water) systems. Equilibrium partitioning of a contaminant between a liquid and gas phase was assumed, with the partitioning given by Henry's Law. The general multiphase transport equations for the air and water phases can then be added to give a single governing equation for contaminant transport (Binning 1994)

$$\frac{\partial (\theta c)}{\partial t} + \nabla \bullet (q c) - \nabla \bullet (D \bullet \nabla c) = 0 \qquad\qquad \text{(Eq. 1)}$$

where:

$$\theta = (R\theta_w + H\theta_a), \quad q = (q_w + Hq_a), \quad D = \theta_w D_w + H\theta_a D_a, \quad c = c_w$$

the subscripts w and a denote the air and water phases respectively, R is the retardation coefficient for sorption of contaminant onto the solid phase, θ_a ($\alpha = a, w$) represents the fluid content of the air and water phases, q_α is the volumetric flux of phase α, $H = c_a/c_w$ is Henry's constant for the partitioning of a liquid into the gaseous phase, D_α is the dispersion tensor for fluid α, and c_α is the volumetric concentration of contaminant in phase α. The governing equation for contaminant transport is written assuming that there are no source/sink terms or chemical reactions in either phase. The equation is defined in the semi-infinite domain Ω_{xt}, and is subject to the boundary conditions:

$$c(x, t) = g_1(x, t) \qquad (x \in \partial\Omega_x^1, t > 0)$$

$$[-D\nabla c(x, t)] \bullet n|_{\partial\Omega} = g_2(x, t) \qquad (x \in \partial\Omega_x^2, t > 0)$$

$$[qc(x, t) - D \bullet \nabla c(x, t)] \bullet n|_{\partial\Omega} = g_3(x, t) \qquad (x \in \partial\Omega_x^3, t > 0)$$

Here the boundary to the spatial domain with boundary condition of type i is denoted $\partial\Omega_x^i$.

For problems with a constant fluid content, the fluid flow and contaminant transport equations are decoupled and the finite volume ELLAM provides computationally efficient and accurate solutions. For variably saturated porous media some care must be taken in the ELLAM solution. In our early work on the finite volume ELLAM in two dimensions, oscillations were found to develop around inflow boundaries where there are sharp gradients of fluid content. The origin of these oscillations appears to be the stringent local mass balance requirements of the finite volume ELLAM. The local contaminant mass balance requirement forces oscillations to develop when there are numerical errors in the fluid flux obtained through the solution of the fluid flow equation. In this paper the origins of the oscillations are explored and an alternate method is developed to overcome them. Comments will be made about the coupling between the multiphase flow and transport equations that have implications for other Lagrangian solution techniques for these equations.

ELLAM for the hyperbolic transport equation

In order to illustrate the importance of the coupling of the fluid flow and contaminant transport equations in cases of variably saturated porous media, a simplified equation system will be considered. The dispersion terms of the contaminant transport will be neglected so that the governing equation is written:

$$\frac{\partial(\theta c)}{\partial t} + \nabla \bullet (qc) = 0 \tag{Eq. 2}$$

The transport equation is coupled to the multiphase fluid flow equations for air and water, which are written

$$\frac{\partial(\rho_\alpha \theta_\alpha)}{\partial t} + \nabla \bullet (\rho_\alpha q_\alpha) = 0 \qquad \alpha = air, water \tag{Eq. 3}$$

The fluid flow equation (Eq. 3) can be combined with the contaminant transport equation (Eq. 2) to find the non-conservative form of the transport equation:

$$\theta\frac{\partial c}{\partial t} + q \bullet \nabla c = \theta\frac{Dc}{Dt} = 0 \tag{Eq. 4}$$

where the material derivative $\dfrac{D}{Dt} = \dfrac{\partial}{\partial t} + \dfrac{q}{\theta} \bullet \nabla$ has been introduced and is defined by

the contaminant velocity $\dfrac{q}{\theta}$. For exact solutions of the governing equations, the conservative (Eq. 2) and non-conservative (Eq. 4) forms of the equation are equivalent. However numerical solutions contain errors so that the solutions based on (Eq. 2) and (Eq. 4) will have different properties and are not equivalent. A finite volume ELLAM employing the conservative form of the equation leads to the stringent local mass balance criterion in the solution. In contrast, the nonconservative form of the equation does not lead to mass conserving solutions. However, an ELLAM based on the the nonconservative form of the equation explicitly incorporates the numerical flow solution in the formulation and so is less sensitive to errors in the fluid velocities.

The ELLAM is developed from the weak form of the conservative governing equation:

$$\int\limits_{0}^{\infty}\int\limits_{\Omega_x} \left[\frac{\partial(\theta c)}{\partial t} + \nabla \bullet (qc) \right] w \ dx \ dt = 0$$

Or after application of Green's theorem:

$$= -\int\limits_{0}^{\infty}\int\limits_{\Omega_x} \left(\theta c \frac{\partial w}{\partial t} + qc \bullet \nabla w \right) dx dt + \int\limits_{0}^{\infty}\int\limits_{\partial\Omega_x} \left[n|_{\partial\Omega_{xt}} \bullet (qc) \, w \right] dS dt$$

$$= -\int\limits_{0}^{\infty}\int\limits_{\Omega_x} \theta c \frac{Dw}{Dt} dx dt + \int\limits_{0}^{\infty}\int\limits_{\partial\Omega_x} \left[n|_{\partial\Omega_{xt}} \bullet (qc) \, w \right] dS dt \qquad \text{(Eq. 5)}$$

The crux of the ELLAM is choice of test function w. In the finite volume ELLAM the test function is chosen to be the constant function satisfying the homogeneous adjoint equation:

$$\theta\frac{\partial w}{\partial t} + q \bullet \nabla w = \theta\frac{Dw}{Dt} = 0 \qquad \text{(Eq. 6)}$$

(Eq. 6) implies that the test function is constant along the characteristics of the governing equation. A finite-dimensional test function space is defined by dividing the space-time domain Ω_{xt} into a set of subdomains Ω_i. The time domain is partitioned into intervals of size $\Delta t^n = t^{n+1} - t^n$ and the spatial domain partitioned into elements that are centred at the nodes of the domain and backtracked in time along the characteristics given by (Eq. 6). The test function for subdomain Ω_i is then defined to be the indicator function for the subdomain having a constant value of one in the subdomain and zero outside the subdomain. An illustration of a typical interior subdomain is shown in Figure 1. Subdomain

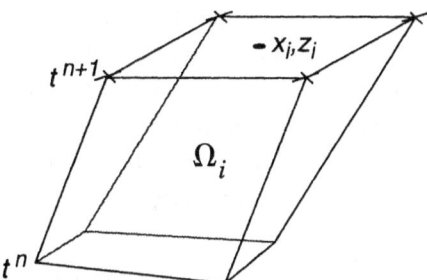

Figure 1 *Definition of the subdomain Ω_i. The domain is a regular rectangle at t^{n+1}, but may be an irregular quadrilateral at t^n.*

definition is altered when the subdomain intersects boundaries and for these details the reader is referred to Binning and Celia (1994).

With this definition of subdomain, the material derivative of the test function is given by:

$$\frac{Dw}{Dt} = -\delta(x - x|_{\partial\Omega_i^{n+1}}) + \delta(x - x|_{\partial\Omega_i^n})$$

where δ is the Dirac delta function. Since the subdomain follows the characteristics $\dfrac{Dw}{Dt}$ is nonzero only on the ends of the subdomain at t^n and t^{n+1}. The interior ELLAM integrals of (Eq. 5) are then evaluated as:

$$\int_{\Omega_i^{n+1}} (\theta c)^{n+1}\, dx - \int_{\Omega_i^n} (\theta c)^n\, dx = 0 \tag{Eq. 7}$$

The approximation given by (Eq. 7) can be viewed as a numerical approximation of the equation:

$$\frac{D(A\theta c)}{Dt} = 0 \tag{Eq. 8}$$

where A is the area of a subdomain. Given the definition of the subdomain, the area A of the subdomain must satisfy the equation:

$$\frac{DA}{Dt} = A\nabla\bullet\frac{q}{\theta} \tag{Eq. 9}$$

These equations demonstrate the nature of the finite volume ELLAM approximation given in (Eq. 7). The change in time of fluid mass in a subdomain is controlled by the area of the subdomain, which deforms as it is tracked along characteristics. The area must satisfy (Eq. 9). Numerical errors introduced in the calculation of fluid fluxes or in the determination of the shape of the subdomain through the backtracking routine result in errors in the determination of the area A and consequently a violation of conservation of fluid mass in the subdomain. Since the ELLAM approximation (Eq. 7) forces conservation of contaminant mass in every subdomain, errors in the amount of fluid mass in the subdomain result in errors in the calculation of contaminant mass. These errors are most apparent in regions of sharp changes in fluid content where the subdomain has greatest deformation. Experience has shown these errors are most likely to occur as oscillations in the vicinity of inflow boundaries.

Two approaches to dealing with this problem can be devised. Either a more accurate calculation of velocities and domain definition is required, or the stringent requirement of local conservation of mass must be relaxed. The second approach is chosen here as it is desirable to develop a numerical method that takes advantage of the information contained in the $D\theta/Dt$ term and is less reliant on extremely accurate velocities.

The ELLAM derives its local mass balance properties from the use of the conservative form of the governing equation. The non-conservative form of the equation (Eq. 4) can be written

$$\theta\frac{\partial c}{\partial t} + q \bullet \nabla c = \theta\frac{Dc}{Dt} = \frac{D(\theta c)}{Dt} - c\frac{D\theta}{Dt} = 0 \qquad \text{(Eq. 10)}$$

The first term on the right hand side is the change of contaminant mass moving along the characteristic and the second term describes the change in time of fluid mass in the subdomain. The inclusion of the second term in the formulation may be viewed as a correction for errors in the approximation of the velocities used to backtrack the subdomains. The weak form of the equation is then based on the non-conservative form of the equation rather than the conservative form of the equation as was done above:

$$\int_0^\infty \int_{\Omega_x} \left[\frac{\partial(\theta c)}{\partial t} + \nabla\bullet(qc)\right] w_i \, dx \, dt = \int_0^\infty \int_{\Omega_x} \left[\frac{D(\theta c)}{Dt} - c\frac{D\theta}{Dt}\right] w_i \, dx dt \qquad \text{(Eq. 11)}$$

The nonconservative ELLAM formulation is completed by evaluation of the integrals on the right-hand side of (Eq. 11). By assuming that the integrand is constant along characteristics, the integral can be approximated as

$$\int_{t^n}^{t^{n+1}} \overline{\left[\frac{D(\theta c)}{Dt} - c\frac{D\theta}{Dt}\right]} dt \int_{\Omega_i^{n+1}} dx$$

where the overbar denotes spatial averaging of the material derivative within the subdomain. The area of the subdomain appearing in the spatial integral is evaluated at t^{n+1}. Thus the change in area does not appear in the nonconservative formulation. Instead the effect of the changing fluid mass in the subdomain appears in the $D\theta/Dt$ term.

To minimize the mass balance errors resulting from the use of the non-conservative

ELLAM, a combined scheme is used with the nonconservative ELLAM only used for subdomains intersecting the boundary and the conservative ELLAM used for interior subdomains.

Results and Discussion

For problems with a constant moisture content the conservative ELLAM gives excellent results. Figure 2 shows a comparison between an analytical solution of (Eq. 1)

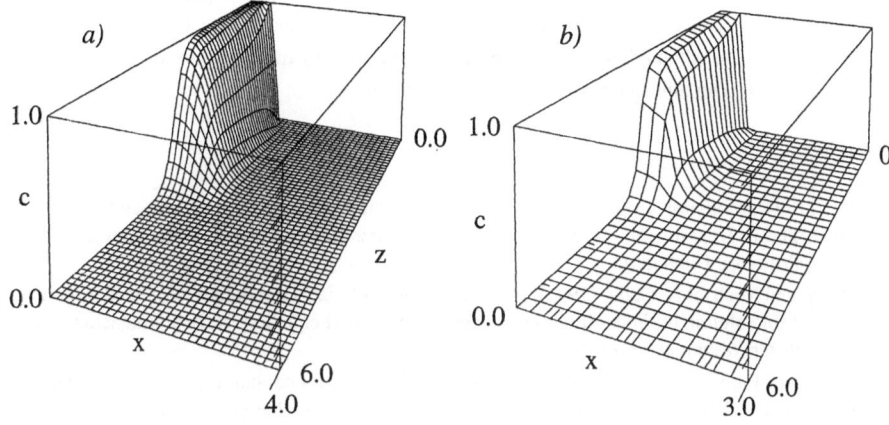

Figure 2 *Comparison of analytical a) and ELLAM solution b) for the contaminant transport equation with a strip source at the top left of the domain, a uniform velocity in the z direction. The dispersivities were set to be $\alpha_L = \alpha_T = 0.01$ and the fluid velocity was constant at 1 in the z direction, with a constant moisture content of 0.37. The solutions are shown at t=3. The ELLAM solution was obtained with a single time step!*

and the solution obtained with the conservative ELLAM using a single time step. A similar solution by the finite element method requires 30 time steps.

For problems with a variable fluid content, the combined conservative/nonconservative ELLAM gives accurate solutions with only small mass balance errors. The conservative/nonconservative ELLAM can utilize far bigger time steps than are possible with the finite element method and still obtain solutions with considerable accuracy. The method is well suited to advection-dominated problems and works for general boundary conditions and heterogeneous media.

The ELLAM is a general framework for deriving Lagrangian solution techniques for the advection-diffusion equation. This work shows that some care must be taken in the implementation of the method, particularly when the transport equation is coupled to the fluid flow equations. The finite volume ELLAM forces local conservation of contaminant mass. Oscillations will result if there are errors present in the fluid velocities used to define the test functions. These oscillations can be controlled by judiciously using a non-conservative ELLAM formulation on the boundaries of the domain. Excellent results are obtained with such a combined scheme.

Acknowledgments

Bill Gray for insight into the changing area of the subdomain, Richard E. Ewing for the suggestion of a combined conservative/nonconservative ELLAM, Helge Dahle and Magne Espedal for discussion. This research was partially funded by the NSF under grant 8657419-CES and by the USNRC under contract number NRC-04-88-074. Although research has been funded in part by these organisations it is not subject to agency review and so no official endorsement should be inferred.

References

Binning, P.J., (1994), Modeling unsaturated zone flow and contaminant transport in the air and water phases, Ph.D. Thesis, Department of Civil Engineering and Operations Research, Water Resources Program, Princeton University.

Binning, Philip and Michael A. Celia, (1994), A finite volume Eulerian-Lagrangian Localized Adjoint method for solution of the multiphase contaminant transport equations in two dimensions, Submitted to *Water Resources Research*.

Celia, M.A. (1994), Eulerian-Lagrangian Localized Adjoint Methods for contaminant transport simulations, Proceedings, Computational Methods in Water Resources, Heidelberg, Germany.

Celia, Michael A., Gang Li, Lin A. Ferrand, Richard E. Ewing (1994), A C^{-1} ELLAM method for simulating reactive and non-reactive transport, Water Resources Program Report, Princeton University.

Celia, Michael A. and Lin A. Ferrand (1993), A comparison of ELLAM formulations for simulation of reactive transport in groundwater, S.Y.Y. Wang (ed.), Proc. Int. Conf. on Hydroscience and Engineering, University of Mississippi, 1829-1836.

Celia, M.A., T.F. Russell, Ismael Herrera, R.E. Ewing (1990), An Eulerian-Lagrangian localized adjoint method for the advection-diffusion equation, *Advances in Water Resources*.

Healy, R.W. and T.F. Russell (1993), A finite-volume Eulerian-Lagrangian Local Adjoint Method for solution of the advection-dispersion equation, *Water Resources Research*, **29**, 7, 2399-2413.

APPLICATION OF AUTOMATIC DIFFERENTIATION TO GROUNDWATER TRANSPORT MODELS

C. H. Bischof[1], G. J. Whiffen[2], C. A. Shoemaker[2], Alan Carle[3], and A. A. Ross[1]

(1) Mathematics and Computer Science Division
Argonne National Laboratory
Argonne, Illinois 60439
(2) School of Civil and Environmental Engineering
Cornell University
Ithaca, New York 14853
(3) Center of Research on Parallel Computing
Rice University
Houston, Texas 77251

Automatic differentiation (AD) is a technique for generating efficient and reliable derivative codes from computer programs with a minimum of human effort. Derivatives of model output with respect to input are obtained exactly. No intrinsic limits to program length or complexity exist for this procedure. Calculation of derivatives of complex numerical models is required in systems optimization, parameter identification, and systems identification. We report on our experiences with the ADIFOR (Automatic Differentiation of Fortran) tool on a two-dimensional groundwater flow and contaminant transport finite-element model, ISOQUAD, and a three-dimensional contaminant transport finite-element model, TLS3D. Derivative values and computational times for the automatic differentiation procedure are compared with values obtained from the divided differences and handwritten analytic approaches. We found that the derivative codes generated by ADIFOR provided accurate derivatives and ran significantly faster than divided-differences approximations, typically in a tenth of the CPU time required for the imprecise divided-differences method for both codes. We also comment on the impact of automatic differentiation technology with respect to accelerating the transfer of general techniques developed for using water resource computer models, such as optimal design, sensitivity analysis, and inverse modeling problems to field problems.

A. Peters et al. (eds.), Computational Methods in Water Resources X, 173–181.
© 1994 Kluwer Academic Publishers. Printed in the Netherlands.

INTRODUCTION

Calculation of derivatives of computer models is necessary for a wide variety of procedures of interest to numerical modelers, for example, sensitivity analysis, optimization, and inverse problems. Several methods have traditionally been used to obtain derivatives of computer model output with respect to input. Analytical calculation can be used whenever it is reasonable to write analytic derivatives and then code them by hand (see, for example, *Chang et al.* [1992]). If a code is extremely complex, this approach may be infeasible. Symbolic differentiation programs often generate very large and inefficient derivative formulas. Writing an adjoint code to calculate derivatives is a well-defined procedure that calculates exact derivatives very efficiently (*Leitmann*, 1981). This method, however, also requires a potentially large amount of human effort.

One of the most common procedures used to obtain model derivatives of complex computer models is the divided-differences method. This method calculates derivatives by perturbing model input by small finite amounts and dividing the resulting model output perturbations by input perturbations (see, for example, *Gorelick et al.*, 1984). This procedure may fail to give accurate derivatives and is inefficient since it requires as many model simulations as there are input parameters of interest. In general, however, the divided-differences method does not require much human effort and, until recently, has been the only universally applicable approach available to obtain derivatives for very complex computer models.

This paper describes the application of automatic differentiation to obtain codes that evaluate derivatives of complex computer models efficiently, exactly, and with a minimum of human effort. Automatic differentiation is a method that produces a derivative code, given the model code and a list of parameters that are considered dependent and independent with respect to differentiation. The method produces a code that will evaluate derivatives exactly (up to machine precision) usually in much less time than the approximate finite-differences method. There are no inherent limits on program size or complexity.

We applied the automatic differentiation tool ADIFOR to a two-dimensional and a three-dimensional groundwater flow and contaminant transport finite-element model to demonstrate the method. The CPU times for automatic differentiation were much less than for the divided-differences method for both models, and somewhat slower than handwritten optimized analytic derivative code (written by *Chang et al* [1992]) for the two-dimensional model. In both cases automatic differentiation produced exact derivatives.

AUTOMATIC DIFFERENTIATION AND ADIFOR

Automatic differentiation techniques rely on the fact that every function, no matter how complicated, is executed on a computer as a (potentially very long) sequence of elementary operations such as additions, multiplications, and elementary functions such as *sin* and *cos*. By applying the chain rule,

$$\frac{\partial}{\partial t}f(g(t))|_{t=t_o} = (\frac{\partial}{\partial s}f(s)|_{s=g(t_o)})(\frac{\partial}{\partial t}g(t)|_{t=t_o}) \qquad (1)$$

over and over again to the composition of those elementary operations, one can compute derivate information of f exactly and in a completely mechanical fashion.

There are two canonical ways of propagating derivatives in automatic differentiation (*Griewank*, 1989). In the forward mode, we propagate derivatives of intermediate values with respect to the input parameters. In the reverse mode, we maintain derivatives of the final result with respect to intermediate quantities. The reverse mode is closely related to the previously mentioned adjoint approach. As a (very rough) rule of thumb, the run-time and storage requirements of the forward mode are linear in the number of model inputs of interest, while the runtime of the reverse mode is linear in the number of model outputs of interest w.r.t. differentiation. The storage requirements of the reverse mode are harder to assess and may be substantial because one must remember or recompute every intermediate value that nonlinearly impacts the final result.

There have been various implementations of automatic differentiation, and an extensive survey can be found in *Juedes* [1991]. Recently, new approaches to automatic differentiation have been explored in the ADIFOR (*Bischof et al.,* 1992a) and ODYSSEE (*Rostaing et al.,* 1993) projects. Both tools transform Fortran code, applying the rules of automatic differentiation, and generating new Fortran code that, when executed, computes derivatives without the overhead associated with "tape interpretation" schemes. We employed the ADIFOR tool in our experiments, which, as we will describe shortly, mainly employs the forward mode. In contrast, ODYSSEE employs the reverse mode.

ADIFOR (Automatic Differentiation in Fortran) (*Bischof et al.,* 1992a) provides automatic differentiation for programs written in Fortran 77. Given a Fortran subroutine (or collection of subroutines) describing a "function," and an indication of which variables in parameter lists or common blocks correspond to "independent" and "dependent" variables with respect to differentiation, ADIFOR produces portable Fortran 77 code that allows the computation of the derivatives of the dependent variables with respect to the independent ones.

ADIFOR accepts almost all of Fortran 77, in particular, arbitrary calling

r\$1 = x(1) * x(2)
r\$2 = r\$1 * x(3)
r\$3 = r\$2 * x(4)
r\$4 = x(5) * x(4)   Reverse Mode for computing
r\$5 = r\$4 * x(3)
r\$1bar = r\$5 * x(2)
r\$2bar = r\$5 * x(1) $$\frac{\partial y}{\partial x(i)}, i = 1, \ldots, 5$$
r\$3bar = r\$4 * r\$1
r\$4bar = x(5) * r\$2

```
do g$i$ = 1, g$p$
  g$y(g$i$) = r$1bar * g$x(g$i$, 1)
            + r$2bar * g$x(g$i$, 2)          Forward Mode:
            + r$3bar * g$x(g$i$, 3)          Assembling ∇y from ∇x(i),
            + r$4bar * g$x(g$i$, 4)          i = 1, ..., 5.
            + r$3 * g$x(g$i$, 5)
enddo
```

Forward Mode:
Assembling ∇y from $\nabla x(i)$, $i = 1, \ldots, 5$.

y = r\$3 * x(5)   } Computing function value

Figure 1: Sample of ADIFOR-generated Code

sequences, nested subroutines, common blocks, and equivalences. The ADIFOR-generated code tries to preserve vectorization and parallelism in the original code and employs a consistent subroutine-naming scheme that allows for code tuning, the exploitation of domain-specific knowledge, and the exploitation of vendor-supplied libraries.

ADIFOR employs a hybrid forward/reverse mode approach to generating derivatives. For each assignment statement, it generates code for computing the partial derivatives of the result with respect to the variables on the right-hand side using the reverse mode approach, and then employs the forward mode to propagate overall derivatives. For example, the statement

$$y = x(1) * x(2) * x(3) * x(4) * x(5)$$

gets transformed into the code shown in Figure 1. Note that none of the common subexpressions $x(i) * x(j)$ is recomputed in the reverse mode section.

ADIFOR-generated code can be used in various ways (*Bischof and Hovland,* 1991): Instead of simply producing code to compute the Jacobian J, ADIFOR produces code to compute $J * S$, where the "seed matrix" S is initialized by the user. So if S is the identity, ADIFOR computes the full Jacobian, and if S is just a vector, ADIFOR computes the product of the Jacobian by a vector. "Compressed" versions of sparse Jacobians can be computed by exploiting the same graph-coloring techniques that are used for divided-differences approximations of sparse Jacobians. In *Bischof et al.* [1994] it is also shown how one can use the flexibility of the ADIFOR interface to "stripmine" Jacobian computations and exploit parallelism to decrease turnaround time for derivative computations.

DIFFERENTIATED MODELS: TWO- AND THREE-DIMENSIONAL FLOW AND TRANSPORT

ISOQUAD: 2D Groundwater Flow and Transport Model

ISOQUAD is a two-dimensional (vertical dimension averaged) Galerkin finite-element model of groundwater transient flow and transport (see *Pinder and Frind*, [1972], and *Pinder and Gray*, [1977]). ISOQUAD is written in Fortran 77 and is on the order of two thousand lines of code. The model assumes the aquifer is confined but does allow for leakage. The model has been used extensively in the optimal groundwater remediation design research (*Chang et al.*, [1992], and *Whiffen and Shoemaker*, [1993]). Analytic expressions that can be coded for the derivatives of ISOQUAD are provided by *Chang et al.* [1992]. These hand-coded, validated derivatives have been optimized for performance and allow a comparison and validation of ADIFOR-generated derivatives and divided-differences derivatives.

The model ISOQUAD is used in its implicit time-stepping mode, on a small mesh of 77 nodes for comparison of derivatives. The model has as input the pumping rates of wells located on m computational nodes. Outputs include contaminant concentrations and hydraulic head values at each of n active nodes for each time step in the simulation. Automatic differentiation was used to obtain the following derivatives:

$$\frac{\partial c_{t+1}}{\partial h_t} \quad \epsilon \; \Re^{n \times n} \tag{2}$$

$$\frac{\partial c_{t+1}}{\partial q_t} \quad \epsilon \; \Re^{n \times m} \tag{3}$$

$$\frac{\partial h_{t+1}}{\partial q_t} \quad \epsilon \; \Re^{n \times m}. \tag{4}$$

The vector $c_t \; \epsilon \; \Re^n$ is the value of the contaminant concentration at time t, the vector $h_t \; \epsilon \; \Re^n$ is the hydraulic head at time t, and the vector $q_t \; \epsilon \; \Re^m$ is the pumping rate at each computational node at time step t. The evaluation of all three derivatives, (2)–(4), together is considered one derivative evaluation. In this example, there are 126 independent variables, corresponding to a hydraulic head value and a contaminant concentration value at each of 63 active nodes.

ADIFOR-generated code produced derivatives that agreed with the validated handwritten code to the order of the machine precision, but executed in much less time than the (imprecise) divided-differences method. In particular, for a 77-element mesh, ADIFOR-generated code calculated derivatives (2)–(4) in about the time it would take to run the original simulation model 17 times. The same derivatives using the divided-differences approach require 126 simulations, one for each independent variable. Automatic differentiation code was somewhat slower than the optimized handwritten code by *Chang et al.* [1992], which requires about the same time as 5 simulations.

TLS3D: 3D Groundwater transport model

TLS3D is a model of the three-dimensional advection diffusion equation. TLS3D employs a Taylor least-squares finite-element procedure to solve the unsteady advection diffusion equation. The model uses a three-dimensional serendipity Hermite element for an eight-node hexahedron. For a complete description of the procedure and three-dimensional model, see *Park and Liggett* [1991]. The code provided to the authors was written in Fortran 77 and is approximately 2300 lines in length.

The model is used on three-dimensional mesh ranging in size from 3 to 32 rectangular box elements to test the scalability of the code generated by ADIFOR. The model has as input the three-dimensional flow field; velocity vectors, v_t, for each simulation time step, t; and an initial contaminant concentration at each node, c_0. Model output is single-species contaminant concentrations at each active node, c_t, for each simulation time step t. Automatic differentiation was used to evaluate the derivative

$$\frac{\partial c_{t+1}}{\partial v_t} \quad \epsilon \quad \Re^{n \times 3n}. \tag{5}$$

Again, automatic differentiation provided a code that calculated model derivatives exactly in much less time than the divided-differences method. Four different-sized discretizations consisting of 4, 7, 16, and 32 three-dimensional elements were used to compare derivative performances.

We obtained the following results on a Sparcstation 10 and a single node of the IBM SP1 parallel computer. The number of independent variables equals three times the number of nodes in the model. The columns labeled $f(x)$ show the runtime (in seconds) of the simulation, the columns labelled $\frac{df(x(i))}{f(x)}$ show the average time required for the derivatives with respect to one independent variable, when the ADIFOR-generated code is used to generate derivatives with respect to 48 independent variables at a time.

# Elements (#independents)	Sparcstation 10 $f(x)$	$\frac{df(x(i))}{f(x)}$	SP1 Node $f(x)$	$\frac{df(x(i))}{f(x)}$
4 (48)	2.0	0.20	0.27	0.17
7 (96)	4.5	0.21	0.51	0.21
16 (243)	30.9	0.20	4.72	0.15
32 (432)	77.9	0.19	13.26	0.14

We see that AD is more than 5 (7) times faster than divided-differences approximations on the Sparcstation 10 and the SP1 node, respectively. No handcoded analytic derivatives of TLS3D are available for comparison.

CONCLUSIONS

We found that automatic differentiation can provide accurate and efficient derivative codes for the complex finite-element codes, ISOQUAD and TLS3D. In particular, we found ADIFOR generated codes that calculated derivatives in approximately 13% of the time required by divided differences for the two-dimensional model ISOQUAD, and between 14% and 21% of the time required by divided differences for the three-dimensional model, TLS3D. A more detailed account of these results will be given in a forthcoming paper.

As a result of our investiations, we believe that automatic differentiation can facilitate model changes by providing a mechanism for generating accurate and efficient derivative codes for models with very little human effort. Automatic differentiation technology will greatly accelerate the transfer of general techniques developed for using water resource computer models, such as optimal design, sensitivity analysis, and inverse modeling problems to field problems. Automatic differentiation can also accelerate the rate at which algorithm and model development and testing can occur by providing exact sensitivity information that may not otherwise be available.

ACKNOWLEDGEMENTS. Bischof's work was supported by the Office of Scientific Computing, U.S. Department of Energy, under Contract W-31-109-Eng-38, and by the National Aerospace Agency under Purchase Order L25935D. Primary support for Whiffen was in the form of a fellowship awarded by the Department of Energy Computational Science Graduate Fellowship Program. Shoemaker was funded in part by a grant from NSF (ASC 8915326) and the IBM Corporation. Carle was supported by the National Aerospace Agency under Cooperative Agreement No. NCCW-0027 and the National Science Foundation through NSF Cooperative Agreement No. CCR-9120008. Ross was supported by the Office of Scientific Computing, U.S. Department of Energy, under Contract

W-31-109-Eng-38, and the National Science Foundation through NSF Cooperative Agreement No. CCR-8809615. TLS3D code and assistance was provided by Dr. Nam-Sic Park and Dr. James A. Liggett. The groundwater remediation optimal control model was modified from Fortran code provided by L. Chang.

REFERENCES

Ahlfeld, D. P., J. M. Mulvey, G. F. Pinder, and E. F. Wood, Contaminated Groundwater Remediation Design Using Simulation, Optimization, and Sensitivity Theory 1. Model Development, *Water Resour. Res.*, *24*(3), 431-441, 1988a.

Ahlfeld, D. P., J. M. Mulvey, G. F. Pinder, and E. F. Wood, Contaminated Groundwater Remediation Design Using Simulation, Optimization, and Sensitivity Theory 2. Analysis of a Field Site, *Water Resour. Res.*, *24*(3), 443-452, 1988b.

Averick B., J. Moré, C. Bischof, A. Carle, and A. Griewank. Computing large sparse Jacobian matrices using automatic differentiation. Technical Report MCS–P348–0193, Mathematics and Computer Science Division, Argonne National Laboratory, 1993. Accepted for publication in SIAM Journal of Scientific Computing.

Bischof, C. , A. Carle, G. Corliss, A. Griewank, and P. Hovland. ADIFOR: Generating derivative codes from Fortran programs. *Scientific Programming*, 1(1):11–29, 1992.

Bischof, C., L. Green, K. Haigler, and T. Knauff. Parallel calculation of sensitivity derivatives for aircraft design using automatic differentiation. Extended Abstract, submitted to the 5th AIAA/NASA/USAF/ISSMO Symposium on Multidisciplinary Analysis and Optimization, 1994.

Bischof, C., and P. Hovland. Using ADIFOR to compute dense and sparse Jacobians. ADIFOR Working Note #2, MCS–TM–158, Mathematics and Computer Science Division, Argonne National Laboratory, 1991.

Chang, L.-C., C. A. Shoemaker, and P. L.-F. Liu, Application of a constrained optimal control algorithm to groundwater remediation, *Water Resour. Res.*, *28*(12), 3157-3173, 1992.

Gorelick, S. M., C. I. Voss, P. E. Grill, W. Murray, M. A. Saunders, and M. H. Wright, Aquifer Reclamation Design: The Use of Contaminant Transport Simulation Combined with Nonlinear Programming, *Water Resour. Res.*, *20*(4), 415-427, 1984.

Griewank, A. On automatic differentiation. In *Mathematical Programming: Recent Developments and Applications*, pages 83–108, Amsterdam, 1989. Kluwer Academic Publishers.

Juedes, D. A taxonomy of automatic differentiation tools. In Andreas Griewank and George Corliss, editors, *Proceedings of the Workshop on Automatic Differentiation of Algorithms: Theory, Implementation, and Application*, Philadelphia, 1991. SIAM.

Leitmann, G., The Calculus of Variations and Optimal Control, Vol. 20 of **Mathematical Concepts and Methods in Science and Engineering**, Plenum Press, New York and London, 1981.

Park, N. -S. and J. A. Liggett, Application of Taylor-Least Squares Finite Element to Three-Dimensional Advection-Diffusion Equation, *Int. J. Numer. Methods Fluids, 13*, 759-773, 1991

Pinder, G. F. and Frind, Application of Galerkin's procedure to aquifer analysis, *Water Resour. Res., 8*(1), 108-120, 1972.

Pinder, G. F. and W. G. Gray, *Finite Element Simulation in Surface and Subsurface Hydrology*, Academic Press, Orlando, Florida, 1977.

Rostaing, N., S. Dalmas, and A. Galligo. Automatic differentiation in Odyssee. *Tellus*, 45a(5):558–568, October 1993.

Whiffen, G. J., and C. A. Shoemaker, Nonlinear Weighted Feedback Control of Groundwater Remediation Under Uncertainty, *Water Resour. Res., 29*(9) 3277-3289, 1993.

USING A NUMERICAL MODEL OF TRANSPORT BASED ON A DETERMINISTIC THEORY TO INFER WELL PROTECTION ZONES IN A CHALKY AQUIFER.

P.Y. BIVER (*) and A. DASSARGUES
Laboratoires de Géologie de l'Ingénieur, d'Hydrogéologie et de Prospection Géophysique (L.G.I.H.), University of Liège, Bat. 19 Sart-Tilman, 4000 LIEGE, BELGIUM.

(*) now at Elf Aquitaine Production, Centre Scientifique et Technique Jean Feger (C.S.T.J.F.), Avenue Larribau, 64018 PAU, FRANCE.

The aim of this paper is to show how tracer test results may be used to determine isochrone lines around a pumping well, for a specified input of solute, in the chalky aquifer of Hesbaye (Belgium).
The use of a deterministic numerical model is interesting as it includes not only convection (as usual), but also the other main transport processes. The influence of the input function is discussed and a methodology to apply this model at a regional scale is given as a conclusion.

INTRODUCTION

The chalky aquifer of Hesbaye is an important water resource in Belgium. It is the main water supply for the town of Liège and its suburbs. In recent years, risks of accidental or diffuse pollutions due to agricultural practices and hazardous waste disposal have increased. For this reason, it is important to define protection zones around pumping wells in order to prevent water pollution and to adapt water production if such a problem occurred.
This paper, divided in three parts, describes a methodology to achieve this goal. In the first part, we give a brief account (see references for details) of tracer tests interpretations performed to obtain transport coefficients of the chalk and to characterize heterogeneities around the well of Crisnée (a production well near a little village of the Meuse valley). In the second part, emphasis is put on the numerical model. The numerical methods used to solve the equations are described. In the third part, we discuss the results obtained by using the same model to quantify the time arrival of a pollutant spill coming from different locations in the neighbourhood of the well.

TRACER TESTS INTERPRETATIONS

Below the site of Crisnée, we observe a typical geological sequence of Hesbaye cretaceous formations: under recent alluvial and colluvial deposits, lays a residual conglomerate with a clayey level at the base. Beneath this level, the chalky aquifer is semi-confined. The aquifer has been tested by a high rate pumping test (125 m^3/h). After stabilisation (18 hours), tracers (uranine and lithium chloride) have been injected in the nearby piezometers (Pz1 and Pz2, see figure 1). These tests have been interpreted in terms of a heterogeneous medium, due to the occurrence of a major fractured zone in the vicinity (figure 1). The geometrical description of heterogeneities has been defined on the basis of a geophysical prospecting on the site.

A. Peters et al. (eds.), Computational Methods in Water Resources X, 183–190.
© 1994 Kluwer Academic Publishers. Printed in the Netherlands.

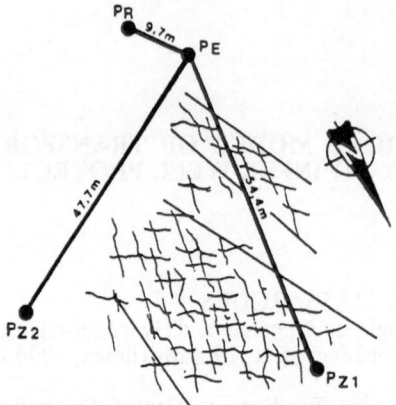

Figure 1: Equipment of the site and location of major fractured zones (Crisnée)

Using the numerical simulator (described below), the following flow and transport characteristics have been obtain by Biver (1993 & and 1994) resulting from detailed calibration procedures.

coefficient	range (from weakly to highly fractured zones)
K (permeability coefficient)	$1\ 10^{-5} \rightarrow 5\ 10^{-5}$ m/s
n_{eff} (effective porosity)	$0.5 \rightarrow 10.0\ \%$
n_{im} ("non effective" porosity)	$10 \rightarrow 40\ \%$
α_d^* (transfer coefficient for dual porosity effect)	$1.0\ 10^{-7} \rightarrow 1.5\ 10^{-6}$ m/s
a_T (transversal dispersivity)	$0.1 \rightarrow 1.9$ m
a_L (longitudinal dispersivity)	$0.5 \rightarrow 9.5$ m

Table 1: Flow and transport coefficients obtained from calibration of Crisnée tests.

Those coefficients correspond to the following mass balance equations:

$$div(-K.gr\bar{a}d\ h) = div(n_{eff}.\vec{v}) = 0 \tag{1a}$$

$$\frac{\partial c_m}{\partial t} + \vec{v}.gr\bar{a}d\ c_m - div(\underline{D}.gr\bar{a}d\ c_m) + \alpha_m.(c_m - c_{im}) = 0 \tag{1b}$$

$$\frac{\partial c_{im}}{\partial t} + \alpha_{im}.(c_{im} - c_m) = 0 \tag{1c}$$

with $\alpha_m = \alpha_d^* / n_{eff}$

$\alpha_{im} = \alpha_d^* / n_{im}$

$(\underline{D})_{ij} = a_T.|\vec{v}|.\delta_{ij} + (a_L - a_T).v_i.v_j / |\vec{v}|$

In this system, one can recognize in (1a) the well-known steady state flow equation, written in terms of piezometric head (h) as the aquifer is considered to be confined. Equation (1b) is the transport equation which describes the variation of concentration c_m in the moving fluid (say in the fractures). It takes into account convective and dispersive fluxes, but also a dual porosity effect involving the so-called "non moving" fluid, located mainly in the pores of the chalk. The mass balance in the pores is given by (1c), describing the variation of concentration in this "non moving" fluid (c_{im}).

NUMERICAL MODEL.

The system defined by equations 1a,b,c has been solved with a finite element method, implemented in the LAGAMINE code (developed in the Civil Engineering Department of the University of Liège). The complete methodology can be summarized as following:

1°) The flow equation (1a) is uncoupled and treated classically with a standard Bubnov Galerkin weighting of the residuals (already implemented in the code).
We write:

$$\int_V div\left[-K.\,gr\bar{a}d\left(N_i(\bar{x}).h_i\right)\right].N_j(\bar{x}).dV = 0 \qquad j = 1,...,N \qquad (2)$$

with $N_i(\bar{x})$ the interpolating function for node i of the finite element mesh,
h_i the nodal piezometric head,
N the total number of nodes,
V the simulation domain.
The solution provides the effective velocities ($\overset{\rho}{v}$), defining the convective flux in (1b)

2°) The mass balance equation in the "non moving" fluid can be solved analytically as it is demonstrated below.
Let us discretize c_m on each time step $\Delta t = t_{n+1} - t_n$. We have:

$$c_m = c_m^{t_n}.N_1(t') + c_m^{t_{n+1}}.N_2(t') \qquad (3)$$

$$\text{with} \quad N_1(t') = 1 - t'/\Delta t$$
$$N_2(t') = t'/\Delta t$$
$$t' = t - t_n$$

Then, (1c) can be written as:

$$\frac{\partial c_{im}}{\partial t'} + \alpha_{im}.c_{im} - \alpha_{im}.\left(N_1(t').c_m^{t_n} + N_2(t').c_m^{t_{n+1}}\right) = 0 \qquad (4)$$

which can be solved analytically versus time.

We obtain:

$$c_{im}(t') = c_{im}^{t_n}.\exp(-\alpha_{im}.t') + \dot{c}_m^{t_n \to t_{n+1}}.t' + \left(c_m^{t_n} - \dot{c}_m^{t_n \to t_{n+1}} / \alpha_{im}\right).\left(1 - \exp(-\alpha_{im}.t')\right) \qquad (5)$$

$$\text{with} \quad \dot{c}_m^{t_n \to t_{n+1}} = \frac{c_m^{t_{n+1}} - c_m^{t_n}}{\Delta t}$$

Result (5) in equation (1b) provides a single transport equation in terms of c_m :

$$L(c_m) = \frac{\partial c_m}{\partial t} + \bar{v}.gr\bar{a}d\ c_m + other\ terms \qquad (6)$$

3°) The single transport equation (6) is the most difficult to solve due to the convective component ($\bar{v}.gr\bar{a}d\ c_m$). If a standard Bubnov Galerkin method is applied, it leads to instabilities which manifest as spurious wiggles on the concentration profiles. To avoid this effect, without introducing artificial dispersion, two new methods have been programmed.

• The first method uses modified weighting functions, in order to obtain more representative weighted residuals. These weighting functions are off-centred in space and time to give more importance to upstream areas without numerical dispersion effect. For this reason, it is called the Full Upwind Petrov Galerkin method (F.U.P.G.). Mathematically, we can write:

$$\int\limits_{\Delta t} \int\limits_{V} L(c_m).W_j(\bar{x},t').dV.dt' = 0 \qquad j = 1,...,N$$

$$\Rightarrow \int\limits_{\Delta t} \int\limits_{V} L\left(N_1(t').c_m^{tn} + N_2(t').c_m^{tn+1}\right).W_j(\bar{x},t').dV.dt' = 0 \quad j = 1,...,N \qquad (7)$$

with $W_j(\bar{x},t') = N_i(\bar{x}).W^t(t') + \Delta W_{j1}(\bar{x}).W^t(t') + \Delta W_{j2}(\bar{x}).\Delta t/2 \,.\partial W^t(t')/\partial t'$

$\Delta W_{j1}(\bar{x}) = \alpha.\Delta l/2 \,.\bar{v}_m/|\bar{v}_m|.gr\bar{a}d\ N_j(\bar{x})$

$\Delta W_{j2}(\bar{x}) = \beta.\Delta l/2 \,.\bar{v}_m/|\bar{v}_m|.gr\bar{a}d\ N_j(\bar{x})$

$W^t(t') = 4.N_1(t').N_2(t')$

$\alpha = \coth(Pe/2) - 2/Pe$

$\beta = Cr/3 - 2.\alpha/(Pe.Cr)$

$Pe = |\bar{v}_m|.\Delta l/D_L =$ Peclet number of the considered finite element

$Cr = |\bar{v}_m|.\Delta t/\Delta l =$ Courant number of the finite element

$\bar{v}_m =$ average convection velocity on the finite element

$\Delta l =$ finite element length in $\bar{v}_m$ direction.

From these expressions one can see that weighting functions are off-centred with two coefficients (α and p) which are optimized to avoid numerical dispersion (Yu and Heinrich 1987).
This numerical scheme is able to treat convection dominated problems as well as dispersion dominated ones if the stability condition $Cr < 1$ is respected. For high convection velocities, this restriction may induce an excessive time discretization refinement. But in our practical case, this limitation is rarely prohibiting.

• The second method, the Hybrid Eulerian Lagrangian Method (H.E.L.M.), treats the convection component with a particle-tracking routine. This allows us to keep

the classical weighting functions in the remaining equilibrium. It can be summarized as following.
Time variation of concentration observed in Frenet axes (moving with convected particles) can be written:

$$\frac{dc_m}{dt} = \frac{\partial c_m}{\partial t} + \vec{v}.\,gr\bar{a}d\,c_m \tag{8}$$

Result (8) in equation (6) gives: $\qquad L'(c_m) = \frac{dc_m}{dt} + other\ terms$ $\qquad$ (9)

This equilibrium leads to the following scheme:

$$\int_{\Delta t}\int_V L'(c_m).\,N_j(\vec{x},t').\,dV.\,dt' = 0 \quad j = 1,...,N$$

$$\Rightarrow \int_{\Delta t}\int_V L'\Big(N_1(t').\,c_m^{tn} + N_2(t').\,c_m^{tn+1}\Big).\,N_j(\vec{x}).\,W^t(t').\,dV.\,dt' \quad j = 1,...,N \tag{10}$$

with $W^t(t') = \delta(t'=\Delta t) =$ collocated function in t_{n+1} (better for large time steps)

$$\frac{dc_m}{dt} = \frac{c_m^{tn+1} - c_m^{*tn}}{\Delta t}$$

$c_m^{*tn} =$ Lagrangian concentration of the particle; it is the concentration which could be observed at time t_{n+1} if the transport was purely convective; it is obtained by reverse particle tracking of global nodes and of intermediate hidden nodes if it is necessary (Yeh 1990).

This new approach is well suited to convection dominated problems but can take into account a non negligible dispersion. It gives better results for larger time steps and requires the condition $Cr > 1$. Time step length is only restricted by dispersion or immobile ("non moving") water effect.

It is clear that those two methods are complementary. Covering all ranges of Cr allows the user to choose time discretization in terms of physical phenomena and not in terms of the numerical scheme's stability. For not purely convective problems, it is even possible to use both methods together with mixed grids (H.E.L.M. and F.U.P.G. finite elements).

Tests have been made in that direction (Biver 1993). However, for previous and subsequent applications, the F.U.P.G. method has been proved to be more efficient.

ASSESSMENT TO WELL PROTECTION ZONES.

Once the numerical model has been calibrated on pumping tests and tracer breakthrough curves, we can keep flow results, transport coefficients, and discretization to infer aquifer behaviour in some prediction situations.

A local pollution with nitrates, at a rate of 1.2 kg/hour, stopped after 5 hours, has been simulated in 20 locations covering all the area around the well. An injection flow rate of 0,25 m³/hour has been considered. The concentration of the spill is supposed to be 10 g/l. The injected surface varies with the effective porosity and the thickness of the fissured zone at the injection point. Its order of magnitude is around 0.5 m². To model the transfer at the injection point, and to quantify the real input function of pollutant in the aquifer, a special finite element has been used. It takes into account the convection velocity variation in the injection area (which leads to a mixed boundary condition for the dispersive flux).

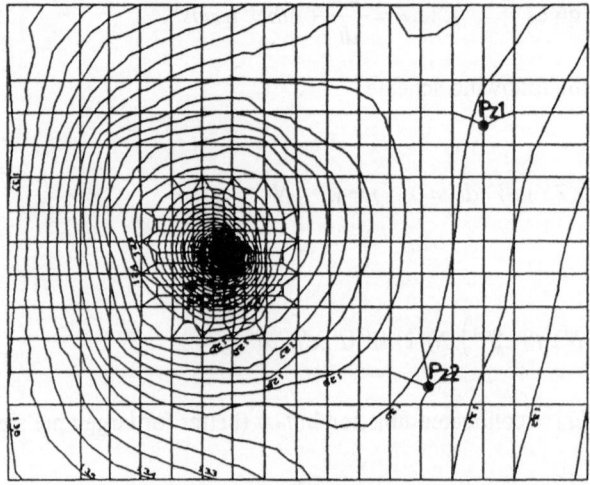

dh = 0.1 m

a) Piezometric head contours (permanent flow).

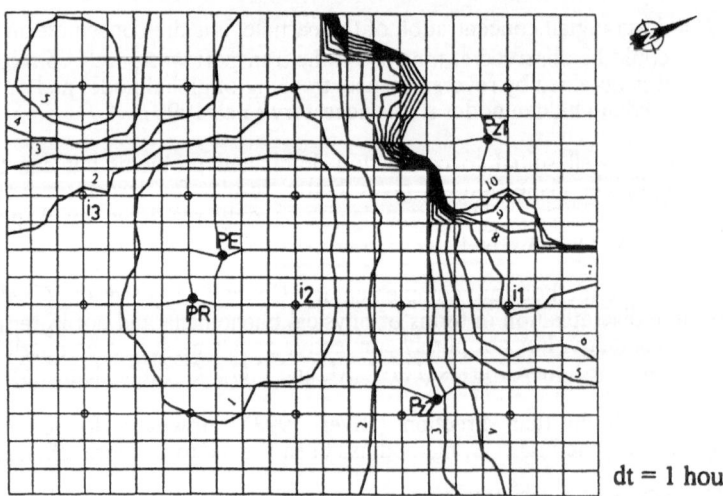

dt = 1 hour

b) Isochrone lines for a concentration of 10 mg/l at the pumping well
(injection of 10 gr/l during 5 hours at a rate of 0,25 m³/h).

Figure 2: Numerical results in pumping conditions (125 m³/h).

With simulation based on the piezometry represented on figure 2a, we can calculate the isochrone lines corresponding to the arrival, at the pumping well, of the very first significant concentrations (i.e. 10 mgr/l or 10 ppm). This result is presented on figure 2b, it is obtained by interpolating the values for the simulated injections. These contour lines are more objective than those which could be obtained by particle tracking. They reflect the three main components of transport in the chalky aquifer of Hesbaye (convection, dispersion, and immobile water effect).

After a quick look, one can see that the zone which is preferentially affected by the pollution around the well is in the direction of piezometer PR. However this zone does not correspond to the known fractured area. This is due to the regional flow of the aquifer which inhibits the pumping flow in this area and which increases it near piezometer PR. For the same reason, a part of the aquifer is totally unaffected by this particular pollution.

If we examine in details the results for 3 injection points (figure 3 a, b, and c,), we can understand that, on the basis of the breakthrough curves, it is possible to generate a great number of isochrone lines associated to different criteria. For instance, if our concentration limit is raised to 50 mgr/l (which is the standard for nitrate pollution in Belgium), the safe zone will be greater including the area represented by injection i_1. This kind of sensitivity analysis is not possible with the particle tracking method.

Moreover, by drawing the isoconcentration curves, we can have an idea of the pollution spreading; it is even possible to define, from those results, the limits of the polluted zone and its evolution in time.

CONCLUSION

We have presented here a methodology, based on a 2D deterministic numerical model, to determine protection zones around a pumping well in a double porosity medium. This method requires more data than the classical particle tracking method but is more rigorous and also flexible to the choices made by the interpreter (concentration limit and pollutant input function).

However, some work still remains:
- the generalisation of the model in 3D to improve our interpretations.
- applying such a methodology to quantify the main movements of nitrates, at the regional scale, in the saturated zone of Hesbaye aquifer.

If the first improvement is relatively obvious, the second is much more difficult. Indeed, nitrate transport in the fractured chalk of Hesbaye is subject to a scale effect problem which is still to be solved. It is more predictable than in a karstic zone but is still due to a fracture network, unlike a simple porous medium. On the available evidence it is unclear whether a continuum exists at high distances; but in our opinion, this may be the case. More tracer experiments are needed to characterize this scale effect. A study of fractures distribution in chalk could also be helpful to define precisely the representative element of volume for transport at regional scale.

ACKNOWLEDGEMENTS

The authors would like to thank the I.R.S.I.A. (Institut pour la Recherche Scientifique dans l'Industrie et l'Agriculture) and the French speaking Community of Belgium which have supported this research. Thanks also to the M.S.M. civil engineering department of the University of Liège for their technical collaboration concerning the LAGAMINE code.

This research was also supported by the Scientific Policy Services of the French speaking community of Belgium. Thanks to IBM Belgium S.A. for the provided hardware support in the framework of a partnership.

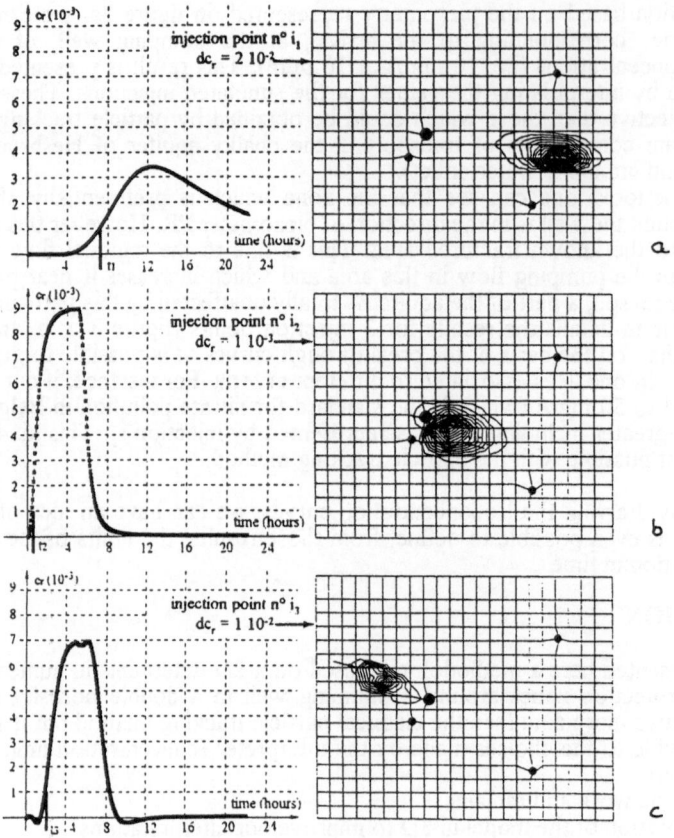

Figure 3: Detailed results of three simulated injections.

REFERENCES

Bear J. & Verruijt A. (1987), "Modelling ground water flow and pollution", Reidel Publ. Company, 415 p.
Biver P. (1994), "Numerical simulation of tracer tests to quantify, at a macroscopic scale, transport processes in a multiporous aquifer", to be published in the proceedings of the I.A.H.R. International Symposium on Transport and Reactive Processes in Aquifers (Zurich, 11-15 April 1994).
Biver P. (1993), "Etude phénoménologique et numérique de la propagation de polluants miscibles dans un milieu à porosité multiple (application au transport des nitrates dans l'aquifère crayeux du Crétacé de Hesbaye)", Ph.D.: University of Liège.
Meus Ph. (1993), "Hydrogéologie d'un aquifère karstique dans les calcaires carbonifères (Néblon Anthisnes, Belgique); apport des traçages à le connaissance des milieux fissurés et karstiques", Ph.D.: University of Liège.
Yeh G.T. (1990), "A Lagrangian-Eulerian method with zoomable hidden fine mesh approach to solving advection-dispersion equations", Water resources research 26: 1133-1144.
Yu C.C. & Heinrich J.C. (1987), "Petrov-Galerkin methods for the multidimensional time dependent convective-diffusion equations", International journal for numerical methods in engineering 24: 2201-2215.

A FOURIER ANALYSIS AND DYNAMIC OPTIMIZATION OF THE PETROV-GALERKIN FINITE ELEMENT METHOD

C. S. CARRANO, JR. (*) and GOUR-TSYH YEH (**)
(*) Department of Aerospace Engineering
The Pennsylvania State University
University Park, PA 16802
USA
(**) Department of Civil Engineering
The Pennsylvania State University
University Park, PA 16802
USA

A Fourier analysis of the linear and quadratic $N+1$ and $N+2$ Petrov-Galerkin finite element methods applied to the one-dimensional transient convective-diffusion equation is performed. The results show that *a priori* optimization of the $N+1$ method is not possible because dissipative errors are introduced as dispersive errors are reduced (any optimization is subjective). However, *a priori* optimization of the $N+2$ Petrov-Galerkin method is possible because the reduction of dispersion errors can be accomplished without the addition of artificial dissipation. The Spectrally Weighted Average Phase Error Method (SWAPEM) for the optimization of the $N+2$ Petrov-Galerkin method is introduced, in which the $N+2$ weighting parameter is chosen at each time step to minimize the integral over wavenumber of the phase error of Fourier modes, weighted by the frequency content of the global solution at the previous time step (obtained via FFT). Optimal values predicted by the method are in excellent agreement with those suggested by the numerical experimentation of others. Simulations of the pure convective transport of a Gaussian plume and a triangle wave are discussed to illustrate the effectiveness of the method.

INTRODUCTION

Insufficient resolution is the primary source of dissipation and dispersion errors in the consistent numerical solution of partial differential equations. Sharp peaks and details of the solution that convect between nodes after each time step are truncated by the interpolation between the nodes. This truncation is termed peak clipping. Conservative schemes will redistribute the displaced mass to nearby nodes so that the total integrated mass remains constant. If this redistribution is symmetric, numerical diffusion results and the details of the solution are smeared. If the redistribution is not symmetric, the shapes of details are distorted and spurious oscillations are generated [4]. More precisely, these errors exhibit themselves as errors in amplitude and phase speed of the Fourier components of the numerical solution, as compared to those of the exact solution. The phase speed errors, which generate spurious oscillations, are generally largest for high wavenumber components because sharp features suffer more clipping than smooth ones. As a result, the numerical description of high gradient regions of the solution is difficult due to their large high frequency content.

A. Peters et al. (eds.), Computational Methods in Water Resources X, 191–198.

Petrov-Galerkin finite element methods can affect changes in the phase and amplitude errors by manipulation of weighting parameters.   In fact, several authors have acknowledged the ability of the N+2 Petrov-Galerkin Method to reduce phase errors without the introduction of dissipation.   The usefulness of these new non-diffusive Petrov-Galerkin methods depends on the analysts's ability to choose appropriate values for the N+2 weighting parameter. Practitioners of this method have therefore resorted to numerical experiments for insight into behavior of the method. Westerink and Shea [3] presented graphically a relationship between the parameter and Courant number for the convection dominated transport of a Gaussian plume. Miller and Cornew [2] have shown experimentally the effects of variation of the standard deviation of the plume on the optimal weighting parameter. Because the optimal parameter depends on the initial conditions of the problem, this type of numerical experimentation is of little practical value.   An analyst trying to compute a previously uninvestigated problem is left completely in the dark. What is needed is a method for the dynamic calculation of the optimal value for the parameter during the computation of an arbitrary problem, without requiring a knowledge of the exact solution to the initial-boundary value problem. This work is in direct response to the aforementioned need. The Spectrally Weighted Average Phase Error Method is introduced, in which the parameter is chosen at each time step during the computation to minimize the integral over wavenumber of the phase error of the Fourier harmonics, weighted by the frequency content of the global solution at the previous time step. Used in conjunction with the N+2 Petrov-Galerkin method, the resulting solution has the property that at each time level, the phase speed of each Fourier mode is as accurate as possible, relative to its importance in describing the solution.

FOURIER PHASE AND AMPLITUDE ERRORS

The exact solution to the convection-diffusion equation with D and u constant can be expressed as the integrated sum of Fourier waves traveling with phase speed u and attenuated at the rate $\exp(-Dk^2t)$ as follows:

$$\phi(x,t) = \frac{1}{2\pi} \int_{-\infty}^{\infty} \hat{\phi}(k,0)\exp(-Dk^2t)\exp[ik(x-ut)]dk \qquad (1)$$

If we impose periodic boundary conditions on the system matrix of the numerical method, the numerical solution at location x_j, and time t^n may be represented in terms of a finite Fourier series as:

$$\phi(x_j,t^n) = \sum_m \bar{\phi}_m^0 \exp(ia_m t^n)\exp(ik_m x_j), \qquad (2)$$

where $\bar{\phi}_m^0$ is the Fourier transform of the initial conditions of the problem. By the algebraic manipulation of this series, we can express it in terms of the magnitude and argument of the numerical eigenvalue spectrum $\xi_m = \exp(ia_m\Delta t)$ as:

$$\phi(x_j,t^n) = \sum_m \bar{\phi}_m^0 \exp(-D_m^* k_m^2 t^n)\exp[ik_m(x_j - u_m^* t^n)], \qquad (3)$$

where we define the numerical phase speed u_m^* and diffusion coefficient D_m^* as

$$u_m^* = \frac{1}{k_m\Delta t}\arg(\xi_m), \qquad D_m^* = -\frac{1}{k_m^2\Delta t}\ln|\xi_m|. \qquad (4)$$

A comparison of equations (1) and (3) reveals that the numerical solution is a sum of

Fourier waves travelling with phase speed $u_m^{\bullet}$ and attenuating at the rate $\exp(-D_m^{\bullet} k_m^2 t^n)$.

Note that the phase speed of the numerical solution is a function of wavenumber, whereas the phase speed of the analytical solution is constant. It is precisely this discrepancy in phase speed that we call numerical dispersion. Dispersion errors cause a non-sinusoidal initial waveform to distort in time, as its component Fourier waves slowly separate in space. The constructive and destructive interference that occurs between the separating Fourier modes produces spurious oscillations in the physical domain. A measure of the dispersive character of the numerical method is the phase error, which is defined as the difference between the radian distance propagated by a numerical Fourier mode and that of the corresponding analytical mode after the latter has propagated one wavelength. We can express the phase error as

$$\Theta_m = 2\pi \left[\frac{1}{ck_m\Delta x} \tan^{-1} \frac{\text{Im}\,\xi_m}{\text{Re}\,\xi_m} - 1 \right] \qquad (5)$$

This equation is will be the cornerstone of the Spectrally Weighted Average Phase Error Method [1].

The ability of the Petrov-Galerkin method to accurately propagate Fourier modes with the correct amplitude is now assessed. We define R_m as the ratio of the amplitude of a component of the numerical solution to that of the analytical solution after the analytic component propagates through 2π radians (one wavelength):

$$R_m = \left[\frac{|\xi_m|}{|\xi^a(k_m)|} \right]^{N_m} \qquad (6)$$

This ratio provides a measure of the numerical dissipation associated with the numerical method. The presence of numerical dissipation damps and smears features of the numerical solution as does analytical diffusion. Therefore, numerical dissipation is often referred to as artificial diffusion. However, the numerical diffusion coefficient $D_m^{\bullet}$ is a function of wavenumber and therefore the damping caused by numerical dissipation need not resemble that resulting from a physical diffusive process in which D is usually wavenumber independent.

Assessment of the Weighting Methods

In this section, the changes in phase and amplitude error of the various weighting methods resulting from adjustments of the weighting parameters are plotted explicitly for all wavenumber components in figures 1 and 2. Examination of these plots shows the ability of each of the weighting methods to correct the causes of spurious oscillation and smearing. In figure 1 the phase and amplitude errors for the N+1 Petrov-Galerkin upstream weighting method with Crank-Nicolson discretization are shown. Examination of figure 1a shows that the severe phase lag of high wavenumber modes characteristic of the Galerkin method can be mostly alleviated with the addition of N+1 bias. It appears that the phase error for any single wavenumber can be brought to zero for some value of α (and for this particular Courant number). However, it is not possible to correct the phase error of all wavenumbers simultaneously. Unfortunately, the correction of phase errors using N+1 weighting is made at the expense of introducing severe amplitude errors, especially at high wavenumbers. These corrections in phase are essentially irrelevant, because the Fourier components whose phase is improved are completely extracted from the solution by artificial diffusion.

In figure 2, the phase and amplitude errors for the N+2 Petrov-Galerkin method with Crank-Nicolson time discretization are shown. Examination of figure 2a shows that the addition of N+2 bias also has the ability to correct phase errors. Once again it appears to be possible to reduce the phase error of any single wavenumber to zero, but not for all wavenumbers simultaneously. However, the phase lag of very high wavenumber modes can be vastly reduced at the expensive of introducing a slight phase lead at midrange wavenumbers. If the solution has a large high frequency content, this may improve the solution further. For this method there are no amplitude errors at any wavenumber and for any value of β. Hence we are free to adjust phase errors without introducing numerical dissipation. It is for this reason that a true optimal value for β exists. There is no tradeoff in this case; there is a clear choice which produces the least numerical error.

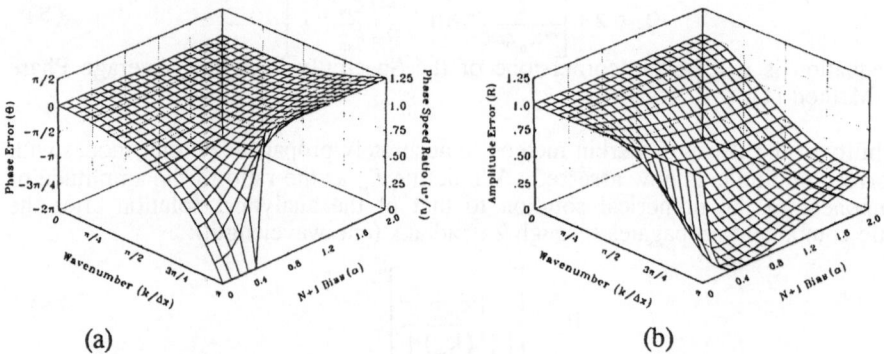

(a)   (b)

Figure 1 Phase error (a) and amplitude error (b) for variations of the N+1 weighting factor α and wavenumber for $c = 0.8$ and $D = 0$.

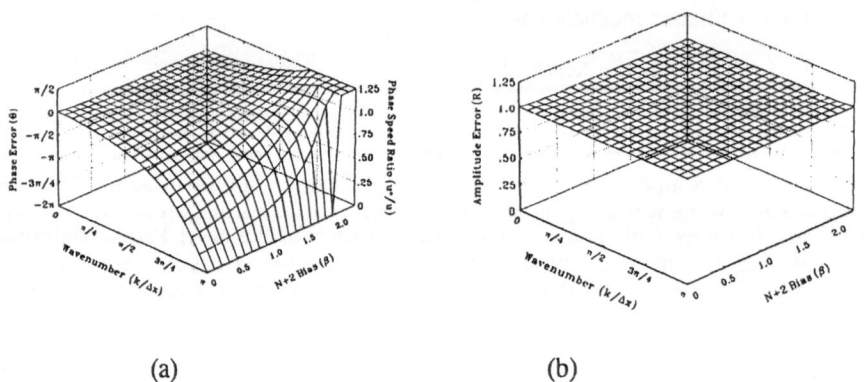

(a)   (b)

Figure 2 Phase error (a) and amplitude error (b) for variations of the N+2 weighting factors β and wavenumber for $c = 0.8$ and $D = 0$.

OPTIMIZATION BY THE SPECTRALLY WEIGHTED AVERAGE PHASE ERROR METHOD

In this paper we introduce the Spectrally Weighted Average Phase Error Method (SWAPEM) for the optimization of the N+2 weighting parameter β. A Crank-Nicolson time discretization is used resulting in a numerical method with perfect amplitude behavior. We define the spectrally weighted average phase error at time t^n as:

$$\overline{\Theta}^n(\beta) = \frac{1}{2\pi} \int_{-\pi/\Delta x}^{\pi/\Delta x} |\, \breve{\phi}^n(k)\, |\, \Theta(k,\beta)\, dk, \qquad (7)$$

where $|\, \breve{\phi}^n(k)\, |$ is the magnitude of the unbounded discrete Fourier transform of the solution at time t^n, and $\Theta(k,\beta)$ is the phase error given in (5), with its functional dependence on k and β emphasized. The spectrally weighted average phase error is an average over all Fourier modes, in which those modes important to the description of the solution are given more weight. This average error is a quantification of the net phase error described in the opening paragraph of this chapter. In the SWAPEM procedure, the optimal N+2 weighting factor for the next time step, β_0^{n+1}, is chosen so that the absolute value of this net phase error is a minimum:

$$\left[\frac{d\,|\,\overline{\Theta}^n(\beta)\,|}{d\beta} \right]_{\beta_0^{n+1}} = 0 \qquad (8)$$

Identically, we can chose β_0^{n+1} such that the square of the net phase error is stationary. This is the recommended approach since numerical minimization algorithms generally converge faster for smoother functions. If quadratic elements are used, the procedure is the same, except that now it is necessary to find the optimal pair (β_c, β_m), which corresponds to the location of the minimum of the surface $\overline{\Theta}^n(\beta_c, \beta_m)$.

The SWAPEM technique will be tested for the pure-convection of a Gaussian plume, and a triangle wave in the unbounded domain. It is well documented that conventional numerical solutions of these problems exhibit severe spurious oscillations or excessive smearing unless very fine meshes are used [2]. For this reason, these problems (among others) have been selected as benchmarks by the convection-diffusion forum of the 7th International Conference on Computational Methods in Water Resources held at the Massachusetts Institute of Technology on June 13-17, 1988. The velocity for the pure convection problems is 0.5, the grid size used is 200. For the case of a Gaussian plume: the initial standard deviation is $\sigma_0 = 264$, the initial location of the peak is $x_0 = 2000$. For the case of a triangular wave pulse: the initial half-width of the wave is $\ell_0 = 1000$ and the initial peak location is $x_0 = 2000$.

At any stage of the time-marching procedure, the concentration $\phi(x_j, t^n)$ is known only at N discrete points (without interpolating with the finite element basis functions), and the eigenvalue spectrum is also discrete. We perform the SWAPEM by approximating the integral in Eq. (7) as:

$$\overline{\Theta}^n(\beta) = \Delta x \sum_m a_m |\, \breve{\phi}_m^n\, |\, \Theta_m(\beta) \, , \qquad (9)$$

where $|\, \breve{\phi}_m^n\, |$ is determined by FFT, and a_m are the coefficients of any numerical integration algorithm requiring function evaluations only at uniform intervals. A closed interval routine may be used despite the removable singularity of $\Theta_m(\beta)$ at $k_m = 0$ since it can be shown that

$$\lim_{k \to 0} \Theta(k, \beta) = 0, \tag{10}$$

for all β. The value of β_0^{n+1} is found by numerically locating the minimum of the square of equation (7) based upon the FFT of the solution at time level n. A simple golden search minimization routine is used in this study.

The resulting scheme has the property that at each time level, the phase speed of each Fourier mode is improved as much as possible relative to its importance in describing the solution waveform. We note that, even if the frequency content of the solution to a particular problem is known *a priori* to be constant in time, selecting a new optimal β at each time step (or every several steps) can still improve the solution. First, roundoff and peak clipping errors redistribute the modal energy throughout the spectrum to some degree. Second, if the waveform does distort during the computation due to residual oscillations or insufficient resolution, this *new* waveform should be propagated as accurately as possible, in order to slow the growth of such deformations.

Figure 3 shows the optimal weighting parameter, β_0^1 determined by SWAPEM for the first time-step ($n=0$). For the first time-step computation, the initial frequency content ϕ_m^0 can be obtained analytically rather than by FFT [1]. Also shown in figure 3 are the curves from figure 7 of the paper by Westerink and Shea [3] showing their model $\beta_0 = 2c^2$ inspired by a local truncation error analysis, and their numerical experiments with a Gaussian plume. The experimental curve (dashed) was reproduced by fitting points on their curve with a straight line, since their original data was not available. No experimental results are currently available for the triangle wave problem.

The SWAPEM gives much better agreement with experiment than the truncation error method. The truncation error method fails in general, because phase errors can be large when the method is of high order. Miller and Cornew [2] have shown that truncation error terms up to the 15th order are amplified by the removal of low order terms such that choosing β_0 to remove these terms can actually increase global errors. Additionally, these authors have noted that optimization by truncation error analysis is fundamentally limited in that the waveform of the solution is not accounted for, whereas numerical experimentation reveals a strong dependence on solution waveform. The Fourier transform of the numerical solution (either obtained analytically, or more likely by FFT) is the ideal way to incorporate this dependence.

The difference between the SWAPEM and experimental results is largest for low Courant numbers. It turns out that in this Courant number range, variations of β have only a small effect on the solution. Hence, the values chosen by numerical experiment are most subject to error in this range. For high Courant numbers, where th solution is sensitive to β, agreement between the SWAPEM and experiment is very good. It turns out that in this region the model $\beta_0 = 2c^2$ also works reasonably well, because β_0 becomes increasingly independent of solution waveform for Courant numbers near one.

Simulation results

In this study, the benchmark problems were simulated at Courant number 0.8 ($u=0.5$, $\Delta x=200$, $\Delta t=320$). Two linear finite element methods of solution were investigated, namely the traditional Galerkin and the N+2 Petrov-Galerkin method where the

SWAPEM is used to determine β_0^{n+1} dynamically. The numerical results for the Gaussian plume and triangle wave benchmark problems produced by these two finite

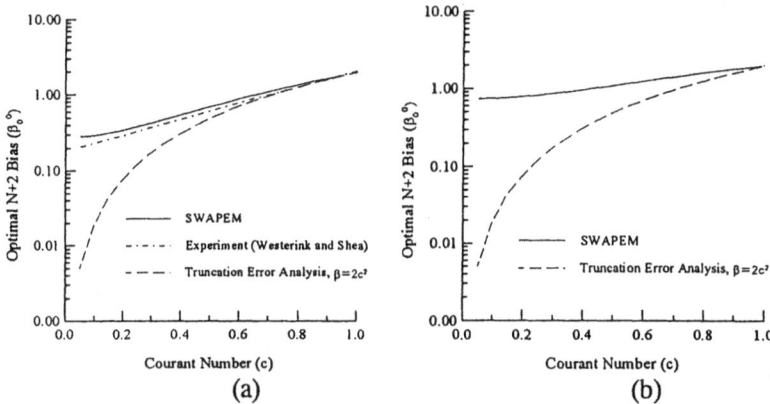

Figure 3 Predictions of the relationship between β_0 and Courant Number for (a) the Gaussian plume problem, (b) the triangle wave problem.

element methods are presented in [1]. Numerical results for the identical Gaussian plume benchmark problems using experimentally determined values for the weighting factors are presented in Westerink and Shea's paper [3]. As discussed earlier, the optimal N+2 weighting factors predicted by the SWAPEM algorithm are quite close to those determined experimentally by Westerink and Shea. Consequently, their numerical results are nearly indistinguishable from those computed with SWAPEM for identical benchmarks problems;   therefore they are not reproduced here to save space.

The simulation results for the Triangle wave problem is shown in figure 4. Both plots have been made at the time $t=9600$. Notice that the addition of N+2 bias whose magnitude is determined by the SWAPEM, (b), significantly reduces spurious oscillations, especially for larger Courant numbers. Also, whereas N+1 bias smears features and lowers peaks of the solution, N+2 bias sharpens features and raises peaks closer to their true values. The addition of optimal N+2 bias significantly increases the rate of convergence, or accuracy on any given grid, of the Galerkin solution.

CONCLUSIONS

The N+2 Petrov-Galerkin method used in conjunction with the SWAPEM algorithm for the dynamic calculation of the optimal N+2 bias has been shown to produce improved solutions to the transient convective-diffusion equation over the oscillatory solutions of the Galerkin method. Only the linear element scheme has been explicitly tested with SWAPEM, but the quadratic scheme is expected to perform similarly, because the behavior of eigenvalue spectra of both methods are similar. The SWAPEM algorithm is easy to incorporate into existing finite element codes, completely automatic, and capable of predicting the N+2 bias for general waveforms, removing the need for tedious problem-dependent numerical experimentation. The fast Fourier Transform is the ideal way to incorporate the dependence of β_0 on the solution without a knowledge of the analytical solution to the initial-boundary value problem.

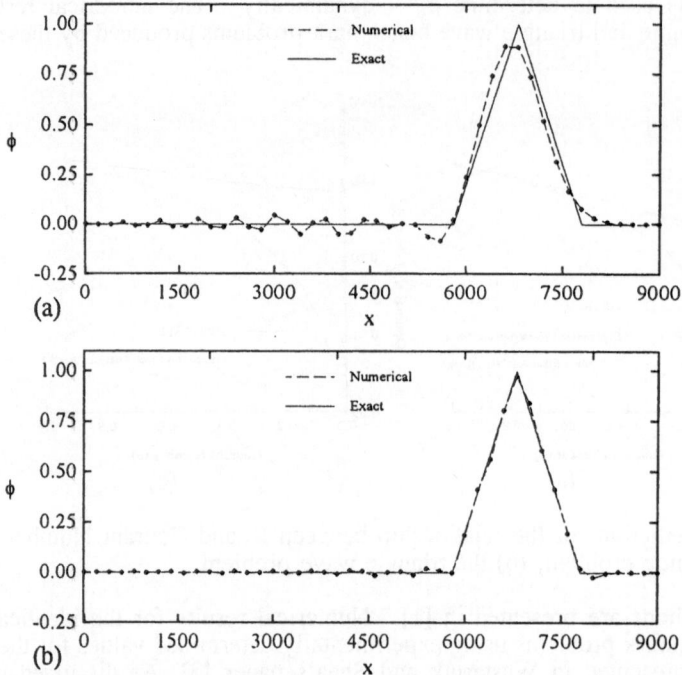

Figure 4 Numerical solution of the pure convection of a triangle wave pulse by (a) the Galerkin method, (b) the N+2 Petrov-Galerkin method with the SWAPEM algorithm.

ACKNOWLEDGEMENTS

This work was funded by an Exploratory and Foundational research grant by the Applied Research Laboratory at the Pennsylvania State University.

REFERENCES

1. Carrano, C., A Fourier Analysis and Optimization of the Petrov-Galerkin Finite Element Method. Masters thesis, Depart. Aerospace Eng., The Pennsylvania State University, 1993. 102 pp.
2. Miller, C., T., and Cornew, F., H., A Petrov-Galerkin Method for Resolving Advective-Dominated Transport, *Computational Methods in Water Resources IX Vol. 1: Numerical Methods in Water Resources*, Computational Mechanics Publications, Boston, and Elsevier Applied Science, New York, 157-164, 1992.
3. Westerink, J., J., and Shea, D., Consistent Higher Degree Petrov-Galerkin Methods for the Solution of the Transient Convection-Diffusion Equation, *Int. J. Numer. Methods Eng.*, Vol. 28, 1077-1101, 1989.
4. Yeh, G. T., Chang, J. R., and Short, T. E., An Exact Peak Capturing and Oscillation-Free Scheme to Solve Advection-Dispersion Transport Equations, *Water Resources Research*, Vol. 28, No. 11, 2937-2951, 1992.

AN IMPROVED FORM OF ADJOINT-STATE EQUATIONS FOR TRANSIENT PROBLEMS

JESUS CARRERA and AGUSTÍN MEDINA
ETSECCPB. Universitat Politècnica de Catalunya
C/ Gran Capitán s/n. 08034 Barcelona

ABSTRACT

Adjoint state equations are useful for computing derivatives of scalar functions of state variables with respect to the parameters of state equations. These derivatives can be used for performing sensitivity and uncertainty analysis. If the scalar function is taken as the objective function of an optimization problem, then they are used for gradient search optimization algorithms. In the framework of parameter estimation, they are now used less often than in the past because Gauss-Newton methods have proven to be much more powerful than gradient search methods. Gauss-Newton methods require computing the jacobian, derivatives of state variables with respect to model parameters. Since conventional adjoint-state equations would require defining one adjoint-state for each observation in space and time, their cost becomes prohibitive whenever the number of observations exceeds the number of parameters. Hence, direct derivation state equations with respect to model parameters is used instead. We reexamine the recursive nature of adjoint-state equations, which allows us to use a simple adjoint-state to compute the derivatives of state variables at one point, but all times, with respect to all model parameters. The resulting equations are applicable both for transient linear flow and for transient linear transport with steady flow. They are only approximate for varying time step, but exact for constant time steps.

1. INTRODUCTION

In comparing different methods for solving the inverse problem, Cooley (1985) concluded that Marquardt's optimization method is the most efficient for steady-state groundwater flow problems. We have found Cooley's conclusions to be valid for transient problems as well. Difficulties with this method, as with all Gauss-Newton methods, arise when computing the derivatives of state variables with respect to model parameters. Two families of methods are available: direct derivatives and adjoint state method. The former consists of taking derivatives of the state equations, which leads to a linear system of equations where the unknowns are precisely the columns of the jacobian matrix. This approach has been used by a large number of authors (e.g., Cooley, 1977; Gavalas et al., 1976; Yoon and Yeh, 1976; Yeh and Yoon, 1981). On the other hand, the adjoint state method can be viewed as a particular case of Lagrange multipliers. However, in its original formulation, it was proposed for computing derivatives of scalar functions. While it could be applied to computing the jacobian matrix (Neuman, 1980a; Sun and Yeh, 1992), it is not competitive whenever the number of observations (in general, the number of scalars whose derivatives are sought) is larger than the number of

A. Peters et al. (eds.), Computational Methods in Water Resources X, 199–206.

parameters. The aim of our work is to present a revised formulation of the adjoint state equations which is competitive for transient problems whenever the number of observation points (regardless of the number of measurements in each point) is smaller than the number of parameters.

Adjoint state equations were originally proposed by Chavent (1971) in a variational framework. Neuman (1980b) extended these equations, which were applied succesfully by Carrera and Neuman (1986). We will use an alternative approach based in formulating the direct problem in its matrix form (Townley and Wilson, 1985). A thorough presentation of the two methods of computing derivatives (direct and adjoint state) both for linear and non-linear problems is given by Carrera et al. (1990).

2. GENERAL ADJOINT STATE EQUATIONS

General adjoint state equations are thoroughly described by Carrera et al. (1990). For the sake of completeness we present here a summary of their equations. Let $f(\boldsymbol{h}, \boldsymbol{c}, \boldsymbol{p})$ be a scalar function (in general, f is the objective function one wants to minimize); let $\boldsymbol{\psi}_i$ be the state equations in step i, which allow obtaining state variables at time i from those at time $i-1$. It can be shown that the derivatives of f with respect to $\boldsymbol{p}$, while recognizing that both $\boldsymbol{h}$ and $\boldsymbol{c}$ depend on $\boldsymbol{p}$, are given by:

$$\frac{df}{d\boldsymbol{p}} = \frac{\partial f}{\partial \boldsymbol{p}} + \sum_{i=0}^{N} \lambda_i \frac{\partial \boldsymbol{\psi}_i}{\partial \boldsymbol{p}} \tag{1}$$

where N is the total number of time steps (notice that $i = 0$ has been included to account for steady-state) and λ_i is the i-th adjoint state. This is the solution of:

$$\frac{\partial f}{\partial \boldsymbol{x}_j} + \sum_{i=0}^{N} \lambda_i \frac{\partial \boldsymbol{\psi}_i}{\partial \boldsymbol{x}_j} = \boldsymbol{0} \qquad j = 0, \ldots, N \tag{2}$$

where $\boldsymbol{x}_j$ is the vector of state variables (either $\boldsymbol{h}_j$ or $\boldsymbol{c}_j$, depending on the meaning of $\boldsymbol{\psi}$). Therefore, computing the gradient of f involves, first, solving (2) for λ_i and, then, substituting it in (1). The adjoint state λ_i ($i=0,\ldots$, N) is linked to f, which appears on both (1) and (2). When obtaining derivatives of a single scalar function $df/d\boldsymbol{p}$, the computational effort is limited to solving $N + 1$ ($i = 0,\ldots,N$) linear systems of NN equations (NN being the dimension of $\boldsymbol{\psi}_i$ and $\boldsymbol{x}_j$). Computational cost is further reduced by the fact that the matrices involved in solving for λ_i (i.e., $\partial \boldsymbol{\psi}_i/\partial \boldsymbol{x}_j$) are identical to those required solving for $\boldsymbol{x}_i$ in the state equations.

However, if $\boldsymbol{f}$ is vectorial (of dimension NO, number of observations), the above computations would have to be repeated NO times. It is clear that the corresponding computer cost can become unacceptably large for a large number of observations. In fact, such cost is larger than that of the alternative methods (direct derivation) whenever NO is larger than NP (dimension of $\boldsymbol{p}$). Since, traditionally, one would need a certain degree of observations redundancy (i.e., $NO >> NP$), the above condition would be rarely met and adjoint-state equations were virtually abandoned for the computation of sensitivity matrices. More recently, however, the revitalization of the geostatistical inverse problem has brought back to fashion problems in which the number of parameters can be larger than the number of observations. Hence, it is not surprising that Sun and Yeh (1992) have proposed using adjoint state methods for computing the jacobian (sensitivity) matrix.

The work of Sun and Yeh or others is constrained to steady-state problems because the number of observations for transient problems can be very large. The remaining of this paper is devoted to examining how adjoint-state equations can still be competitive for, at least, two cases (1) transient flow, possibly with steady-state initial conditions and (2) transient transport with steady-state flow.

3. CASE 1. TRANSIENT FLOW

In this case, the state variables, $\boldsymbol{x}_j$, are the vector of nodal heads $\boldsymbol{h}_j$, $(j = 0, ..., N)$, $\boldsymbol{h}_0$ represents initial heads, which may be assumed to represent steady-state flow. State equations are the numerical version of the flow equation:

$$\boldsymbol{\psi}_0 = \boldsymbol{A} h_0 - \boldsymbol{b}_0 = \boldsymbol{0} \tag{3}$$

$$\boldsymbol{\psi}_j = \left(\theta\boldsymbol{A} + \frac{\boldsymbol{D}}{\Delta t}\right)\boldsymbol{h}_j - \left[(\theta - 1)\boldsymbol{A} + \frac{\boldsymbol{D}}{\Delta t}\right]\boldsymbol{h}_{j-1} - \boldsymbol{b}_{j-1+\theta} = \boldsymbol{0} \tag{4}$$

where $\boldsymbol{A}$ is the $NN \times NN$ symmetric "conductance" matrix, which depends on transmissivity and leakance factors; $\boldsymbol{D}$ is the storage matrix, which depends on storativity and is often taken as diagonal; $\boldsymbol{b}_j$ is the vector of nodal sinks and sources at time j. Readers may notice that a θ-weighted time integration scheme is implied in (4). The exact form of $\boldsymbol{A}$, $\boldsymbol{D}$ and $\boldsymbol{b}$ depends on the method for spatial approximation of the flow equation (that is, finite elements, finite differences, etc). The choice of solution method is unimportant at this stage.

Let us now assume that we want to obtain derivatives of $\boldsymbol{a}^t\boldsymbol{h}_k$ with respect to $\boldsymbol{p}$. We will assume, without loss of generality, that $\boldsymbol{a}^t$ is not a function of $\boldsymbol{p}$ ($\boldsymbol{a}$ may be a function of transmissivities, for example, if $\boldsymbol{a}^t\boldsymbol{h}_k$ represents measurable boundary fluxes). Adjoint state equations are obtained by substituting $f = \boldsymbol{a}^t\boldsymbol{h}^k$ in (2), which leads to:

$$\boldsymbol{a}^t\delta_j^k + \boldsymbol{\lambda}_j^k\left(\theta\boldsymbol{A} + \frac{\boldsymbol{D}}{\Delta t}\right) - \boldsymbol{\lambda}_{j+1}^k\left(\frac{\boldsymbol{D}}{\Delta t} + (\theta - 1)\boldsymbol{A}\right) = \boldsymbol{0} \qquad j = 0, ..., N \tag{5}$$

where the superindex k is being used to denote that $\boldsymbol{\lambda}_j^k$ is the adjoint state of $\boldsymbol{a}^t\boldsymbol{h}_k$ (i.e., k is a constant, though it will vary when computing the derivatives of heads at a different time); δ_j^k is Kronecker's delta. Eq. (5) is solved backwards in time, from $j = N$ through $j = 0$. It should be noticed that $\boldsymbol{\lambda}_{N+1}^k = \boldsymbol{0}$. Also, $\boldsymbol{\lambda}_j^k = \boldsymbol{0}$ for $j > k$, which is a consequence of $\boldsymbol{\lambda}_{N+1}^k = \boldsymbol{0}$.

The most important property of (5) is that k is just a dummy variable. Therefore, adjoint states for different times are identical, though shifted in time:

$$\boldsymbol{\lambda}_j^k = \boldsymbol{\lambda}_{j+m}^{k+m} = \boldsymbol{\lambda}_{j+N-k}^N \qquad \text{or} \qquad \boldsymbol{\lambda}_j^N = \boldsymbol{\lambda}_{j+k-N}^k \tag{6}$$

The importance of (6) stems from the fact that one does not need to compute a different adjoint state for each k value. A single adjoint state is sufficient. Actually the above needs an additional correction for the case in which initial heads are given by a steady-state. In such case, the corresponding adjoint state equation would be:

$$\lambda_0^k A = \lambda_1^k \Big[\frac{D}{\Delta t} + (\theta - 1) \frac{A}{\Delta t} \Big] \tag{7}$$

That is, a different λ_0^k is needed for each k value. This is a minor problem, but can be solved by adding eq. (5) from $j = 0$ through N, which leads to:

$$\sum_{j=0}^{N} \lambda_j^k A = -a^t \tag{8}$$

If (8) is substracted from eq. (5), for $k = 0$, $j = 0$,

$$\lambda_0^0 A = -a^t \tag{9}$$

one obtains the adjoint state for $j = 0$ as:

$$\lambda_0^k = \lambda_0^0 - \sum_{j=1}^{k} \lambda_j^k \tag{10}$$

These equations can now be plugged in (1) to compute the derivatives of f with respect to p for varying k. Let f_k be $a^t h_k$. The algorithm for computing df_k/dp for $k = 0, ..., N$ can be summarized as follows:

1. Solve direct problem eqs [(3) and (4)].
2. Solve (9) for λ_0^0.
3. For $j = N$ through $j = 0$ perform the following steps:
   3.1. Solve (5) for λ_j^N (starting with $\lambda_{N+1}^N = 0$).
   3.2. For all observation points between $k = N - j + 1$ and $k = N$, update df_k/dp $\Big(f_k = a^t h_k \Big)$ by sequential use of (1), i.e.

$$\frac{df_k}{dp} = \frac{df_k}{dp} + \lambda_{j+k-N}^k \frac{\partial \psi_{j+k-N}}{\partial p} \tag{11}$$

   where $\lambda_{j+k-N}^k = \lambda_j^N$, except for $k = 0$ (in which case, it equals λ_0^0).
4. Repeat steps 2 and 3 for varying f_k (i.e., by changing a wherever several observation points are available).

Details on this procedure will be given for case 2, which is more complex.

4. CASE 2. TRANSIENT TRANSPORT ON A STEADY-STATE FLOW FIELD

State variables are heads for steady-state [x_0 in (1) and (2)] and concentrations for transient [x_j for $j > 0$ in (1) and (2)]. It would be easy to incorporate a steady-state initial concentration. It would not be possible, however, to take advantage of the simplifications discussed here if transport were taking place on a transient flow field. Hence, state equation is (3) for steady-state, while the transient is given by:

$$\boldsymbol{\Psi}_j = \left(\theta \boldsymbol{E} + \frac{\boldsymbol{F}}{\Delta t}\right)\boldsymbol{c}_j - \boldsymbol{g}_{j+\theta-1} - \left(\frac{\boldsymbol{F}}{\Delta t} + (\theta - 1)\boldsymbol{E}\right)\boldsymbol{c}_{j-1} = \boldsymbol{0} \qquad (12)$$

where $\boldsymbol{c}_j$ is the vector of nodal concentrations at time j, $\boldsymbol{E}$ is a matrix incorporating dispersive and advective terms, $\boldsymbol{F}$ accounts for storage (hence, it is a function of porosity) and $\boldsymbol{g}_j$ includes sinks and sources. $\left(g_{j+\theta-1} = \theta g_{j-0} + \left(1 - \theta\right)g_{j-1}\right)$

As in case 1, f_k is of the form $\boldsymbol{a}^t \boldsymbol{c}_k$. Hence, eq. (5) can be rewritten as:

$$\boldsymbol{a}^t \delta_j^k + \boldsymbol{\mu}_j^k \left(\theta \boldsymbol{E} + \frac{\boldsymbol{F}}{\Delta t}\right) - \boldsymbol{\mu}_{j+1}^k \left(\frac{\boldsymbol{F}}{\Delta t} + (\theta - 1)\boldsymbol{E}\right) = \boldsymbol{0} \qquad (13)$$

where we have used $\boldsymbol{\mu}$ instead of $\boldsymbol{\lambda}$ to express that $\boldsymbol{\mu}$ is the adjoint state of a transport problem. As before, eq. (13) is solved sequentially in time, starting with $\boldsymbol{\mu}_N$ ($\boldsymbol{\mu}_{N+1} = \boldsymbol{0}$). The main difference with case 1 is that the steady state adjoint vector $\boldsymbol{\lambda}_0^k$ is given by (from (2)):

$$\boldsymbol{\lambda}_0^k \boldsymbol{A} + \boldsymbol{a}^t \delta_0^k + \sum_{j=1}^{k} \boldsymbol{\mu}_j^k \left(\frac{\partial \boldsymbol{E}}{\partial \boldsymbol{h}_0} c_{j+\theta-1} + \frac{\partial \boldsymbol{F}}{\partial \boldsymbol{h}_0} \frac{c_j - c_{j-1}}{\Delta t} + \frac{\partial g_{j+\theta-1}}{\partial \boldsymbol{h}_0}\right) = 0 \qquad (14)$$

Actually this expression can be formally simplified. It is not hard to show that the last term in the left hand side is just $\partial f_k / \partial \boldsymbol{h}_0$, so that (14) becomes:

$$\boldsymbol{\lambda}_0^k \boldsymbol{A} - \frac{\partial f_k}{\partial \boldsymbol{h}_0} = 0 \qquad k > 0 \quad ; \quad \boldsymbol{\lambda}_0^0 \boldsymbol{A} = -\boldsymbol{a}^t \qquad k = 0 \qquad (15)$$

These equations do not really simplify the computation because obtaining $\partial f_k / \partial \boldsymbol{h}_0$ is by no means simple. Hence, we keep (14). Notice, however, that computing $\boldsymbol{\lambda}_0^k$ is not as difficult as it might look, because the coefficients matrix remains constant and the right hand side can be updated iteratively.

In summary, the algorithm to compute the derivatives of concentrations (or linear functions of them) at all times is given by:

1. Solve direct problem [(3) and (12)] and store state variables
2. Solve (9) for $\boldsymbol{\lambda}_0^0$
3. For $j = N$ through $j = 0$, perform the following steps:
   3.1. Solve (13) for $\boldsymbol{\mu}_j^N$ (starting with $\boldsymbol{\mu}_{N+1}^N = \boldsymbol{0}$)
   3.2. Update right hand side of (14) for all observations times between $k = N-j+1$ and $k = N$ and solve (14) for $\boldsymbol{\lambda}_0^k$ only if $k = N - j + 1$ is an observation time
   3.3. Use (1) to update $df_k/d\boldsymbol{p}$.
4. Repeat steps 2 and 3 for varying f (i.e., by changing $\boldsymbol{a}$ wherever several observation points are available).

Program implementation details of this algorithm are discussed below. Since the details are somewhat involved, we suggest the reader to proceed to the conclusions, unless he(she) is actually implementing the algorithm.

Implementation hints

The above algorithm is somewhat difficult to implement because one has to keep track simultaneously of adjoint state and state variables time dependence, together with the updating of eqs. 1 (actually, N equations, one for each observation time) and 14 (also N equations). To make things worse, one has to solve the adjoint state equations reversely in time. Furthermore, numerical accuracy may require a number of solution time increments between sequential observation times. This section is devoted to discuss how to overcome such difficulties.

Table 1: Structure of adjoint states and state variables to be used in the computation of $\partial f_k / \partial p$ (eq. 1, $f_k = \boldsymbol{a}^t \boldsymbol{c}_k$), for varying k. For each observation time k, we display the adjoint state vector and the related state variables for computing each term of the sum in eq. (1)

Obs time j	k=N Adj. State	k=N State Var	k=N-1 Adj. State	k=N-1 State Var		k=1 Adj. State	k=1 State Var	k=0 Adj. State	k=0 State Var
N	μ_N^N	c_N, c_{N-1}	–	–	$\cdots$	—	—	—	—
N-1	μ_{N-1}^N	c_{N-1}, c_{N-2}	μ_{N-1}^{N-1}	c_{N-1}, c_{N-2}	$\cdots$	—	—	—	—
N-2	μ_{N-2}^N	c_{N-2}, c_{N-3}	μ_{N-2}^{N-1}	c_{N-2}, c_{N-3}	$\cdots$	—	—	—	—
.	.	.	.	.		.	.	.	.
.	.	.	.	.		.	.	.	.
1	μ_1^N	c_1, c_0	μ_1^{N-1}	c_1, c_0	$\cdots$	μ_1^1	c_1, c_0	—	—
Flow (s.s.)	λ_0^N	$\boldsymbol{h}_0$	λ_0^{N-1}	$\boldsymbol{h}_0$		λ_0^1	$\boldsymbol{h}_0$	λ_0^0	$\boldsymbol{h}_0$

NOTE: Observe that state variables remain unaltered in each row, while adjoint state vectors remain unaltered along diagonals (recall that $\mu_N^N = \mu_{N-1}^{N-1} = ... = \mu_1^1$; $\mu_{N-1}^N = \mu_{N-2}^{N-1} = \mu_1^2$; ...; $\mu_2^N = \mu_1^{N-1}$).

The structure of adjoint state vectors and state variables is shown in Table 1. Each column contains the variables to be used for computing $\partial f_k / \partial p$ by means of eq. (1). In fact, each component in Table 1 contains the variables required by the corresponding term in the sum in (1). For example:

$$\frac{df_{N-1}}{d\boldsymbol{p}} = \frac{\partial f_{N-1}}{\partial \boldsymbol{p}} + \lambda_0^{N-1} \frac{\partial \boldsymbol{\psi}_0}{\partial \boldsymbol{p}} + \mu_1^{N-1} \frac{\partial \boldsymbol{\psi}_{N_1}}{\partial \boldsymbol{p}} + \cdots + \mu_{N-1}^{N-1} \frac{\partial \boldsymbol{\psi}_{N-1}}{\partial \boldsymbol{p}} \qquad (16)$$

Notice that $\boldsymbol{\psi}_0$ is given by eq. (3) which involves $\boldsymbol{h}_0$, and $\boldsymbol{\psi}_1$ through $\boldsymbol{\psi}_{N-1}$ are given by eq. (12), which involve $\boldsymbol{c}_1$ and $\boldsymbol{c}_0$ through $\boldsymbol{c}_{N-1}$ and $\boldsymbol{c}_{N-2}$, as shown in Table 1. It should also be noticed that one may not need all columns of Table 1. In fact, only the columns corresponding to solution times which coincide with observation time need to be stored.

The above paragraph might suggest that a huge storage is required. Actually, storage needs are moderate. In addition to the standard variables of any finite element code, one needs to store the following arrays:

- state variables: $h_0, c_1, ..., c_N$ (NTIM × NUMNP, where NTIM is the number of solution times and NUMNP is the number of nodes). This is the largest array, but can be stored in disk because it is going to be used sequentially.

- $\partial f_k/\partial p$ (NINT×NPAR, where NINT is the number of observation times and NPAR is the number of parameters to be estimated). Notice that often NINT << NTIM. Notice also that, in most cases, one does not need simultaneously the derivatives of state variables for all observation points (see Medina et al., 1990 for the use of the jacobian matrix $\partial f/\partial p$, $f = (f_1, ..., f_{NINT})$, when computing the hessian matrix).

- $\partial E/\partial h_0$. This term, which is needed for updating the right hand side of (14), might be enormous. However, it can be stored element-wise, in which case the number of scalars to be stored is NUMEL×LMXNDL×LMAX, where NUMEL is the number of elements, LMXNDL is the maximum number of nodes per element and LMAX equals LMXNDL×(LMXNDL-1)/2.

With these definitions in mind the procedure outlined above can now be explained in more detail. After having solved the direct problem and having stored the state variables (step 1), one may proceed to solve eq. (9) for λ_0^0 for the current observation point (step 2). Obviously, this computation can be skipped if a head measurement is not available at this point. Otherwise, one may use eq. (1) to compute the derivatives of heads with respect to model parameters.

Step 3 is aimed at performing the column-wise sums of Table 1 required for computing eq. (1), as explained in eq. (16). This requires 3 substeps: first, compute the adjoint state (μ_j^k); second, update the right hand side of eq. (14) for λ_0^k and, third update $\partial f_k/\partial p$. In performing these computations one should keep in mind the following remarks. First, δ_j^k is taken as 1 only for $j = N$ because $\mu_N^N = \mu_{N-1}^{N-1} = \cdots = \mu_1^1$ ($\mu_j^k = 0$ for $j > k$). Second, eq. (14) is identical to (9) for $k > 0$. Also, the term $\partial F/\partial h_0$ equals zero for linear flow problems (In fact, we have only kept it for possible generalizations of this algorithm. Third, only one term of the sum in eq. (14) is added for every observation time k during each sweep of step 3 (that is, for each j value). This implies that the right hand side of eq. (14) is also updated column-wise. Finally, the actual solution of eq. (14) needs not be performed at each observation time, but one may wait until the end of the step 3 loop, which may allow one to take advantage of vectorization. Once step 3 is finished for one observation point, one may use $\partial f_k/\partial p$ to update the gradient and the approximate hessian of the objective function (see Medina et al., 1990 or Carrera et al., 1990 for details). Steps 2 and 3 are repeated for every observation point.

CONCLUSIONS

The equations we have presented in this work allow one to compute the jacobian matrix after solving (NTIM+ NINT)×NO linear systems of equations (NTIM is the number of solution time increments, NINT is the number of time dependent measurements at each of the NO observation points). In geostatistical inverse problems, this number can be much smaller than the more conventional direct derivation equations, which require solving NTIM×NPAR systems.

ACKNOWLEDGEMENTS

This work has been performed in the framework of a project jointly funded by ENRESA and the European Union on the framework of the RADWAS program.

REFERENCES

Carrera, J. and S.P. Neuman (1986) "Estimation of aquifer parameters under steady-state and transient conditions: I. Background and Statistical framework", Wat. Resour. Res., 22, 2, 199-210.

Carrera J. , F. Navarrina, L. Vives, J. Heredia and A. Medina. (1990) "Computational Aspects of the Inverse Problem". Comp. Meth. in Wat. Resour., 513-523.

Chavent, G. (1971 "Analyse fonctionelle et identification de coefficients repartis dans les equations aux derivees partielles" These de docteur en sciences, Univ. de Paris, Paris.

Cooley, R.L. (1977) "A method of estimating parameters and assessing reliability for models of steady state groundwater flow. 1. Theory and numerical properties". Wat. Resour Res., 133 (2), 318-324

Cooley R. and R. Naff. (1985) "Regressión Modeling of Groundwater flow". U.G.S.S. Open-file Report 85-180, 450 pp.

Gavalas G.R., P.C. Shah and T.H. Seinfeld. (1976) "Reservoir history matching by Bayesian estimation", Soc. Pet. Eng. J., 261, 337-350.

Medina, A., J. Carrera and G. Galarza. (1990) "Inverse modelling of coupled flow and solute transport problems". ModelCare 90: Calibration and Reability in Groundwater Modelling, IAHS Publ. no. 195, 185-194.

Neuman, S.P. (1980a) "A statistical approach to the inverse problem of aquifer hydrology, 3, Improved solution methods and added perspective". Water Resour. Res., 16(2), 331-346,

Neuman, S.P. (1980b) Adjoint state finite elements equations for parameter estimation. Finite Elements in Water Resources, edited by S.Y. Wang, 2.66-2.75, U.S. Department of Agriculture, New Orleans, La.

Sun, N.-Z. and W.W.-G. Yeh (1992) "A stochastic inverse solution for transient groundwater flow: parameter identification and reliability analysis". Wat. Resour. Res., 28(12), 3269-3280.

Townley, L.R. and J.L. Wilson (1985) "Computationally Efficient Algorithms for Parameter Estimation and Uncertainty Propagation in Numerical Models". Wat. Resour. Res., 21(12), 1851-1860.

Yeh, W.W.G and Y.S. Yoon (1981) "Aquifer parameter identification with optimum dimension in parametrization", Water Resour. Res., 17(3), 664-672.

Yoon, Y. S. and W.W.G. Yeh (1976) "Parameters identification in an homogeneous medium with the finite element method". Soc. Pet. Eng. J., 16(4), 217-226.

EULERIAN-LAGRANGIAN LOCALIZED ADJOINT METHODS FOR CONTAMINANT TRANSPORT SIMULATIONS

M. A. CELIA
Water Resources Program, Department of Civil Engineering
Princeton University
Princeton, NJ 08544
USA

The Eulerian-Lagrangian Localized Adjoint Method (ELLAM) provides a framework within which general characteristic-based approximations may be developed for a variety of contaminant transport problems. The ELLAM formalism provides a systematic framework for implementation of boundary conditions, leading to mass-conservative numerical schemes. The computational advantages of ELLAM approximations have been demonstrated for a number of one-dimensional transport systems. In addition, the ELLAM formalism provides a platform on which many characteristic-based methods can be compared and analyzed. Practical implementation of ELLAM algorithms in multiple spatial dimensions requires careful algorithmic development for evaluation of certain space-time integrals that arise in the formulation. The most promising three-dimensional formulations combine attractive features from one-dimensional ELLAM approximations with certain aspects of other characteristic-based methods.

INTRODUCTION

The Eulerian-Lagrangian Localized Adjoint Method (ELLAM) is a general characteristic-based numerical solution procedure that applies to a variety of transport equations. It is a member of the general family of numerical approximations referred to as Localized Adjoint Methods (LAMs). LAM approximations follow from the original work of Herrera (Herrera, 1984; Herrera et al., 1985), and include the Optimal Test Function (OTF) methods of Celia et al. (1989) and Neuman (1990). In these methods, test functions are chosen as part of the solution procedure by requiring the test functions to satisfy locally the homogeneous adjoint equation associated with the governing equation. OTF methods use this philosophy in a semi-discrete sense; this leads to Optimal Spatial Approximations (OSAs) that are characterized by Courant-number limitations and excessive numerical diffusion. This is characteristic of all OSA's (Bouloutas and Celia, 1988). The advantages of the ELLAM approach are due to the incorporation of space-time test functions, within the Localized Adjoint framework, that respect the hyperbolic nature of the secondary characteristics of the transport equation. This alleviates Courant-number limitations and produces a general characteristic-based approximation algorithm.

A. Peters et al. (eds.), Computational Methods in Water Resources X, 207–216.
© 1994 Kluwer Academic Publishers. Printed in the Netherlands.

Details of the ELLAM approach have been presented by Celia et al. (1990), Russell (1990), Celia and Zisman (1990), Ewing and Celia (1992), Herrera et al. (1993), Healy and Russell (1993), and Binning (1994). All of these works have been restricted to one spatial dimension, with the exception of Binning (1994), who developed a two-dimensional ELLAM approximation to model contaminant transport in unsaturated soils. This paper focuses on issues important to the development of practical three-dimensional ELLAM simulators. It also presents comparisons of ELLAM with more traditional characteristic-based solution methods such as finite difference-based Eulerian Lagrangian Methods (ELMs), the Method of Characteristics (MOC) as used in USGS codes, the Modified Method of Characteristics (MMOC), and particle methods.

FUNDAMENTALS OF ELLAM

To illustrate the underlying concept of ELLAM, consider the following transport equation for subsurface contaminants,

$$Lc \equiv \frac{\partial}{\partial t}(\theta c) + \nabla \bullet (qc) - \nabla \bullet \theta D \bullet c + \theta \lambda c = \theta F \tag{1}$$

where θ is moisture content, c is solute concentraion, q is the volumetric flux vector, D is the hydrodynamic disprsion tensor, λ is a decay constant, and F represents a source/sink term. Let the space-time domain over which this equation applies be denoted by Ω_x, with spatial domain Ω_x and temporal domain $[0, T]$. Assume that appropriate boundary conditions are specified; these boundary conditions are abitrary in terms of type and spatial distribution along the boundary $\partial \Omega_x$. To derive the general ELLAM approximation to this equation, the conservative form of the Equation (1) is maintained and a weak solution is sought such that

$$\int_{\Omega_{xt}} \left[\frac{\partial}{\partial t}(\theta c) + \nabla \bullet (qc) - \nabla \bullet \theta D \bullet \nabla c + \lambda \theta c \right] w_i(x, t) \, dx dt = \int_{\Omega_{xt}} \theta F w_i dx dt \tag{2}$$

In this equation, the test function $w_i(x,t)$ represents a space-time function associated with node number i. The functional form of the test function is determined as part of the ELLAM procedure. This is consistent with the general philosophy of LAM algorithms (Herrera et al., 1993).

Choice of Test Functions

A cornerstone of LAM approximations is the systematic choice of test functions as part of the overall numerical development. The criterion for choice of test functions is derived by rewriting Equation (2) such that the adjoint operator L^* acts on the test function w_i. For Equation (1), the formal adjoint of L is

$$L^* w_i = -\theta \frac{\partial w_i}{\partial t} - q \bullet \nabla w_i - \nabla \bullet \theta D \bullet \nabla w_i + \theta \lambda w_i \tag{3}$$

For test functions that are defined piecewise, such that each function $w_i(x,t)$ is defined on the set of (space-time) subintervals $\{\Omega_e\}$ $(e=1,...,E)$, application of integration-by-parts results in integrals along the boundaries of the subintervals and integrals over the interior of each subinterval. The latter integrals have integrands cL^*w_i. Localized Adjoint Mehods define the test functions so that the interior integrals are indentically zero, that is,

$$L^*w_i = 0 \qquad \text{within each } \Omega_e \tag{4}$$

This provides the criterion for definition of the test functions $\{w_i\}$. Note that the solution space for Equations (3) and (4) is infinite-dimensional, such that an infinite number of test functions may be chosen that satisfy the LAM requirement. Therefore, while the LAM approach provides a systematic formalism for development of numerical approximations, it also allows considerable freedom of choice. Judicious choice of the finite number of test functions to be used in the approximation requires a combination of careful numerical analysis and thorough understanding of the dominant physical and chemical processes.

To illustrate the concepts involved in a general ELLAM development, consider the case where Equation (1) is defined in one spatial dimension and time. Further, assume that the numerical solution is known through discrete time level t^n. The problem at hand is to determine the solution at time level $t^{n+1} = t^n + \Delta t$. As described in Celia et al. (1990) and Celia and Zisman (1990), the Eulerian-Lagrangian Localized Adjoint Methods (ELLAM) involves definition of test functions that satisfy the the advective and diffusive parts of the equations separately. The test functions are required to satisfy the following,

$$\theta \frac{\partial w_i}{\partial t} + q \bullet \nabla w_i - \theta \lambda w_i = 0$$

$$\nabla \bullet \theta D \bullet \nabla w_i = 0 \tag{5}$$

For the one-dimensional case, the first of these equations implies that w_i varies exponentially $(\lambda \neq 0)$ or is constant $(\lambda = 0)$ along the characteristic directions $dx/dt = q/\theta$, while the second equation describes the spatial variation of the test function. For simplicity of representation, let θ, q, and D be constant. Then a simple choice for test function $w_i(x,t)$ is the standard piecewise linear (in space) "hat" function that is associated with the spatial node i at time $t = t^{n+1}$ and that "travels" in space-time along the characteristics. Such a function, illustrated in Figure 1, was originally used by Celia et al. (1990). Note that alignment of the test function with nodes at time t^{n+1} implies that a backward characteristic-type of approximation will result. Also, when node i is sufficiently close to an inflow boundary, part (or all) of the function w_i intersects the spatial boundary before it reaches time level t^n. This boundary intersection has important implications for boundary condition implementation and associated mass-conservation properties (see Celia et al. (1990) for discussions and associated figures).

Piecewise linear test functions that are C^0-continuous in space are not the only choice for test functions. Following Healy and Russell (1993) and Celia et al. (1994), piecewise constant (in space), C^{-1}-continuous test functions may be defined. A typical C^{-1} test function is shown in Figure 2. Notice that the piecewise linear test functions are defined on a "corner-centered" grid while the piecewise constant functions are defined on a block-centered grid.

The ELLAM Approximating Equation

To illustrate the general features of ELLAM approximations, consider the one-dimensional version of Equation (1) with the test functions illustrated in Figure 2. To derive the ELLAM approximation, two approaches may be followed. The first uses the equation in which L^*w_i appears explicitly, such that all other terms in the equation are evaluated along the boundaries of each subdomain Ω_e. This approach was followed in the original work of Celia et al. (1990). A second approach is to use the test functions derived from the adjoint equation, but apply them to Equation (2) and integrate the equation by parts only once in space. This approach, used originally by Russell (1990), is analogous to standard finite element approaches. It has the advantage that only physical quantities arise as boundary evaluations. Because of this property, the second approach appears to be preferable and is used herein. Following this approach, with subdomain definitions as per Figure 2, the resulting equation has the following form,

$$\int_0^l \theta^{n+1} c^{n+1} w_i^{n+1} dx - \int_0^l \theta^n c^n w_i^n dx + \int_{t^n}^{t^{n+1}} \left(qc - \theta D \frac{\partial c}{\partial x} \right)\bigg|_{x=l} w_i(l,t)\, dt$$

$$- \int_{t^n}^{t^{n+1}} \left(qc - \theta D \frac{\partial c}{\partial x} \right)\bigg|_{x=0} w_i(0,t)\, dt + \int_{t^*_{i-1/2}}^{t^{**}_{i-1/2}} \left(\theta D \frac{\partial c}{\partial x} \right)\bigg|_{B_{i-1/2}} dt - \int_{t^*_{i+1/2}}^{t^{**}_{i+1/2}} \left(\theta D \frac{\partial c}{\partial x} \right)\bigg|_{B_{i+1/2}} dt = 0 \tag{6}$$

This equation may be viewed as an explicit statement of mass balance along the "traveling control volume" defined by $w_i(x,t)$. Celia and Ferrand (1993) demonstrated that approximations based on Equation (6) produce results similar to those using the piecewise linear test functions of Celia et al. (1990). Detailed computational considerations using this one-dimensional equation can be found in Healy and Russell (1993).

While the one-dimensional equation is useful to understand the underlying ideas of ELLAM, and to illustrate graphically the subdomains and test functions, general practical application of the method requires extension to three spatial dimensions. The ELLAM concept extends easily to three dimensions. The equation analogous to the one-dimensional equation above may be written as follows, where the test function $w_i(x,t)$ is the three-dimensional analog of the one-dimensional function illustrated in Figure 2.

$$
\int_{\Omega_x} \theta^{n+1} c^{n+1} w_i^{n+1} dx - \int_{\Omega_x} \theta^n c^n w_i^n dx
$$

$$
+ \int_{t^n}^{t^{n+1}} \int_{\partial\Omega_x} (qc - \theta D \bullet \nabla c) \bullet n_{\partial\Omega_x} w_i \big|_{\partial\Omega_x} da\, dt + \int_{\partial\Omega_i} (\theta D \bullet \nabla c) \bullet n_{\partial\Omega_i} dv = 0
$$

(7)

In this equation, $\partial\Omega_x$ represents the boundary of the total spatial domain whose outward normal direction is $n_{\partial\Omega_x}$, while $\partial\Omega_i$ denotes the boundary of the subregion associated with test function i, exclusive of the boundary $\partial\Omega_x$, whose outward normal is $n_{\partial\Omega_i}$. Note that the subdomain Ω_i is defined in three-dimensional space and time, so the boundary $\partial\Omega_i$ is a hypersurface through space-time whose outward normal is also a function of space-time. For the simple one-dimensional case, $\partial\Omega_i$ corresponds to the lines $B_{i \pm 1/2}$. Summation of Equation (7) for all test functions w_i demonstrates the mass-conservative nature of the approximation, independent of boundary condition type. Also, an equation analogous to Equation (7) results for the three-dimensional analog of the piecewise linear test functions illustrated in Figure 1.

Choice of Trial Function and Grid Design

To complete the ELLAM approximation, a trial function must be chosen to represent the unknown c. To date, one-dimensional approximations have used piecewise linear interpolation on fixed (in time) spatial grids (Celia et al., 1990; Celia and Zisman, 1990; Russell, 1990; Healy and Russell, 1993), while the two-dimensional algorithm of Binning (1994) used a piecewise constant approximation. Trial functions for the three-dimensional approximations defined by Equation (7) may use both piecewise constant and piecewise linear functions. The flow solution is assumed to be defined on a three-dimensional block-centered finite difference grid, with volumetric fluxes calculated along the faces of each block. For piecewise constant test functions, the subdomains Ω_e coincide with the finite difference block at time t^{n+1}. If piecewise constant trial functions are used, then block-centered node locations are defined at which discrete approximations to c are determined. For piecewise linear trial functions, the finite difference blocks are used as finite elements for the transport solution, such that nodes are placed at the corners of the blocks. If piecewise linear test functions are used, they are defined similarly to the piecewise linear trial functions.

Evaluation of Integrals

Equations (6) and (7) provide integral equations for the ELLAM approximations. The integrals in these equations can be evaluated after specific choices are made for trial and test functions, coupled with choices for discrete time evaluations. Some of these integrals are easily evaluated, while others require significant effort. The spatial integral at time t^{n+1} (the first integral in Equations (6) and (7)) is the easiest to evaluate, because both the trial and test functions are defined relative to the spatial grid at time t^{n+1}. Evaluation of this integral leads to standard finite element mass matrices. The diffusion integrals (the fifth and sixth integrals in Equation (6), the fourth integral in Equation (7)) are usually simplified by using an implicit, one-point approximation in time, using time interval $\Delta t^* = t^{**} - t^*$, with the appropriate time levels illustrated in Figure 2. The spatial integrals evaluated at time t^n (second integral in Equations (6) and (7)) are relatively easy to evaluate in one dimension but become much more challenging in multiple dimensions. This is due to the multi-dimensional deformation of each finite element (or finite volume) on which the test functions are defined as the geometry is back-tracked from time t^{n+1} to time t^n. The computational difficulties are due to the need to define the geometry at time t^n, which requires mapping of points along the boundary of the element and subsequent interpolation and mapping onto the fixed spatial grid at the old time level. Binning (1994) used such a mapping in two dimensions in a procedure that was very computationally intensive. This is especially true when part or all of the volume being mapped intersects a spatial boundary before reaching time level t^n. Based on the experience of Binning (1994) for two-dimensional problems, use of this explicit mapping of test functions is considered impractical in two and three dimensions. For one-dimensional problems, this step is relatively simple, because the boundaries of the spatial subdomains are points rather than lines or surfaces. Indeed, analytical evaluation of the integrals is possible in one dimension.

The most practical approach for evaluation of the spatial integrals at time t^n as well as the integrals evaluated along inflow boundaries is to use numerical integration, as proposed for the one-dimensional case by Healy and Russell (1993). On the fixed spatial grid, θ^n and c^n are easliy defined; the difficult evaluation is the test function w_i^n. Rather than back-tracking the geometry and estimating the test function by mapping the deformed geometry onto the fixed grid, discrete integration points defined on the fixed grid at t^n can be forward-tracked to time t^{n+1}, where evaluation of w_i is straight-forward. Then the integral is estimated as

$$\int_{\Omega_x} \theta^n c^n w_i^n dx \approx \sum_{e=1}^{E} \sum_{p=1}^{P} \Gamma_p \theta_p^n c_p^n w_i^n \Big|_{x_p} \qquad (8)$$

where P interpolation points, located at $\{x_p\}$, are used in each element e, with Γ_p an appropriate weighting factor. Algorithmically, this is implemented by evaluating θ and c at integration point x_p and time t^n, then tracking the point x_p from t^n to t^{n+1} and determining which test functions are nonzero at $x_p(t^{n+1})$. Tracking is most easily accomplished by using analytical integration within each cell, based on the analysis of Goode (1990). Analogous calculations apply to the boundary integrals, where discrete integration points are distributed in two-space and time.

COMPARISON TO OTHER METHODS

Finite difference-based Eulerian-Lagrangian Methods (ELM's) have been widely used in modeling transport in surface waters (see, for example, Baptista (1987) and references therein). ELM's may be viewed in the ELLAM context as follows: The spatial integrals at time t^{n+1} are evaluated using one-point numerical integration, with the integration point corresponding to a node point. This produces a lumped mass matrix. At time t^n, the integral is again calculated with a one-point integration rule, with the location of the integration point determined by back-tracking the point used to evaluate the integral at the new time level. To compensate for the crude approximation to the spatial integral, higher-order spatial interpolation is often used for c^n. Boundary conditions are usually restricted to fixed-concentration inflow conditions and zero-diffusive-flux outflow conditions, and the diffusion terms are treated implicitly (in time) using the equivalent of low-order spatial interpolation and reduced integration. Therefore, in the framework of ELLAM, ELM's may be seen as using low-order interpolation at t^{n+1}, higher-order interpolation at t^n, a one-point approximation to spatial integrals, and a simplified treatment of boundary conditions. Different combinations of the trial and test functions described above (piecewise constant, piecewise linear) lead to ELLAM expressions for the integrals at t^n that are analogous to linear, quadratic, and cubic interpolation in ELM's.

Particle methods may also be viewed in the ELLAM framework. Forward tracking of discrete-mass particles is analogous to the tracking of discrete integration points whose "mass" is determined by the product of concentration, moisture content, and the weighting factor Γ_p (which acts as a "volume"). The ELLAM test functions may be viewed as analogs to the projection functions used to determine concentrations from discrete particle distributions (see Tompson and Dougherty, 1992). Key differences between the methods are the way diffusion is handled (particle-based random walks versus concentration-based flux calculations) and the continuous tracking of particle locations. As it is currently constructed, the ELLAM algorithm re-initializes the spatial location of each integration point after every time step, with uniform coverage over the entire domain. "Adaptive" numbers of integration points, based on concentrations, is a logical extension of ELLAM that would more closely tie ELLAM to particle methods.

Other methods of interest include the Modified Method of Characteristics (MMOC - see, for example, Russell, 1985) and the Method of Characteristics (MOC) approach of Konikow and Bredehoeft (1978). The relationship between MMOC and ELLAM is relatively straight-forward; it has been discussed elsewhere (Celia et al., 1990; Healy and Russell, 1993) and will not be considered here. The MOC algorithm is similar to particle methods in that discrete quantities are tracked forward in time, with spatial positions continuously recorded. While concentrations are assigned to each tracked point, a fixed volume is also implicitly assigned to each point so that the result is equivalent to tracking a discrete-mass particle. The diffusion calculation is done explicitly, with the concentrations modified by a relatively ad-hoc rule. The MOC algorithm is most closely related to an ELLAM approximation in which integration points are distributed in space based on concentration, and for which test functions are defined on the spatial grid at time t^n and then tracked forward to time t^{n+1}. In ELLAM, the test functions would be deformed in space at t^{n+1}; in MOC, projection functions analogous to those used in particle methods are applied, which is equivalent to non-deforming test functions.

DISCUSSION

When considering practical three-dimensional ELLAM algorithms, several aspects of the problem must be considered. First, certain computational conveniences that are useful in one-dimensional problems are not feasible in three dimensions. For example, Healy and Russell (1993) used "strategic" integration points in space and time to locate the boundaries of w_i at t^n and at inflow boundaries. Boundaries of w_i are points in one dimension and are determined easily by backtracking. In two dimensions, the boundaries become deforming lines, and in three dimensions they are deforming surfaces. As demonstrated by Binning (1994), identification of these deforming boundaries is computationally intensive in two dimensions and likely to be infeasible in three dimensions. Rather than tracking the boundaries, it is more plausible to use particle-tracking concepts to determine how many integration points to use in each element at each time level. Such an approach gives an "adaptive" strategy for defining the number of integration points used to evaluate integrals.

The "adaptive" integration-points concept does not overcome the need for spatial (grid) resolution in the vicinity of steep concentration fronts. This can be accomplished by following the adaptive gridding procedures of, for example, Dahle et al. (1990). Additional computational observations include the fact that the major computational components in ELLAM can be performed in parallel, and some additional analysis is necessary to deal effectively with problems that have significant reactions. Overall, the ELLAM approach offers a foundation on which general characteristic-based algorithms may be developed and analyzed. It also provides a platform on which different approximations can be viewed in a consistent framework.

ACKNOWLEDGEMENTS

Much of this work is a result of ongoing collaborations with other members of the ELLAM group: Ismael Herrera, Tom Russell, and Dick Ewing. Their contributions are gratefully acknowledged. I have also had many helpful discussions with Philip Binning, Magne Espedal, Helge Dahle, and Lin Ferrand.

REFERENCES

Baptista, A.M., *Solution of Advection-Dominated Transport by Eulerian-Lagrangian Methods using the Backward Method of Characteristics*, Ph.D. Thesis, Dept. of Civil Engineering, M.I.T., 1987.

Binning, P.J., *Modeling Unsaturated Zone Flow and Contaminant Transport in the Air and Water Phases*, Ph.D. Thesis, Dept. of Civil Engineering and Operations Research, Princeton University, 1994.

Bouloutas, E.T. and M.A. Celia, "An Analysis of a Class of Petrov-Galerkin and Optimal Test Function Methods," *Proc. VII Intl. Conf. Comp. Meth. in Water Res.*, Celia et al., eds., 15-20, 1988.

Celia, M.A., G. Li, R.E. Ewing, and L.A. Ferrand, "A C-1 ELLAM Approximation for Simulating Reactive and Nonreactive Transport," Water Resources Program Report, Princeton University, 1994..

Celia, M.A., J.S. Kindred, and I. Herrera, "Contaminant Transport and Biodegradation: I. A Numerical Model for Reactive Transport in Porous Media," *Water Resources Research,* 25, 1141-1148, 1989.

Celia, M.A. and S. Zisman, "An Eulerian-Lagrangian Localized Adjoint Method for Reactive Transport in Groundwaters," *Proc. VIII Intl. Conf. Comp. Meth. in Water Res., Vol. I: Comp. Meth. in Subsurface Hydrology*, Gambolati et al., eds., Springer, 383-392, 1990.

Celia, M.A., T.F. Russell, I. Herrera, and R.E. Ewing, "An Eulerian-Lagrangian Localized Adjoint Method for the Advection-Diffusion Equation," *Advances in Water Resources, 13*(4), 187-206, 1990.

Celia, M.A. and L.A. Ferrand, "A Comparison of ELLAM Formulations for Simulation of Reactive Transport in Groundwater," *Proc. 1993 Intl. Conf on Hydrosc. and Engrg.*, Wang, S.Y.Y., ed., Univ. of Mississippi, 1829-1836, 1993.

Dahle, H., M. Espedal, R.E. Ewing, and O. Saevareid, "Characteristic Adaptive Subdomain Methods for Reservoir Flow Problems," *Numerical Methods for PDE's, 6*(4), 279-309, 1990.

Ewing, R.E. and M.A. Celia, "Numerical Methods for Reactive Transport and Biodegradation," *Proc. IX Intl. Conf. Comp. Meth. in Water Res., Vol 1: Num. Meth. in Water Res.*, Russell et al., eds., Comp. Mech. Publ., 51-58, 1992.

Goode, D.J., "Particle Velocity Interpolation in Block-Centered Finite Difference Groundwater Flow Models," *Water Resources Research, 26*(5), 925-940, 1990.

Healy, R.W. and T.F. Russell, "A Finite-Volume Eulerian-Lagrangian Localized Adjoint Method for Solution of the Advection-Dispersion Equation," *Water Resources Research, 29*(7), 2399-2413, 1993.

Herrera, I., *Boundary Methods: An Algebraic Theory*, Pitman Publ. Co., London, 1984.

Herrera, I., L. Chargoy, and G. Alduncin, "Unified Approach to Numerical Methods, III. Finite Differences and Ordinary Differential Equations," *Numerical Methods for PDE's, 1*(4), 241-258, 1985.

Herrera, I., R.E. Ewing, M.A. Celia, and T.F. Russell, "Eulerian-Lagrangian Localized Adjoint Methods: The Theoretical Framework," *Numerical Methods for PDE's, 9*(4), 431-458, 1993.1993.

Konikow, L.F. and J.D. Bredehoeft, "Computer Model of Two-Dimensional Solute Transport and Dispersion in Ground Water," *USGS Tech. Water Resour. Invest. Book 7*, Chap. C2, 90 pp., 1978.

Neuman, S.P., "Adjoint Petrov-Galerkin Method with Optimal Weight and Interpolation Functions defined on Multi-dimensional Nested Grids," *Proc. VIII Intl. Conf. on Comp. Meth. in Water Res., Vol. 2: Comp. Meth. in Surface Hydrology*, Gambolati et al., eds., Comp. Mech. Publ., 347-356, 1990.

Russell, T.F., "Time Stepping along Characteristics with Incomplete Iteration for a Galerkin Approximation of Miscible Displacement in Porous Media," *SIAM J. Num. Anal., 22*, 970-1013, 1985.

Russell, T.F., "Eulerian-Lagrangian Localized Adjoint Methods for Advection-dominated Problems, *Numerical Analysis 1989*, Griffith and Watson, eds., Pitman Research Notes in Math. Series, 228, Longman Scientific and Technical, U.K., 206-228, 1990.

Tompson, A.F.B. and D.E. Dougherty, "Highly Resolved Simulations of Chemical Migration in Physically and Chemically Heterogeneous Porous Media," *Proc. IX Intl. Conf. Comp. Meth. in Water Res., Vol 1: Num. Meth. in Water Res.*, Russell et al., eds., Comp. Mech. Publ., 671-686, 1992.

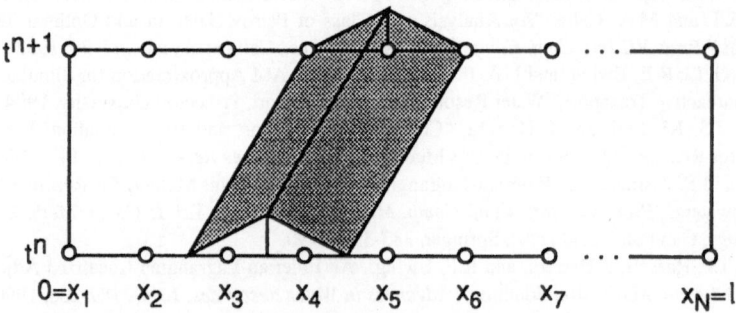

Figure 1: Piecewise linear, C^0-continuous ELLAM test function $w_4(x,t)$ for the case of $\lambda = 0$.

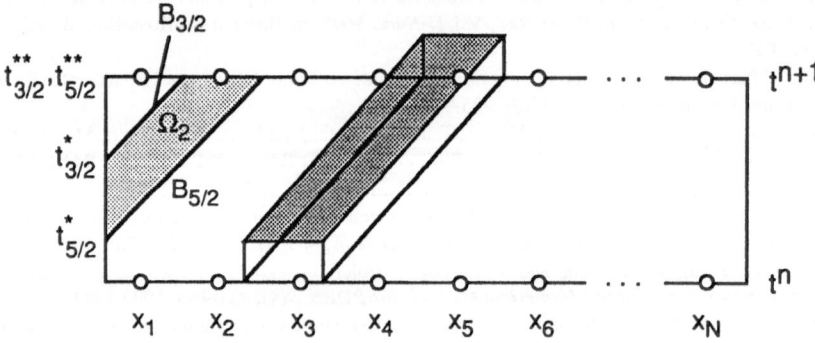

Figure 2: Piecewise constant, C^{-1}-continuous test function $w_4(x,t)$ for the case of $\lambda = 0$. Subdomain Ω_2 is shown, with boundaries $B_{3/2}$ and $B_{5/2}$. All $t_{j/2}^* = t^n$ and $t_{j/2}^{**} = t^{n+1}$ for $5<j<2N+3$.

MODELING THREE-DIMENSIONAL SUBSURFACE FLOW, FATE AND TRANSPORT OF MICROBES AND CHEMICALS (3DFATMIC)

J. R. CHENG (*), G. T. YEH (*), and T. E. Short (**)
(*) Department of Civil and Environmental Engineering
The Pennsylvania State University
University Park, PA 16802
U.S.A.
(**) Robert Kerr Environmental Research Laboratory
Environmental Protection Agency
Ada, OK 74820
U.S.A.

A three-dimensional model simulating the subsurface flow, microbial growth and degradation, microbial-chemical reaction, and transport of microbes and chemicals has been developed. The model is designed to solve the coupled flow and transport equations. Basically, the saturated-unsaturated flow field is described by the well-known Richards' equation with variant hydraulic conductivity changing with microbial and chemical concentrations. Seven components, namely one substrate, two electron acceptors, one nutrient, and three types of microbes, might exist in subsurface systems and compose seven simultaneous advective-dispersive-reactive transport equations. Since numerical problems may be introduced by using the conventional finite element method, the modified Lagrangian-Eulerian numerical scheme with adapted zooming and peak capturing is applied to solving these seven nonlinear partial differential equations accurately. The results from the flow and transport parts affect each other. Therefore, the coupling loop must be solved iteratively. To verify the developed model, this paper presents the simulation results of the published examples in the literature.

INTRODUCTION

The processes, which are the growth and decay of microbes and the biodegradation and transport of chemicals in heterogeneous subsurface environments, have been widely recognized as important phenomena in contaminant transport in groundwater. The ongoing research in subsurface flow and transport is toward combining these processes with advective-dispersive processes in transport models. However, not only chemical and biochemical processes but also flow is included in the system. Baek et al. (1989) presented the idea that the coupling of flow and transport can overcome the difficulties of microbial and chemical transport. Among the many recent developed models in the literature, Corapcioglu and Haridas (1984, 1985), Borden and Bedient (1986), Kindred and Celia (1989), Kommalapati et al. (1991), and Baek et al. (1989), all developed their models by using the same conceptual framework, which are based on the assumption of neglect of microscopic configuration. Most of systems only include substrate and microorganisms in one-spatial dimensional regions except Borden and Bedient considered three components: substrate, microorganisms, and oxygen, while Corapcioglu and Haridas (1985) solved the system in a two-dimensional domain.

A. Peters et al. (eds.), Computational Methods in Water Resources X, 217–224.
© 1994 Kluwer Academic Publishers. Printed in the Netherlands.

These studies also only focused on aerobic microbial degradation, and oxygen is the limited factor. Widdowson et al. (1988) pointed out that not only aerobic respiration and substrate but also anaerobic respiration and nutrients play important roles in subsurface environments. Therefore, all the above models have (at least) two simultaneous partial differential equations and (at most) nine PDEs which are solved analytically or numerically. With assumptions made for these governing equations, we can obtain an analytical solution; otherwise, numerical methods are the only way to solve these mixed parabolic-hyperbolic transport equations. However, most present numerical schemes are limited by the Peclet number, that means, it introduces numerical dispersion or numerical oscillation when Pe is greater than a limited number. Therefore, the numerical solution of transport equations has been the subject of this research.

The presented three-dimensional model, 3DFATMIC, is designed to simulate water flow through saturated-unsaturated media and the fate and transport of seven components (one substrate, two electron acceptors, one nutrient, and three microbial populations). The Galerkin finite element method is used to discretize the Richards' equation for obtaining the flow field and a new Lagrangian-Eulerian finite element numerical scheme, LEZOOMPC [Yeh et al., 1993], is implemented to solve seven simultaneous nonlinear partial differential equations without numerical dispersion and oscillation, even with a large mesh Peclet number. To demonstrate the model's capability and accuracy, five examples presented by Widdowson et al. (1988) are examined with the developed model.

MATHEMATICAL MODEL

The designed model is able to solve the following system of governing equations along with initial and boundary conditions, which describe flow and transport through saturated-unsaturated media (Yeh et al., 1993). The governing equation for a fluid density-dependent flow is basically the modified Richards' equation. The governing equations for transport are derived based on the continuity of mass and flux laws. The major processes are advection, dispersion/diffusion, adsorption, decay, source/sink, and microbial-chemical interactions. The governing equations are given as follows.

$$\frac{\rho}{\rho_w}\frac{d\theta}{dh}\frac{\partial h}{\partial t} = \nabla \cdot [K_s K_r \cdot (\nabla h + \frac{\rho}{\rho_w}\nabla z)] + \frac{\rho^*}{\rho_w}q\,(or\ -\frac{\rho}{\rho_w}q)$$

$$R_d \cdot \frac{\partial C}{\partial t} + V \cdot \nabla C = \nabla \cdot \theta D \cdot \nabla C - \Lambda \cdot R_d \cdot C + \frac{\rho_w}{\rho}V \cdot \nabla(\frac{\rho}{\rho_w})C + q_{in}(C_{in} - \frac{\rho^*}{\rho}C) + F(C)C \tag{1}$$

The saturated hydraulic conductivity tensor K_s (T/L) is given by

$$K_s = K_{sw}\frac{(\rho/\rho_w)}{(\mu/\mu_w)} \tag{2}$$

where K_{sw} is the referenced saturated hydraulic conductivity tensor (L/T), K_r is the relative hydraulic conductivity or relative permeability, h is the referenced pressure head defined as p/ρ_w in which p is pressure (M/LT^2), t is time (T), z is the potential head (L), q is the source and/or sink (L^3/T), and θ is the moisture content; $\rho(C)$, ρ_w and ρ^* are the density (M/L^3) of fluid, fresh water, and injected fluid, respectively. $C = \{ C_1, C_2, C_3, C_s, C_o, C_n, C_p \}^T$ where $C_s, C_o, C_n, C_p, C_1, C_2,$ and C_3 are dissolved concentration (M/L^3) of substrate, oxygen, nitrate, nutrient, microbe #1 (aerobic respiration), microbe #2 (anaerobic respiration), and microbe #3 (aerobic and anaerobic respirations). The density and dynamic viscosity of fluid are functions of microbial and chemical concentrations and are assumed to take the following form

$$\frac{\rho}{\rho_w} = 1 + \sum_i \left(\frac{1}{\rho_w} - \frac{1}{\rho_i} \right) C_i \tag{3}$$

$$i = 1, 2, 3, s, o, n, p$$
$$\mu = \mu_w + \beta_s C_s + \beta_o C_o + \beta_n C_n + \beta_p C_p + \beta_1 C_1 + \beta_2 C_2 + \beta_3 C_3 \tag{4}$$

where ρ_i are the intrinsic density of chemical i and β_s, β_o, β_n, β_p, β_1, β_2, and β_3 are the viscosity effecting factor of associated species (L^2/T). The tensors $\mathbf{R_d}$, Λ, and $\mathbf{D}$ are the retardation factor standing for ($\theta + \rho_b K_d$), transformation rate constants, and dispersion coefficient, respectively. The velocity field ($\mathbf{V}$) obtained from the flow equation is then input to the transport equation. Function F depends on the concentration distribution in the system and is formulated by the modified Monod relationship. It is assumed that microbe #1 utilizes substrate under aerobic conditions, microbe #2 utilizes substrate under anaerobic conditions, and microbe #3 utilizes substrate under both aerobic and anaerobic conditions.

The flow chart in Fig. 1 depicts the procedure in which 3DFATMIC solves the above governing equations.

EXAMPLES

The following five examples (Widdowson et al., 1988) are presented for model verification. The hydrodynamic properties and numerical parameters are identical to those listed in Widdowson et al.'s (1988) paper. These test cases consider a column discretized by $100 \times 1 \times 1$ elements in which the size is 0.5 cm $\times$ 1 cm $\times$ 1 cm. The dispersion coefficient and flow velocity were chosen to be 5 cm^2/day and 10 cm/day, respectively.

Example 1: Oxygen-Based Respiration

This simulation considers the fate and transport of substrate, oxygen, and microbe #1 solely

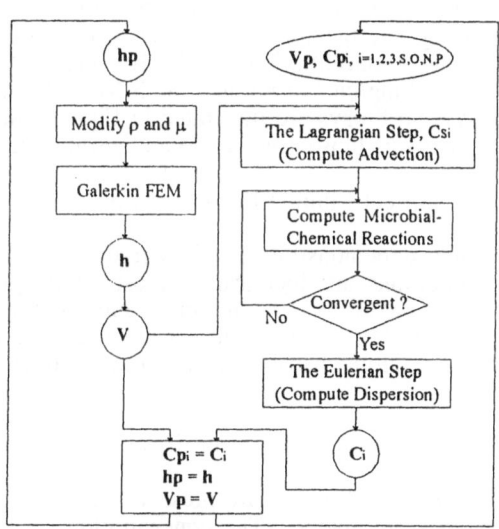

hp, Vp, Cp : pressure head, velocity, and concentration at previous time step

h, V, C : pressure head, velocity, and concentration at current time step

Cs : the Lagrangian concentration at current time step

Figure 1 The flow chart of 3DFATMIC

through oxygen-based respiration, which means the concentration of nitrate is zero and the nutrient is in excess. Initial concentrations of substrate, oxygen, and microbe #1 are 1.0×10^{-3}, 2.0×10^{-3} and 5.65×10^{-4} mg/cm^3, respectively. The influent concentrations are 1.0×10^{-2} mg/cm^3 of substrate, 2.0×10^{-3} mg/cm^3 of oxygen, and 5.65×10^{-4} mg/cm^3 of microbes.

Example 2: Nitrate-Based Respiration

Anaerobic (nitrate-based) respiration was considered to be the sole biodegradation mechanism in this simulation. The initial and boundary conditions are the same as the above example except nitrate and microbe #2 were involved instead of oxygen and microbe #1.

The results of these two examples simulated at t = 4 and 8 days are depicted in Figs. 2 and 3, respectively. They show a similar trend to those given by Widdowson et al. (1988).

Example 3: Multiple Electron Acceptor Respiration

Both oxygen and nitrate are important electron acceptors in aquifers and some kinds of microbes can perform aerobic and anaerobic respirations, i.e., oxygen and nitrate act as electron acceptors for oxidizing substrate. This example simulates the system including substrate, oxygen, nitrate, nutrient and microbe #3, and the associated concentrations are as in the previous examples, which are 1×10^{-3}, 2×10^{-3}, 5×10^{-3}, and 3.64×10^{-3} mg/cm^3, respectively. The microbial kinetic parameters are the same as those in the above two examples except the maximum specific growth rates and microbial decay coefficients are reduced to one-fifth of Widdowson's values. The boundary influent concentrations are 2×10^{-2} mg/cm^3 of substrate, 2×10^{-3} mg/cm^3 of oxygen , 5×10^{-3} mg/cm^3 of nitrate, and 3.64×10^{-3} mg/cm^3 of microbes. The results at t = 4 days and 8 days are illustrated in Figs. 4 and 5. Again, these results show similar behaviors as those given by Widdowson et al. (1988).

Example 4: Nutrient Limited Biodegradation

This example is to demonstrate the biodegradation also limited by nutrients. Initial and boundary conditions are identical to those of oxygen-based respiration except the nutrient concentration is 10^{-3} mg/cm^3 initially the influent concentration is 10^{-3} mg/cm^3. After 8 days, the nutrient boundary condition is increased to 6×10^{-3} mg/cm^3. The resulting distributions at 4 days, 8 days, and 12 days are shown in Figs. 6, 7, and 8.

Example 5: Substrate Limited Denitrification

Initially, substrate and nitrate are present at 8×10^{-3} mg/cm^3 and 10^{-3} mg/cm^3 in the whole region, respectively. The concentration of nitrate is introduced through the boundary at 10^{-2} mg/cm^3. The concentration profiles at 4 days and 8 days are plotted in Figs. 9 and 10.

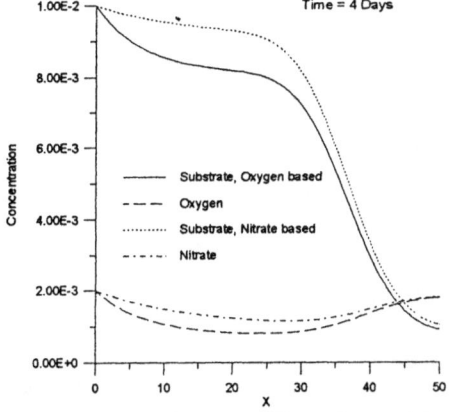

Fig. 2   Simulated results of Example 1 and 2

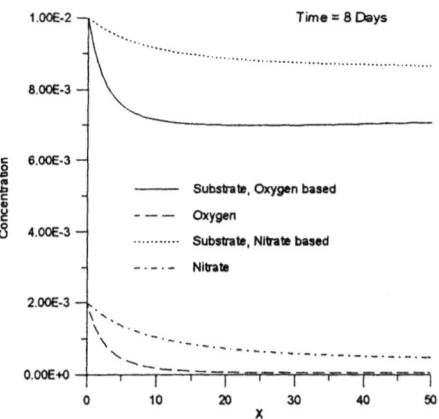

Fig. 3   Simulated results of Example 1 and 2

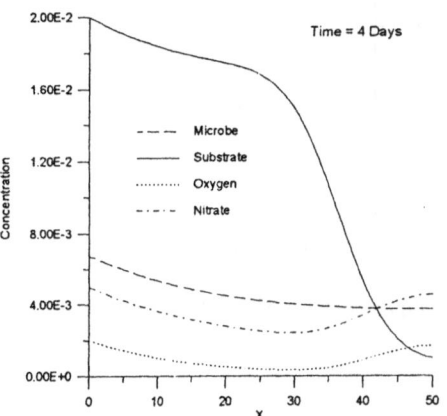

Fig. 4   Simulated results of Example 3

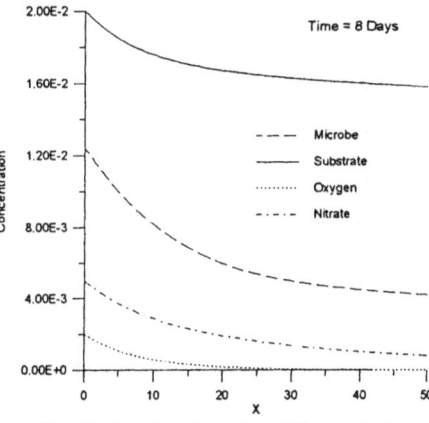

Fig. 5   Simulated results of Example 3

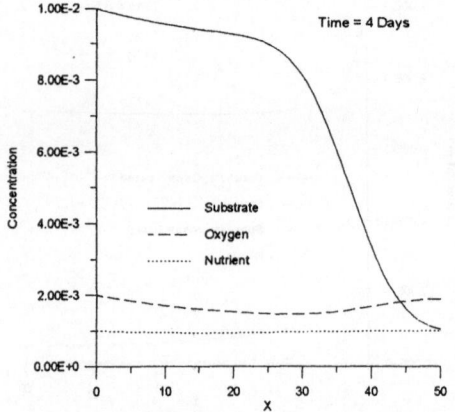

Fig. 6 Simulation results of Example 4

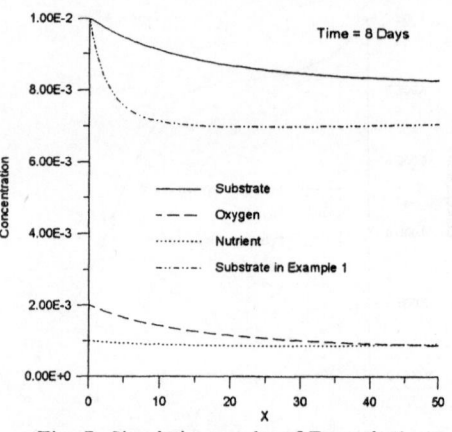

Fig. 7 Simulation results of Example 4

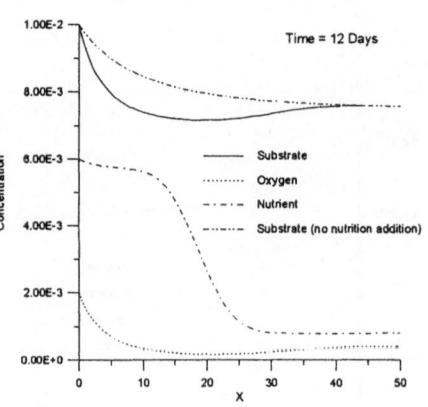

Fig. 8 Simulation results of Example 4

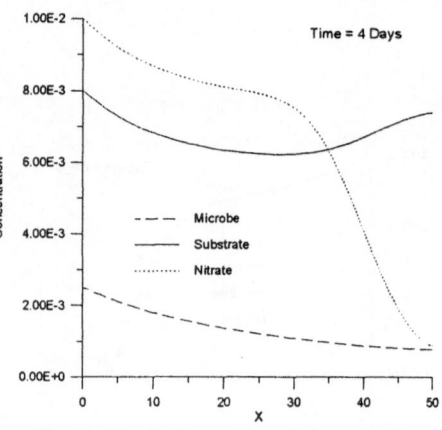

Fig. 9 Simulation results of Example 5

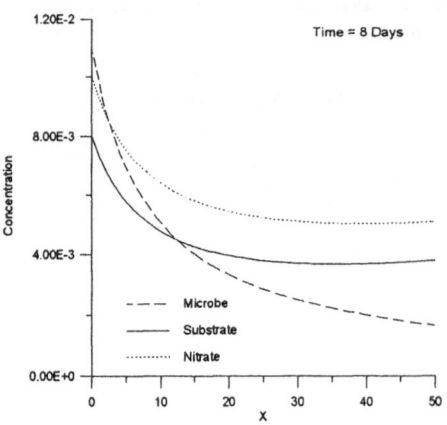

Fig. 10 Simulation results of Example 5

DISCUSSIONS

The presented model is designed to solve complicated flow and transport problems in a 3-dimensional domain at the same time. However, the verification cases to test the model capabilities are hard to find and the presented examples are the most completed ones in the literature. Although the above five examples are applied in a one-dimensional column, they still provide enough evidences to show the model's correctness.

The first two simulations (Examples 1 and 2) show the differences (Figs. 2 and 3) between oxygen-based and nitrate-based respirations. Since nitrate-based respiration is less efficient than oxygen-based respiration, the associated microbes can not utilize as much substrate under anaerobic conditions as those under aerobic conditions. In other words, this phenomena also reflects on the consumption of oxygen and nitrate.

The third example (Example 3) simulates multiple electron acceptors respiration which can occur in real aquifers. The results (Fig. 4) show that ample nitrate exists at the beginning and end of the column at $t = 4$ days so that the anaerobic respiration around these regions is inhibited. Excluding these areas, the rest column is characterized as an anaerobic zone because of almost depletion of oxygen. At $t = 8$ days, the substrate front passes through the end of the column and oxygen is completely depleted through most of the column. A large amount of nitrate is also consumed in anaerobic zone.

As mentioned earlier, nutrient supplementation is essential in biodegradation in the system. Example 4 demonstrates the importance of nutrients. Figs. 6 and 7 show that the substrate consumption and oxygen depletion are relatively less than those of Example 1 because of insufficient nutrients in the system. After sufficient time has elapsed, the change of substrate concentration profile is significant with additional nutrient added into the system. All in all, it proves that nutrients really play a very important role in the presented model.

The last example (Example 5) is to present that denitrification of nitrate is limited by the availability of substrate. From Figs. 9 and 10, the results indicate the substrate apparently controls the consumption of nitrate.

The model developed herein is intended to be applied to solve real world problems with interaction between flow and chemical transport coupling microbial growth and biodegradation. This model also provides many features and capabilities to overcome numerical problems when dealing with multidimensional and complicated biorestoration cases. Validation with field data would be the next topic to assure the model's feasibility.

ACKNOWLEDGEMENT

This research is supported by Robert S. Kerr Environmental Research Laboratory, U.S. Environmental Protection Agency, under cooperative agreement CR-818322. This article has not been subjected to the Agency's peer and administrative review and therefore does not necessarily reflect the view of the Agency and no official endorsement should be inferred.

REFERENCES

Baek, N. H., L. S. Clesceri, and N. L. Clesceri (1989) "Modelling of Enhanced Biodegradation in Unsaturated Soil Zone", Journal of Environmental Engineering, 115(1), 151-172.

Borden, R. C. and P. B. Bedient (1986) "Transport of Dissolved Hydrocarbons Influenced by Oxygen-Limited Biodegradation: 1. Theoretical Development", Water Resour. Res., 22(13), 1973-1982.

Corapcioglu, M. Y. and A. Haridas (1984) "Transport and Fate of Microorganisms in Porous Media: A Theoretical Investigation", Journal of Hydrology, Vol. 72, 149-169.

Corapcioglu, M. Y. and A. Haridas (1985) "Microbial Transport in Soils and Groundwater: A numerical Model", Adv. Water Resour., Vol. 8, 188-200.

Kommalapati, R. R., G. Wang, D. Roy, and D. D. Adrian (1991-1992) "Transport and Retention of Microorganisms in Porous Media: Comparison of Numerical Techniques and Parameter Estimation", J. of Environmental System, Vol. 21(2), 121-142.

Kindred, J. S. and M. A. Celia (1989) "Contaminant Transport and Biodegradation: 2. A Conceptual Model and Test Simulations", Water Resour. Res., 25(6), 1141-1148.

Yeh, G. T., J. R. Chang, and T. E. Short (1993) "2DFATMIC: User's Manual of a Two-Dimensional Subsurface Flow, Fate and Transport of Microbes and Chemicals", Technical Report to EPA. Department of Civil Engineering, Penn State University, University Park, PA 16802. March, , 106 pp.

Yeh, G. T., H. P. Cheng, and J. R. Chang (1993) "A two-dimensional Lagrangian-Eulerian approach with adapted zooming and peak capturing algorithm (2DLEZOOMPC)". EOS Transaction American Geophysical Union 74(16): 156.

CAN WE COMPUTE EXACT PATHLINES IN 3-D GROUNDWATER FLOW MODELS?

C. CORDES[*] and W. KINZELBACH[**]
[*] Department of Civil Engineering, Kassel University, 34109 Kassel
[**] Institute of Environmental Physics, Heidelberg University, 69120 Heidelberg

Numerical 3D models for pathlines and transport in groundwater are usually only verified by the comparison to two-dimensional or trivial three-dimensional solutions. In fact, only few closed solutions exist and they do not allow to test models comprehensively. Therefore it is not surprising to find that erroneous pathlines in three dimensions as one may obtain with customary numerical models are not easily recognized. Yet, correct pathlines are the prerequisite for catchment and pollutant arrival studies as they are carried out in groundwater protection and waste isolation. In two dimensions each pathline represents a water divide and, if boundary fluxes of a domain are known, the correspondence between entrance point and exit point is a priori known and independent from the flow details in the interior. In three dimensions the situation is fundamentally different. It is shown that

- although the potential flow in a simply connected domain is irrotational, streamlines may twist and a coordinate trihedral may rotate as it moves with the flow.
- when a FD or FE flow model is used as the basis for computation of pathlines or transport, it should be noted that either irrotationality or continuity is satisfied in a weak sense only. The error in the velocity field is small, sufficient discretization provided, yet it may cause large errors in the pathlines.
- the error can be limited, depending on the location of pathlines. Different models are compared and criteria are presented which allow to check the accuracy of pathlines.

The theory of pathlines in potential flow

The fundamentals of potential flow theory have already been formulated in the last century, e.g. Helmholtz' theorems about vortex motion (1858). In classical applications, however, only pressure and velocity were of interest, but "not the individual destination of single particles" (Prandtl, 1949). In groundwater flow the question concerning the whereabouts of a water particle arises in connection with the localization and remediation of contaminated groundwater.

The task seems trivial because once the velocity field is known, pathlines from arbitrary starting points can be calculated by particle tracking. Nevertheless, cases

225

A. Peters et al. (eds.), Computational Methods in Water Resources X, 225–232.
© 1994 *Kluwer Academic Publishers. Printed in the Netherlands.*

can occur where this post-processing is not satisfactory, i.e. at an impervious boundary or at the surface of a solid body, where velocities become singular.

In the two-dimensional case, singular points cause no problems since zero divergence ensures that pathlines never intersect and an impervious boundary always coincides with a pathline. Which law helps us in three dimensions? Does the irrotationality in a simply connected domain prevent pathlines from twisting? Does it avoid that fluid particles retain a transversal displacement after flowing around an obstacle?

In two dimensions irrotationality means that at any point the change of v_x in y-direction is identical to the change of v_y in x-direction:

$$\frac{\partial v_x}{\partial y} = \frac{\partial v_y}{\partial x} \tag{1}$$

Therefore, it is often assumed that a cross, consisting of water particles, does not rotate as it moves with the flow. However, this is not true. Consider, for example, the flow around an impervious cylinder with radius R:

$$v_x = v_0 \left[1 + R^2 \frac{y^2 - x^2}{(x^2 + y^2)^2} \right] \qquad v_y = -v_0 R^2 \frac{2xy}{(x^2 + y^2)^2} \tag{2}$$

Isochrones are shown in Figure 1. All pathlines have been started at the same time on the left hand side. It can be seen that a cross at the upstream side is rotated while it passes the cylinder.

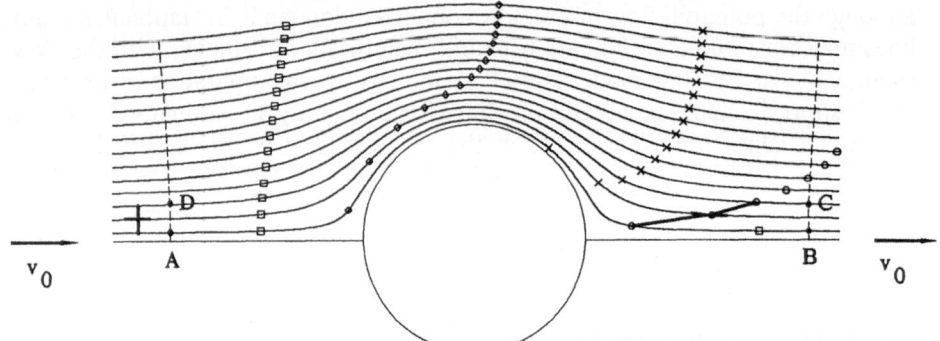

Fig. 1 Flow around an impervious cylinder

In a simply connected flow field without rotation, the circulation

$$\Gamma = \int v \, dS \tag{3}$$

along any closed curve S is zero. v denotes the velocity component tangential to the curve. Choose S as the path ABCDA in Figure 1 with the parts AB and CD identical to the pathlines between points A and B and between C and D. From B towards C and from D towards A the curve coincides with the dotted equipotential lines. Since there is no circulation along an equipotential line, the circulation from A to B must be equal to the circulation from D to C. In contrast, the traveltimes

$$t = \int \frac{1}{v} \, dS \tag{4}$$

are different. Thus, all displacements show no rotation after an infinitesimal time step, but after a finite traveltime the cross does turn.

Does it also twist in three dimensions? Let a constant vertical velocity be superimposed with the two-dimensional velocity field (eq. 2). This involves no rotation or divergence since the derivative of a constant is zero. The traveltime on the path between A and B is longer than between D and C. Thus after the same horizontal displacement a water particle on the path between A and B is subject to a larger vertical displacement and the pathlines shear. Generally, cases can occur in which pathlines twist around each other several times.

Nevertheless, three theorems can be formulated concerning pathlines in potential flow (Cordes, 1994):

(1)   If two particles have an infinitesimal distance at one moment, then they have an infinitesimal distance all the time unless they pass a stagnation point.
(2)   All particles on the surface of an impervious obstacle pass at least one stagnation point at the upstream side and at least one stagnation point at the downstream side.
(3)   At an edge, either one pathline coincides with the edge or all particles flow round the edge exactly at a right angle (but never obliquely).

In the first case of theorem 3 there is flow towards the edge and all velocity components except the one along the edge become zero, leading to one pathline along the edge. If water does not flow towards but around the edge, all components except the one along the edge become infinite. In reality, of course, the edge is not infinitely sharp. Yet velocity components around the edge are considerably larger and pathlines cross the edge at a right angle.

Some of the phenomena can be illustrated by the flow around a banana-shaped obstacle (Figure 2). Since no analytical solution exists, pathlines covering the surface are drawn qualitatively (In the next paragraph it will be proven that there is also no satisfactory numerical solution). The obstacle shows four stagnation points. Only one of these points lies on the upstream side and thus is passed by all pathlines on the surface. Furthermore, it can be seen that if an aggregate of particles passes such an obstacle, some particles retain a large transversal displacement.

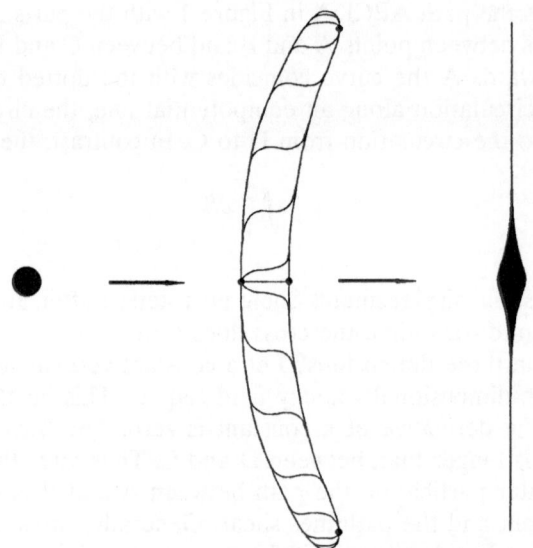

Fig. 2 Pathlines on the surface of a banana-shaped obstacle and the projection of an aggregate of particles before and after passing it

Numerical Models

In a two-dimensional domain with known boundary fluxes there is a unique correspondence between entrance point and exit point of a pathline. This is not the case in three dimensions where the exit point additionally depends on the geometry and the distribution of hydraulic conductivity in the interior of the domain.

The uniqueness in two dimensions is ensured by continuity while in three dimensions irrotationality has to be satisfied too. The latter requirement yields problems when discretized models are used since either zero divergence or irrotationality (or both) are fulfilled in a weak sense only (e.g. in head models using linear finite elements continuity is satisfied at the patch around each node but not at element edges).

In any case, zero divergence is to be preferred because the shortcomings of discontinuous fluxes which we documented in two dimensions (Cordes and Kinzelbach, 1992) show even worse in the third dimension which is usually discretized by a few elements only. By means of a simple post-processing Pollock (1988) converted the irrotational solution of a FD-head-model into a velocity field without divergence. In a similar way this can be achieved in finite elements (Cordes and Kinzelbach, 1992).

Streamfunction models in two dimensions have the advantage that boundaries with prescribed flux (including impervious boundaries) represent Dirichlet-type boundary conditions for a Laplace problem. In other words: values of the streamfunction are directly known. In three dimensions values of the two streamfunctions λ and χ can be prescribed only on one inflow or one outflow boundary (not on impervious boundaries) since the extent of streamline twist depends on the structure of the interior and is not known a priori. Therefore, the streamfunction approach loses its advantages in three dimensions.

Under certain conditions numerical flow models converge, meaning that the solution for heads and velocities tends to the exact solution with increasing mesh refinement. This is not always the case for pathlines based on the same flow field.
According to theorem 3 the flow towards an edge always involves one pathline along this edge because normal velocity components are zero. Using the above-mentioned models, normal velocities are only approximately zero and this is not sufficient: The particle will immediately leave the edge and take a completely different path.
Figure 3 shows the x-y-projection of three pathlines surrounding a cube in parallel flow ($v_x = v_y = v_z$). The cube is discretized into 20 * 20 * 20 FD-cells and cells in the vicinity have the same size. Three particles are started exactly upstream of the stagnation point in such a way that either of them strikes one of the three upstream faces. At the downstream side, according to theory, the particles have to flow along the edges and meet in the stagnation point. As can be seen, the rotation, which cannot be avoided when velocities are interpolated from continuous cell fluxes (e.g. after Pollock), prevents pathlines to follow the edges, although a very fine discretization is used.

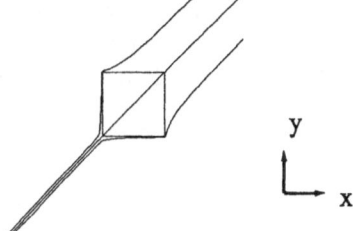

Fig. 3 x-y-projection of pathlines round a cube in parallel flow ($v_x = v_y = v_z$)

A streamfunction model would lead to similar results since not velocities but only values of streamfunction are computed directly on the edge. Directly calculating velocities (at nodes) is only possible using the velocity field approach (Zijl, 1984). Zijl proved that with sufficient boundary conditions (well posed problem) the potential flow problem can also be described by three coupled Poisson-type equations for v_x, v_y, and v_z. This involves irrotationality as well as zero divergence in a weak sense only. In contrast to the advantages in the case of flow towards an edge, the velocity field approach does not allow to simulate the flow around an edge correctly. Zijl recommends (and applied) his model only in horizontally stratified domains (conductivity does not change in the x-y-plane but only in the vertical direction) because then the three differential equations become decoupled.

Problems also occur in absolutely homogeneous parts of the domain, far from any boundary if flow is not parallel. Pollock's cell-wise velocity interpolation based on continuous cell fluxes yields zero rotation and divergence inside the cells. On the cell faces the normal velocity components are continuous but the tangential ones are not, leading to shearing of pathlines.

Figure 4 shows a block-shaped flow model with edgelengths 4L, 3L, and 2L. In its interior four impervious cubes with edgelength L are situated. The boundary conditions are an homogeneously distributed inflow on the left-hand-side, and the corresponding equally distributed outflow on the right-hand-side (due to symmetry), while the remaining faces are impervious.

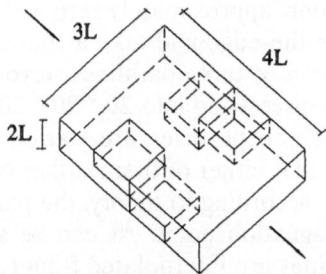

Fig. 4 Block-shaped model containing 4 impervious cubes

The model is discretized into 20*15*10 cubic cells and velocities are calculated according to Pollock. 12 Pathlines are started in the middle of the left boundary, 6 of them just above and the other 6 just below the boundary between the 5th and 6th layer of cells (Figure 5). It can be seen that the pathlines shear widely.

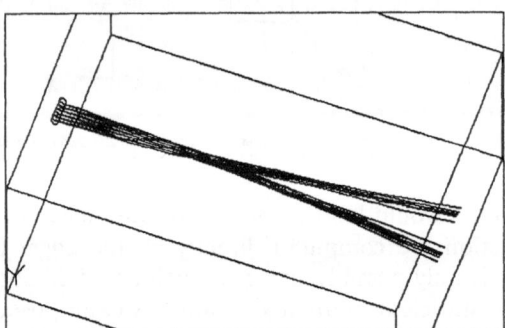

Fig. 5 Shearing pathlines in an FD-model with 20*15*10 cubic cells

The tank-and-tube-model

The error in the pathlines is always caused by the error in the approximate solution of the velocity field, and since hardly any analytical solutions exist its size is difficult to estimate. The only values we know for sure are the waterbalance fluxes through the cell faces. If we try to interpolate velocities we always introduce rotation or divergence.

We suggest to hold on to the fluxes and interpret them as fluxes through the tubes of a tank-and-tube-model. Since any assumption on the velocity distribution in the tank (cell) must be avoided the only reasonable approach here is: Full mixing. Thus, the aim of tracing a single particle has to be given up. Instead, we can only state the probability that a particle appears in a tank or a tube depending on the location of the tank or tube in which it started. The probability of appearance can be interpreted as a concentration.

The procedure of "concentration tracking" is quite simple. In each tube there is one water flux and one concentration, and as their product a solute flux. In a tank the concentration is calculated by the sum of all incoming solute fluxes divided by the sum of the incoming water fluxes. The concentration in a tube is always the same as in its upstream tank.

From its results, the tracking-procedure is nothing but a steady state upwind transport scheme without hydrodynamic dispersion. Since the procedure is explicit, no global equation system has to be solved. An approach similar to central differences makes no sense, because physically the concentration in a tube must never depend on the concentration in its downstream tank but only on the one in the upstream.

The mixing in the nodes leads to a phenomenon which in upwind differences is called numerical dispersion. The extent of spreading depends on the degree of discretization. The dispersion caused by a coarse grid corresponds with the uncertainty, because information is provided about few cell water balances only. Thus, the so-called numerical dispersion is nothing but a measure for the uncertainty of pathlines due to the discretization.

Unfortunately, the uncertainty depends on the grid orientation too. Even if finite elements are used it will be difficult in space to dicretize the mesh in "principal directions" as Frind (1982) did for the two-dimensional case.

Convergence criteria for pathlines

Is there any alternative to concentration tracking? The answer is negative if one seeks a solution excluding any hidden errors. However, unless no extremely fine discretization is available this means that advective transport of a point-like input does no longer take the one-dimensional shape of a pathline.

In which cases can we use pathlines? The discretization in Figure 3 is already very fine and even a further refinement will not force the particles to flow along the edges. In Figure 5 the situation is essentially different. Although one might not have expected such a large shearing of pathlines in a model which is already discretized into 10 layers, further refinement of the grid reduces the shearing and the pathlines converge to the exact solution. In the case shown in Figure 3 a similar convergence behaviour only occurs if the obstacle is not absolutely impermeable because then no stagnation points exist and no pathline must flow exactly along an edge.

In general, pathlines at the interface between two areas of different hydraulic conductivity do converge with increasing degree of discretization but compared to cases where pathlines stay away from such heterogeneities (e.g. Figure 5) a finer mesh is required. Note that whether or not an obstacle is absolutely impermeable mainly influences the velocity near its surface while, for example, the pathline shearing in Figure 5 would be qualitatively the same even if the cubes in Figure 4 had a small non-zero permeability.

Stagnation points do not always lie on impervious surfaces but can also be caused by the operation of a well. In a discretized model such a stagnation point is difficult to locate. The point may lie in the interior of a cell or an element where a continuous velocity distribution exists. But it might also be located on the common edge or corner of neighbouring cells or elements where velocities are discontinuous. Then pathlines started on a small circle around the stagnation point are not able to cover the surface of the capture zone completely, as they should do from theory.

In a practical approach the extent of pathline change due to grid refinement can always be used as a criterion for the accuracy - as long as three conditions are satisfied. Pathlines must stay away from:
- the surface of impervious obstacles (since it can not be ensured that pathlines pass at least one stagnation point at the upstream and the downstream side respectively),
- impervious boundaries unless the boundary is absolutely plane (because pathlines must not pass any edge),
- stagnation points of any kind.

References

Cordes, C. and Kinzelbach, W. (1992) "Continuous Groundwater Velocity Fields and Path Lines in Linear, Bilinear, and Trilinear Finite Elements", Water Res. Res., Vol. 28, No. 11, 2903-2911.

Cordes, C. (1994) "Bahnkurven in Potentialströmungen", Dissertation, Fachgebiet Technische Hydraulik, FB 14, Universität - Gesamthochschule Kassel, 34109 Kassel.

Frind, E. O. (1982) "The principal direction technique: A new approach to ground water contaminent transport modelling", 4th International Conference on Finite Elements in Water Resources, Hannover, Springer-Verlag Berlin, 13/25-42.

Pollock, D. W. (1988) "Semianalytical computation of pathlines for finite difference models", Ground Water, Vol. 26, No. 6, 743-750.

Prandtl, L. (1949) Strömungslehre, Friedr. Vieweg & Sohn Verlag, Braunschweig.

Zijl, W. (1984) "Finite-Element Methods Based on a Transport Velocity Representation for Groundwater Motion", Water Res. Res., Vol. 20, No. 1, 137-145.

MODELING OF NONLINEAR ADSORPTION IN CONTAMINANT TRANSPORT

CLINT N. DAWSON
Department of Computational and Applied Mathematics
Rice University
Houston, Texas 77251
U.S.A.

A numerical procedure for modeling equilibrium and non-equilibrium adsorption kinetics in contaminant transport is outlined. The procedure is based on upwinding the advective term and incorporating diffusion using a mixed finite element method. General, nonlinear adsorption isotherms may be incorporated in the model. Numerical results for equilibrium adsorption with a nonlinear isotherm of Freundlich type are given for one and two-dimensional test cases.

INTRODUCTION

Solutes in groundwater may undergo chemical reactions (adsorption) with the surrounding porous skeleton during the transport process. These reactions involve the adhesion of the solute in an extremely thin layer of molecules to the pore surface, and in effect, represent mass transfer between the fluid and the solid phases. Knowledge of the influence of adsorption on the transport of the solutes is of fundamental importance in understanding how pollutants spread through the soil.

Adsorption of dissolved solute onto the pore surface may be fast (equilibrium) or slow (non-equilibrium), depending on the rate of adsorption compared to the rate of bulk fluid flow. Let c $(M/L^3) \geq 0$ denote the concentration of the solute in the groundwater, and let s $(1/L^3)$ denote the mass of contaminant adsorbed on the solid matrix per unit mass of solid. In the case of non-equilibrium adsorption reactions, the mass transfer between fluid and solid is assumed to satisfy an ordinary differential equation of the form

$$s_t = k(f(c) - s), \tag{1}$$

where k is a rate parameter. The equilibrium case is obtained in the limit as $k \to \infty$, resulting in

$$s = f(c). \tag{2}$$

A. Peters et al. (eds.), Computational Methods in Water Resources X, 233–240.

The function f appearing in (1) and (2) is an adsorption isotherm. Common isotherms are due to Langmuir and Freundlich [13]. We will consider only the Freundlich isotherm, where

$$f(c) = Kc^p, \quad K \, (1/M) \geq 0. \tag{3}$$

In (3), $p > 0$ and in most cases of physical interest, $p \leq 1$. The choice $p = 1$ is commonly used in groundwater models because it results in a linear transport equation. However, as noted in [13], a linear model is not adequate to accurately describe adsorption in many situations, and nonlinear models are needed. Therefore, in this paper, we will concentrate on stable and accurate methods for modeling both linear and nonlinear adsorption phenomena. For more details on the fundamentals of sorption processes in porous media, see, for example, [9, 13].

Below, we outline a numerical procedure for modeling advection, diffusion, and adsorption of a single chemical species dissolved in groundwater. Both equilibrium and non-equilibrium adsorption will be considered. The Freundlich isotherm will be assumed with varying exponent p. The numerical method outlined here summarizes and extends methods which have been discussed previously in [4, 5]. In these methods, the underlying algorithm is the mixed finite element method for diffusion equations with the lowest-order Raviart-Thomas approximating spaces. Advection is incorporated into the procedure by upwinding the advective flux using, for example, a higher-order Godunov procedure. The higher-order upwinding methods which we employ are well-suited for advection-dominated flow with possibly nonlinear advection terms. Error estimates for these upwind-mixed methods were first derived in [6, 7], and estimates for the nonlinear problems considered here can be found in [5].

Recently, the upwind-mixed method described below has been used to study the long-term behavior of solutions to contaminant transport equations in one [10] and two [8] space dimensions, assuming equilibrium adsorption and $f(c)$ of Freundlich type. In these studies, the behavior of the solution for large time was predicted mathematically using matched asymptotic analysis. From this analysis, an asymptotic form was obtained to which the solution is expected to converge as time increases. The particular form depends on the exponent p in (3). As the analysis is heuristic, numerical experiments have been used to validate the mathematical assumptions. In all cases, the numerical solution obtained by our method showed precisely the same behavior as that predicted by the analysis.

The rest of this paper is outlined as follows. In the next section, we describe the mathematical model and the upwind-mixed finite element method. In Section 3, we give numerical results for one and two-dimensional test cases, and conclusions are given in Section 4.

THE MATHEMATICAL AND NUMERICAL MODELS

Though, in general, flow and transport in porous media may involve multiple species partitioned among several flowing phases, we will concentrate on the case of a single flowing phase (water) with a single chemical solute, moving through a homogeneous, saturated porous medium. We will assume flow is at steady state, and that transport is described by advection, molecular diffusion and mechanical dispersion, and adsorption. Other types of chemical reactions, such as biodegradation, will be neglected, though they can easily be incorporated into the mathematical and numerical models. We will also neglect source terms such as injection and production wells.

Under these assumptions, the transport of a chemical species is described by [1]

$$\phi c_t + \rho_b s_t + \nabla \cdot (\mathbf{u}c - \mathbf{D}\nabla c) = 0, \quad \text{on } \Omega \times (0, T]. \tag{4}$$

In (4), ϕ represents porosity, ρ_b $(M) > 0$ the bulk mass density of the soil, $\mathbf{u}$ (L/T) the Darcy velocity, and $\mathbf{D}$ (L^2/T) the hydrodynamic diffusion/dispersion tensor. We assume Ω is a polygonal domain in $\mathbb{R}^d$ and time $T > 0$. Assume the boundary of Ω, $\partial\Omega = \Gamma_1 \cup \Gamma_2$, where Γ_1 is an inflow boundary, where $\mathbf{u} \cdot \eta < 0$, and on Γ_2 is an outflow or noflow boundary, where $\mathbf{u} \cdot \eta \geq 0$; here η is the unit outward normal to $\partial\Omega$. We supplement (4) with the initial and boundary conditions

$$c(\mathbf{x}, 0) = c^0(\mathbf{x}), \qquad \mathbf{x} \in \Omega, \tag{5}$$

$$(\mathbf{u}c - \mathbf{D}\nabla c) \cdot \eta = (\mathbf{u}c_1) \cdot \eta, \quad \text{on } \Gamma_1 \times (0, T], \tag{6}$$

$$(\mathbf{D}\nabla c) \cdot \eta = 0, \quad \text{on } \Gamma_2 \times (0, T], \tag{7}$$

where c^0 and c_1 are specified. In the case of non-equilibrium adsorption, we must also have an initial condition on s,

$$s(\mathbf{x}, 0) = s^0(\mathbf{x}), \quad \mathbf{x} \in \Omega. \tag{8}$$

Denote $\mathbf{v}c$ by $\mathbf{a}$ and $\mathbf{D}\nabla c$ by $\mathbf{z}$. Let $(\cdot, \cdot)$ represent the $L^2(\Omega)$ inner product, and let $H(div; \Omega) = \{\mathbf{v} \in (L^2(\Omega))^d : \nabla \cdot \mathbf{v} \in L^2(\Omega)\}$. Let $W = L^2(\Omega)$, $\mathbf{V} = H(div; \Omega)$, and $\mathbf{V}^0 = \mathbf{V} \cap \{\mathbf{v} : \mathbf{v} \cdot \eta = 0\}$. Then, the mixed weak form of (4) is given by

$$(\phi c_t + \rho_b s_t + \nabla \cdot (\mathbf{a} + \mathbf{z}), w) = 0, \quad w \in W, \tag{9}$$

$$(\mathbf{D}^{-1}\mathbf{z}, \mathbf{v}) = (c, \nabla \cdot \mathbf{v}), \quad \mathbf{v} \in \mathbf{V}^0. \tag{10}$$

In the case of non-equilibrium adsorption we add the equation

$$(\rho_b s_t, w) = (\rho_b k(f(c) - s), w), \quad w \in W. \tag{11}$$

Let $W_h \subset W$ and $\mathbf{V}_h \subset \mathbf{V}$ be the lowest-order Raviart-Thomas [12] approximating spaces defined on a triangulation of Ω into elements with maximum diameter $h > 0$. Let $\mathbf{V}_h^0 = \mathbf{V}_h \cap \{\mathbf{v} : \mathbf{v} \cdot \eta = 0\}$. We note that W_h consists of functions which are constant on each element in the triangulation. The space $\mathbf{V}_h$ is more complicated, and is discussed below. Let $\Delta t > T/N$ for some positive integer N, and $t^n = n\Delta t$, $n = 0, \dots, N$. Let $\partial_t \phi^n = (\phi^n - \phi^{n-1})/\Delta t$. Approximate $c(\cdot, t^n)$, $s(\cdot, t^n)$ by C^n, $S^n \in W_h$, respectively. We also approximate $\mathbf{a}(\cdot, t^n)$, $\mathbf{z}(\cdot, t^n)$ by $\mathbf{A}^n \in \mathbf{V}_h$, $\mathbf{Z}^n \in \mathbf{V}_h^0$, respectively. These approximating functions are assumed to satisfy the following equations. For $n = 0$,

$$(C^0, w) = (c^0, w), \quad w \in W_h, \tag{12}$$

$$(S^0, w) = (s^0, w), \quad w \in W_h. \tag{13}$$

For $n = 1, \dots, N$,

$$(\phi \partial_t C^n + \rho_b \partial_t S^n + \nabla \cdot (\mathbf{A}^{n-1} + \mathbf{Z}^n), w) = 0, \quad w \in W_h, \tag{14}$$

$$(D^{-1} \mathbf{Z}^n, \mathbf{v}) = (C^n, \nabla \cdot \mathbf{v}), \quad \mathbf{v} \in \mathbf{V}_h^0. \tag{15}$$

In the case of non-equilibrium adsorption, we add

$$(\rho_b \partial_t S^n, w) = (\rho_b k(f(C^n) - S^n), w), \quad w \in W_h, \tag{16}$$

while in the case of equilibrium adsorption,

$$S^n = f(C^n). \tag{17}$$

The only unknown quantity in (14) is the advective flux $\mathbf{A}^{n-1} \approx (\mathbf{u}c)(\cdot, t^{n-1})$. In our method $\mathbf{A}^{n-1} \in \mathbf{V}_h$ is calculated explicitly from the solution given at t^{n-1}. In the lowest-order spaces, a function $\mathbf{v} \in \mathbf{V}_h$ is completely determined by the values of $\mathbf{v} \cdot \eta_K$, where η_K is the normal vector to the boundary of an element K. For example, suppose $\Omega \subset \mathbb{R}^2$ is divided into triangles, and suppose the triangle edges are numbered $k = 1, \dots, N_e$. Let η_k denote a unit vector normal to edge k. We associate with each edge a basis function $\mathbf{v}_k \in \mathbf{V}_h$ which satisfies (i) each component of $\mathbf{v}_k$ is linear on each triangle, and (ii) $\mathbf{v}_k \cdot \eta_l = \delta_{kl}$, where δ_{kl} is one if $k = l$ and zero otherwise. Then, $\mathbf{v} \in \mathbf{V}_h$ can be written as

$$\mathbf{v}(\mathbf{x}) = \sum_{k=1}^{N_e} (\mathbf{v} \cdot \eta_k) \mathbf{v}_k(\mathbf{x}).$$

We remark that $\mathbf{v}_k$ is nonzero only on the triangles which share edge k.

On inflow boundary edges, corresponding to the region Γ_1, we set $\mathbf{A}^{n-1} \cdot \eta_k = c_1(\mathbf{v} \cdot \eta_k)$. Since $\mathbf{Z}^{n-1} \cdot \eta_k = 0$, the inflow boundary condition (6) is satisfied. There are many ways to calculate $\mathbf{A}^{n-1} \cdot \eta_k$ on internal and outflow element edges. One trivial way is by upwinding. Let C_L^{n-1} denote the value of C^{n-1} on the element to the "left" of edge k (relative to the direction of the normal vector η_k), and C_R^{n-1} the value to the "right" of edge k. Then set

$$\mathbf{A}^{n-1} \cdot \eta_k = \begin{cases} (\mathbf{v}C_L^{n-1}) \cdot \eta_k, & \text{if } \mathbf{v} \cdot \eta_k > 0, \\ (\mathbf{v}C_R^{n-1}) \cdot \eta_k, & \text{if } \mathbf{v} \cdot \eta_k < 0. \end{cases} \tag{18}$$

This simple type of upwinding is usually insufficiently accurate. However, accuracy can be improved by calculating better approximations to the left and right states C_L and C_R. For rectangular elements, such approximations are discussed in [2, 7]. These higher-order approximations are constructed by a two-step process. First, the constant values C^{n-1} are postprocessed to obtain a linear function on each element. This improves the spatial accuracy. To improve the temporal accuracy, characteristic tracing can be used to predict the advective flux across edge k from time t^{n-1} to time t^n. Using these ideas, the formal accuracy of the advective flux can be improved to second-order in space and time, see [6]. Similar types of advective flux approximations for triangular elements can be found in [3].

The system (14)-(16) is a nonlinear system of equations in C^n, S^n, and $\mathbf{Z}^n$. Using (15) and (16), $\mathbf{Z}^n$ and S^n can be eliminated, giving a system in C^n only. In the case of equilibrium adsorption, we also obtain a system in C^n only. The existence and uniqueness (assuming Γ_1 is nonempty) of C^n follows from the fact that $f(c)$ is a continuous, monotone function of c, and by invoking standard theory for continuous, monotone operators [11].

NUMERICAL RESULTS

In this section, we give numerical results for the large-time behavior of solutions for one and two-dimensional test cases. We will concentrate only on the equilibrium case with a single exponent $p = 1/2$. In one-dimension, consider the equation

$$c_t + (\sqrt{c})_t + c_x - 10^{-2}c_{xx} = 0. \tag{19}$$

In two dimensions, consider

$$c_t + (\sqrt{c})_t + c_x - \Delta c = 0. \tag{20}$$

In (19) and (20) the quantities have been scaled so that c, x, and t are dimensionless. In both cases, the initial condition was a pulse centered at the origin and the computational domain Ω was chosen sufficiently large so that no-flow boundary conditions could be assumed.

In [10], the large-time behavior of solutions to (19) was studied using matched asymptotic expansion. There it was shown that, up to $o(1)$ terms, $tc(x,t)$ converges as $t \to \infty$ to a profile $\bar{c}(x,t)$ where

$$\bar{c}(x,t) = \begin{cases} 0, & x < 0, \\ \frac{x^2}{4t}, & 0 < x < 4\sqrt{t}, \\ 0, & x > 4\sqrt{t}. \end{cases} \tag{21}$$

In Figure 1, we plot $\bar{c}$ at $t = 1500$ versus $\nu = x/\sqrt{t}$ as computed by our numerical procedure. This result was obtained using a computational domain of length 2500 with $h = 1$ and $\Delta t = 1.5$. As seen in the figure, the computed solution is very close to the asymptotic profile given in (21). As t increases, the convergence improves. Plots for $t = 1000$ and $t = 2000$ are given in [10].

Results for the two-dimensional equation (20) are still in progress. Preliminary analysis shows that $c(x, y, t) \approx t^{-1.2} v(X, Y)$ as $t \to \infty$, where $X = x/t^{.4}$ and $Y = y/t^{1.2}$. A numerical approximation to v is given in Figure 3. Here we have plotted a three-dimensional perspective of $t^{1.2} C$ versus X and Y at $t = 3000$. Figure 2 represents the view along the X axis. As demonstrated in the figure, a shock develops as the solution propagates, just as in one space dimension. The matched asymptotic analysis indicates that the behavior observed in the figure is very close to the predicted behavior. In this simulation, the computational domain $\Omega = [-200, 200] \times [-200, 200]$ and 160,000 uniform grid blocks were used. The time step $\Delta t = 1$.

In the above simulations, the method used for calculating the advective flux in the one-dimensional case is explained in detail in [4]. In the two-dimensional case, the method used is an extension of ideas given in [4, 7, 2].

CONCLUSIONS

Adsorption is an important phenomenon in the transport of contaminants in groundwater. The type of adsorption, whether linear, nonlinear, equilibrium, or non-equilibrium, can have a significant impact on the spread of contaminants. We have outlined a numerical method suitable for modeling a wide-range of adsorption processes. This method has been analyzed theoretically and applied to a number of test cases, and has been shown to accurately model adsorption processes even when they are highly nonlinear.

Acknowledgments: This work was supported by grants from the U. S. Department of Energy.

References

[1] J. Bear, *Dynamics of fluids in porous media*, Elsevier, New York, 1972.

[2] J. B. Bell, C. N. Dawson, and G. R. Shubin, *An unsplit higher-order Godunov scheme for scalar conservation laws in two dimensions*, J. Comput. Phys., **74**, pp. 1-24, 1988.

[3] B. Cockburn, S. Hou, and C.-W. Shu, *The Runge-Kutta local projection discontinuous Galerkin finite element method for conservation laws IV: The multidimensional case*, Math. Comp., **54**, pp. 545-581, 1990.

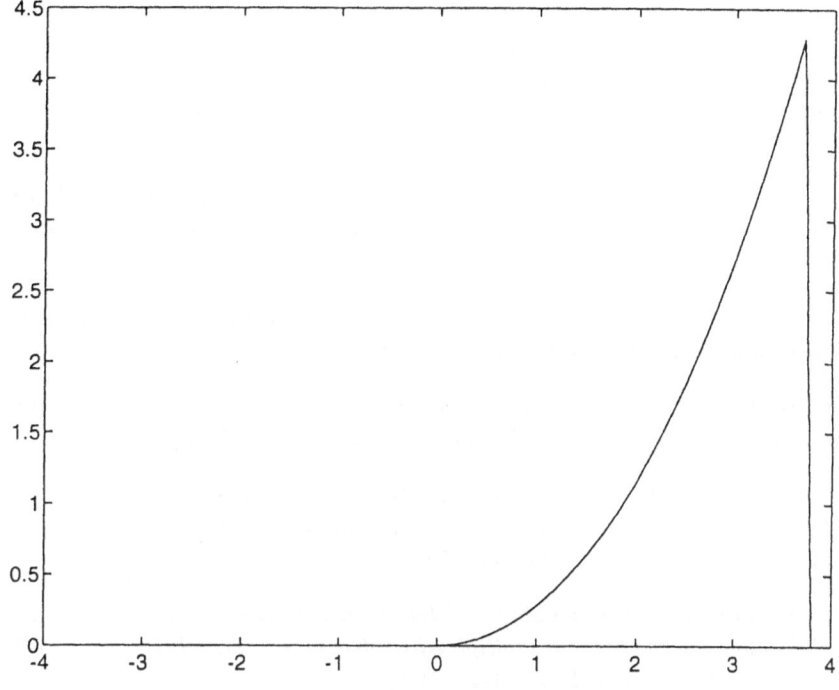

Figure 1: $\bar{c}$ at $t = 1500$ versus $\nu = xt^{-.5}$.

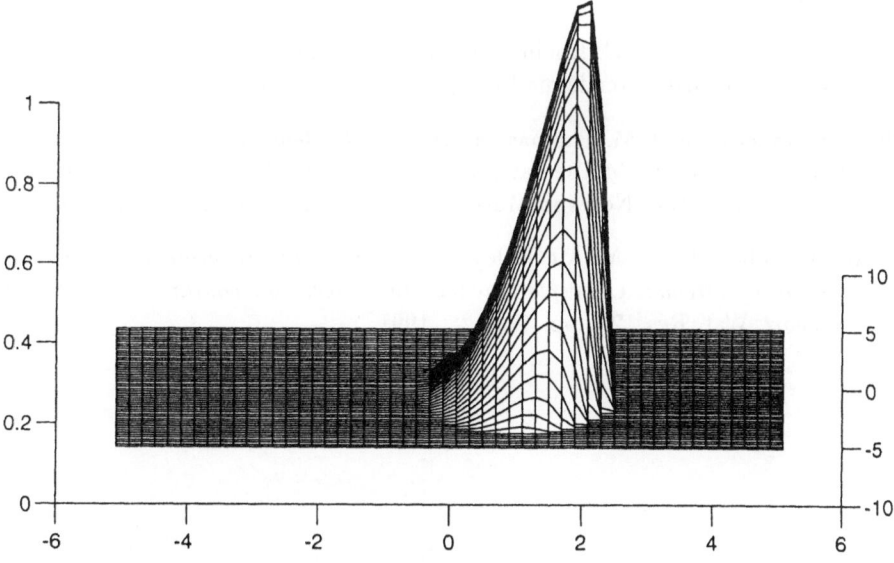

Figure 2: $t^{1.2}C$ versus X and Y at $t = 3000$. View is along the X axis.

[4] C. N. Dawson, *Simulation of nonlinear contaminant transport in groundwater by a higher order Godunov-mixed finite element method*, Applications of Supercomputers in Engineering II, C. A. Brebbia, D. Howard, and A. Peters, eds., Computational Mechanics Publications, Southampton, UK, pp. 419-433, 1991.

[5] C. N. Dawson, *Analysis of upwind-mixed methods for a nonlinear contaminant transport equations*, Technical Report TR93-57, Dept. of Comp. and Appl. Math., Rice University, Houston, TX.

[6] C. N. Dawson, *Godunov-mixed methods for advective flow problems in one space dimension*, SIAM J. Numer. Anal., **28**, pp. 1282-1309, 1991.

[7] C. N. Dawson, *Godunov-mixed methods for advection-diffusion equations in multidimensions*, SIAM J. Numer. Anal., **30**, pp. 1315-1332, 1993.

[8] C. N. Dawson, C. J. van Duijn, and R. E. Grundy, *Asymptotic profiles with finite mass in two-dimensional contaminant transport through porous media: the fast reaction case*, in preparation.

[9] C. J. van Duijn and P. Knabner, *Solute transport in porous media with equilibrium and non-equilibrium multiple-site adsorption: Travelling waves*, J. Reine Angewandte Math. **415**, pp. 1-49, 1991.

[10] R. E. Grundy, C. J. van Duijn, and C. N. Dawson, *Asymptotic profiles with finite mass in one-dimensional contaminant transport through porous media: the fast reaction case*, to appear in the Quarterly Journal of Mechanics and Applied Math.

[11] J. M. Ortega and W. C. Rheinboldt, *Iterative solution of nonlinear equations in several variables*, Academic Press, New York, 1970.

[12] P. A. Raviart and J. M. Thomas, *A mixed finite element method for 2nd order elliptic problems*, in Mathematical Aspects of the Finite Element Method, Rome 1975, Lecture Notes in Mathematics, Springer-Verlag, Berlin, 1977.

[13] W. J. Weber, Jr., P. M. McGinley, and L. E. Katz, *Sorption phenomena in subsurface systems: Concepts, models and effects on contaminant fate and transport*, Wat. Res., **25**, pp. 499-528, 1991.

EULERIAN-LAGRANGIAN LOCALIZED ADJOINT METHODS FOR TRANSPORT OF NUCLEAR-WASTE CONTAMINATION IN POROUS MEDIA

R.E. EWING
Institute for Scientific Computation
Texas A&M University
College Station, Texas 77843–3404

H. WANG and R.C. SHARPLEY
Department of Mathematics
University of South Carolina
Columbia, South Carolina 29208

The contamination of the groundwater by a variety of hazardous pollutants makes a proper description and understanding of contaminant transport in porous media very important to predict and plan groundwater management programs and remediation strategies. The resulting mathematical models involve a transient fully coupled system of equations for transport of total fluid, heat, brine, and trace-species radionuclides. The numerical simulation of this system is often plagued with various numerical difficulties. In this paper, we apply Eulerian-Lagrangian localized adjoint methods (ELLAM) to provide improved numerical approximations to the solution of this system. ELLAM are formulated to systematically adapt to the changing features of governing partial differential equations. The time-stepping along characteristics of the advection part of the flow in ELLAM allows the use of large time-steps in a stable and accurate fashion. Moreover, ELLAM can treat boundary conditions systematically and maintain mass conservation, which makes ELLAM superior to many Eulerian-Lagrangian methods (ELM) while maintaining their numerical advantages. Specific linearization techniques based on the inherent physics of the system are discussed.

1. INTRODUCTION

Groundwater, on which many people depend for their water supply, has been increasingly threatened by organic, inorganic, and radioactive pollutants as well as high-level nuclear waste introduced into the environment by improper disposal and accidental release of these hazardous materials. While remediation methods are extremely expensive and unpredictable in their success, mathematically based nu-

A. Peters et al. (eds.), Computational Methods in Water Resources X, 241–248.

merical models have proven to be a very feasible, practical, and efficient approach to plan groundwater management programs and remediation strategies. A mathematical model has been developed for use by the Nuclear Regulatory Commission to analyze deep geologic nuclear waste disposal facilities, which involves a fully coupled transient system of equations for transport of total fluid, heat, brine, and trace-species radionuclides. The equations used here are obtained by combining appropriate continuity and constitutive relations and have been derived by several authors [1,3]. The resulting relations for the total fluid, heat, brine, and the ith component of radionuclides may be stated as follows:

$$-\nabla \cdot \underline{u} - q + R'_s = 0, \tag{1}$$

$$-\nabla \cdot (H\underline{u}) + \nabla \cdot (\mathbf{E}_H \nabla T) - q_L - qH - q_H = [\phi c_p + (1 - \phi)\overline{\rho}_R c_{pR}]\frac{\partial T}{\partial t}, \tag{2}$$

$$-\nabla \cdot (\hat{c}\underline{u}) + \nabla \cdot (\mathbf{E}_c \nabla \hat{c}) - q\hat{c} - q_c + R_s = \phi\frac{\partial \hat{c}}{\partial t}, \tag{3}$$

$$\nabla \cdot (c_i \underline{u}) + \nabla \cdot (\mathbf{E}_c \nabla c_i) - qc_i - q_{c_i} + q_{0i} + \sum_{j=1}^{N} k_{ij}\lambda_j K_j \phi c_j$$
$$-\lambda_i K_i \phi c_i = \phi K_i \frac{\partial c_i}{\partial t}, \quad i = 1, 2, \cdots, N. \tag{4}$$

The quantity $q = q(\mathbf{x}, t)$ is a production term, $R'_s = R'_s(\hat{c}) = \frac{c_s \phi k_s f_s}{1+c_s}(1 - \hat{c})$ is a salt-dissolution term for fluid equation. $q_L = q_L(T)$ is a heat loss to under/overburden term, qH is an injected enthalpy term, q_H is a produced enthalpy term, $q\hat{c}$ is an injected brine term, $q_c = q_c(\hat{c})$ is a produced brine term, and $R_s = R_s(\hat{c}) = \phi k_s f_s(1-\hat{c})$ is a salt-dissolution term for brine equation. qc_i are injected components terms, $q_{c_i} = q_{c_i}(c_i)$ are produced component terms, $q_{0i} = q_{0i}(c_i)$ are waste leach terms, $\sum_{j=1}^{N} k_{ij}\lambda_j K_j \phi c_j$ describe generation of component i by decay of j, and $\lambda_i K_i \phi c_i$ is the decay for component i.

Equations (1)–(4) are coupled by a mixing rule for viscosity, $\mu = \mu(\hat{c})$, and three auxiliary relations for Darcy flux, fluid enthalpy and fluid internal energy given, respectively, by

$$\underline{u} = -(\mathbf{k}/\mu)\left(\nabla p - \rho_0 \frac{g}{g_c}\nabla z\right), \tag{5}$$

$$H = U_0 + U + p/\rho_0, \tag{6}$$

$$U = c_p(T - T_0), \tag{7}$$

where $U_0 = U_0(\mathbf{x})$ and $T_0 = T(\mathbf{x}, 0)$.

The tensors E_j $(j = H, c)$, in Equations (2)–(4) are sums of molecular diffusion and fluid dispersion terms and are small compared to the fluid velocity $\underline{u}$, creating advection dominated equations. The nonlinearities in the dispersion terms significantly complicate the related computational schemes. Also, absorption of radionuclides is included via an assumption of a linear equilibrium isotherm. This yields the distribution coefficient k_{di} and the retardation factor for each component

$$K_i = 1 + \rho_R k_{di}(1 - \phi)/\phi. \tag{8}$$

In the system above, our primary unknowns are the pressure p, the Darcy velocity $\underline{u}$, the concentration $\hat{c}$ of the dominant species (brine in our case), the concentrations c_i $(i = 1, 2, \ldots, N)$ of the trace-species (radionuclides in our case) and the temperature T. The formulation of the problem is closed with appropriate boundary conditions, initial conditions and possibly some compatibility conditions between the source and sink terms depending on the types of boundary conditions specified.

2. OVERVIEW OF SOLUTION PROCEDURES

The system (1)–(5) presented in Section 1 is a large coupled system of strongly nonlinear partial differential equations. In fact, the coefficient $\mu(\hat{c})$ of Equations (1) and (5), which determine the pressure p and the Darcy velocity $\underline{u}$, depends on the concentration $\hat{c}$ of the dominant species given by Equation (3). In this manner, Equations (1) and (5) are coupled to Equation (3). On the other hand, Equation (3) is coupled to Equations (1) and (5) through the Darcy velocity $\underline{u}$. Also, Equations (2) and (4) are also coupled to Equations (1) and (5) through the Darcy velocity $\underline{u}$. But the concentrations c_i $(i = 1, 2, \ldots, N)$ of the trace-species do not affect the coefficient μ in Equation (5) and so do not affect the Darcy velocity $\underline{u}$. Moreover, Equations (2)–(4) are usually strongly advection-dominated transport equations whose numerical simulation often causes various numerical difficulties, which we will discuss more in Section 2.1.

It is the combination of strong nonlinearities and close couplings between the equations that causes difficulties in solving the system (1)–(5). Due to the enormous size of the applications, the equations cannot be solved in a fully-coupled, fully-implicit fashion. Thus, some linearization techniques must be used to obtain the numerical solutions. However, a blind linearization with little regard to the properties of the equations or the solutions can result in extremely large, ill-conditioned, nonlinear systems; the accurate solution of these equations can be extremely difficult and expensive. Choices of implicitness and decoupling of the equations are forms of linearization and must be carefully analyzed. The linearization techniques in this paper use a time marching algorithm coupled with the application of Lagrangian coordinates for estimating the nonlinear coefficients. We present the details in Section 4.

2.1. Numerical Methods for Transport Equations

After linearizing the system (1)–(5), we turn to the numerical methods for the individual equations. In [6], we presented finite element methods on the logically rectangular hexahedral meshes to simulate the system (1)–(5) in three-dimensional porous media. Even though the original domain is not rectangular and our finite-element partition is not a three-dimensional rectangular partition, the partition still maintains the same data structure as a rectangular domain with a rectangular partition. This gives us the flexibility to implement the system for real field simulations over three-dimensional nonrectangular domains. However, because Equations (2)–(4) are strongly advection-dominated, the standard finite element/difference methods tend to exhibit excessive oscillation that is unrelated to the underlying physical processes, unless of course, very small time steps are used. Moreover, the derived algebraic system has a very strongly nonsymmetric coefficient matrix that forces any algebraic solver to converge slowly. While the finite element methods with upstream weighting can remove the oscillations, they tend to yield numerical solutions with severe nonphysical damping that smears out important information along fluid fronts where complex and important physics and chemistry take place. Therefore, the Eulerian numerical methods may be inefficient for large three-dimensional applications due to the excessive computational requirements to overcome these difficulties.

In [3], Ewing *et al.* treated Equations (2)–(4) by a combination of a modified method of characteristics (MMOC). The time-stepping along the characteristics of the advection part of the flow allows the use of large time steps in a stable and accurate fashion. Nevertheless, the major drawback of MMOC and many other Eulerian-Lagrangian methods (ELM) is that they do not conserve mass and that they have difficulty in treating boundary conditions. Since small amounts of contamination can cause serious environmental problems, we must accurately model and predict the transport of even small quantities of pollutants. Thus, mass-conservative numerical schemes are essential. In addition, the treatment of boundary conditions also has significant effect on the numerical solutions due to the strong Lagrangian nature of the problems. In this paper, we develop ELLAM [2] to solve Equations (2)–(4). ELLAM maintains all the numerical advantages of ELM while overcoming their drawbacks. We will present our ELLAM in Section 3.

2.2. Numerical Methods for Pressure Equations

As for the pressure p and the Darcy velocity $\underline{u}$, if we substitute Equation (5) into Equation (1), we obtain a second-order elliptic equation for the pressure p

$$-\nabla \cdot \left((\mathbf{k}/\mu) \left(\nabla p - \rho_0 \frac{g}{g_c} \nabla z \right) \right) = -q + R'_s. \tag{9}$$

In [6] we solve Equation (9) by finite element methods over the logically rectangular hexahedral meshes. After we obtain the numerical solution p^h, we put p^h into Equation (5) to compute the Darcy velocity $\underline{u}^h$. However, because of the strong heterogeneity of the subsurface geology, the media property $\mathbf{k}$ and $\hat{c}$-dependent μ are in general not smooth. Therefore, the exact solution p of Equation (9) is not smooth and consequently the numerical solution p^h might not be very accurate. As a result, the numerical Darcy velocity $\underline{u}^h$ obtained by taking the numerical differentiation of p^h and multiplying p^h by a rough coefficient $\mathbf{k}/\mu$ is even less accurate, which in turn affects the accuracy of approximations to the concentrations $\hat{c}$, c_j and the temperature T when it is put into Equations (2)–(4).

While the pressure p might not be smooth, the Darcy velocity $\underline{u}$ is usually very smooth. The mixed finite element method (MFEM) solves for the Darcy velocity $\underline{u}^h$ and the pressure p^h simultaneously, and generates very accurate approximations for both. Therefore, we use MFEM to solve the pressure Equations (1) and (5), referring the readers to the references for details.

3. AN ELLAM SCHEME FOR TRANSPORT OF SINGLE SPECIES

In this section we derive an ELLAM scheme for a single-species, advective-diffusive-reactive transport equation. In Section 4, we will describe the sequential techniques for linearizing and decoupling the system (1)–(5). While we have successfully applied ELLAM to solve multidimensional transport equations [5], for simplicity in this paper we present our ELLAM scheme for the following one-dimensional transport equation

$$\mathcal{L}c \equiv \left(\Phi\, c\right)_t + \left(V\, c - D\, c_x\right)_x + K\, c = f, \qquad x \in (a,b), \;\; t \in (0,T]. \qquad (10)$$

Let J and N be two positive integers. We define the partitions of space and time as $x_j = a + j\Delta x$ $(j = 0, 1, \ldots, J)$ and $t^n = n\Delta t$ $(n = 0, 1, \ldots, N)$, where $\Delta x = \frac{b-a}{J}$ and $\Delta t = \frac{T}{N}$ are spatial mesh and time step, respectively. Since we use a time-marching algorithm to solve the system (1)–(5), we consider space-time test functions w that vanish outside $[a, b] \times (t^n, t^{n+1}]$ and are discontinuous in time at each time level t^n, whose exact form will be given below as part of the ELLAM development. With these test functions, we can write out the space-time variational formulation for Equation (10) as follows:

$$\int_a^b \Phi(x, t^{n+1})u(x, t^{n+1})w(x, t^{n+1})dx + \int_{t^n}^{t^{n+1}} \int_a^b Du_x w_x \, dx dt$$
$$+ \int_{t^n}^{t^{n+1}} \left(Vu - Du_x\right)w \,\big|_a^b dt - \int_{t^n}^{t^{n+1}} \int_a^b u\left(\Phi\, w_t + V\, w_x - K\, w\right)dx dt \qquad (11)$$
$$= \int_a^b \Phi(x, t^n)u(x, t^n)w(x, t_+^n)dx + \int_{t^n}^{t^{n+1}} \int_a^b f\, w dx dt,$$

where $w(x, t_+^n) = \lim\limits_{t \to t_+^n} w(x, t)$.

Based on ideas from the localized adjoint method [2], the test functions w should be chosen from the solution space of the adjoint equation of Equation (10):

$$\mathcal{L}^* w \equiv -\Phi(x, t)w_t - V(x, t)w_x - \left(D(x, t)w_x\right)_x + K(x, t)w = 0. \qquad (12)$$

The solution space of Equation (12) is infinite-dimensional. Since only a finite number of test functions should be chosen, we can split the adjoint operator $\mathcal{L}^*$ in different ways. Different splittings lead to different classes of numerical methods, including optimal test function methods and Eulerian-Lagrangian methods. Notice that the exact solutions of Equation (10) vary exponentially along the characteristics. We define the test functions w by the following splitting:

$$-\Phi w_t - V w_x + K w = 0 \qquad \text{and} \qquad -(D w_x)_x = 0. \tag{13}$$

For our method to be practical, we choose computable test functions w to approximately satisfy Equation (13). For simplicity, we approximate the characteristics by the one-point backward Euler rule (more complicated rules can also be applied). We define the test functions w to be piecewise linear functions at the new time level t^{n+1} (or at the outflow boundary) and to vary exponentially along the approximate characteristics. Substituting these test functions into (11), and dropping the truncation error term, we obtain our ELLAM scheme after delicate treatments. Due to the space limitation, we only present our ELLAM scheme at a typical interior node x_j:

$$
\begin{aligned}
\int_{x_{j-1}}^{x_{j+1}} & \Phi(x, t^{n+1}) c(x, t^{n+1}) w_j(x, t^{n+1}) dx \\
& + \int_{x_{j-1}}^{x_{j+1}} \Phi_3^K(x, t^{n+1}) D(x, t^{n+1}) c_x(x, t^{n+1}) w_{jx}(x, t^{n+1}) dx \\
& - \int_{x_{j-1}}^{x_{j+1}} \Phi_4^K(x, t^{n+1}) D(x, t^{n+1}) c_x(x, t^{n+1}) w_j(x, t^{n+1}) dx \\
= & \int_{x_{j-1}}^{x_{j+1}} \Phi_1^K(x, t^{n+1}) \Phi(x^*, t^n) c(x^*, t^n) w_j(x, t^{n+1}) dx \\
& + \int_{x_{j-1}}^{x_{j+1}} \Phi_2^K(x, t^{n+1}) f(x, t^{n+1}) w_j(x, t^{n+1}) dx,
\end{aligned} \tag{14}
$$

where x^* is the foot of the approximate characteristic emanating backward from x at t^{n+1}, $\Phi_1^K, \dots, \Phi_4^K$ are Jacobian factors due to the spatial and temporal variations of the characteristics [4].

By choosing the test functions w in (11) to the approximate solutions of Equations (13), our ELLAM systematically adapt to the changing features of governing partial differential equations. Relative importance of retardation, advection, diffusion and reaction is directly incorporated into the numerical methods by the use of these test functions. The time-stepping along characteristics of the advection part of the flow in ELLAM allows the use of large time-steps in a stable and accurate fashion. Moreover, ELLAM can treat boundary conditions systematically and maintain mass conservation. Therefore, ELLAM overcome the major drawback of many ELM while maintaining their numerical advantages. Moreover, our theoretical and numerical results also confirm the strength of ELLAM. We refer interested readers to [4,5].

4. SEQUENTIAL SOLUTION PROCEDURES

Having developed our ELLAM scheme for a single-species transport Equation (10), we describe the time-marching sequential solution procedures for system (1)–(5). Suppose that we know the numerical approximations to the concentrations $\hat{c}(\mathbf{x}, t^n)$, $c_i(\mathbf{x}, t^n)$ $(i = 1, 2, \ldots, N)$ and the temperature $T(\mathbf{x}, t^n)$ at the time level t^n; the algorithm can be stated as follows:

1. Putting $\hat{c}(\mathbf{x}, t^n)$ into $\mu(\hat{c})$ in Equations (1) and (5), we can solve these two equations for $\underline{u}^h(\mathbf{x}, t^n)$ and $p^h(\mathbf{x}, t^n)$ by MFEM.

2. Since the Darcy velocity $\underline{u}$ usually varies less rapidly than the concentrations, we use $\underline{u}^h(\mathbf{x}, t^n)$ as the projected value for $\underline{u}^h(\mathbf{x}, t^{n+1})$. Then, Equation (3) is reduced to a linear advective-diffusive-reactive transport equation (10) with unknown $\hat{c}^h(\mathbf{x}, t^{n+1})$. Then, we can apply the ELLAM scheme derived in the last section to solve Equation (3) for $\hat{c}^h(\mathbf{x}, t^{n+1})$.

3. Next we solve Equations (4) for $c_i(\mathbf{x}, t^{n+1})$ $(i = 1, 2, \ldots, N)$. Note that Equations (4) are coupled together for all c_i through the reaction terms. Thus, we need to solve these equations iteratively. We can solve these equations for $c_i(\mathbf{x}, t^{n+1})$ in the Gauss-Seidel manner. Namely, when we solve the ith equation in Equations (4), we choose the estimated values for $c_i(\mathbf{x}, t^{n+1})$ $(i = 1, 2, \ldots, N)$ in the coefficients of these equations to be the known values $c_j^h(\mathbf{x}, t^{n+1})$ at the current time level t^{n+1} and to be the known values $c_j^h(\mathbf{x}^*, t^n)$ at the foot of the characteristics at the previous time level t^n for $j > i$. Hence, the ith equation in Equations (4) is reduced to a linear variable-coefficient advective-diffusive-reactive transport equation (10). Applying our ELLAM in Section 3 and repeating this process for $i = 1, 2, \ldots, N$, finishes the solution of Equations (4).

4. Then we can solve Equation (2) for the temperature $T(\mathbf{x}, t^{n+1})$ by our ELLAM in Section 3 since Equation (2) is also an advective-diffusive-reactive transport equation.

The above procedure yields numerical solutions $\underline{u}^h(\mathbf{x}, t^n)$, $p^h(\mathbf{x}, t^n)$, $\hat{c}^h(\mathbf{x}, t^{n+1})$, $c_i^h(\mathbf{x}, t^{n+1})$ $(i = 1, 2, \ldots, N)$, and the $T^h(\mathbf{x}, t^{n+1})$. We can repeat this procedure to march in time until we obtain the numerical solutions at all the time steps.

In the above, we decoupled Equations (4) by updating the reaction terms with the Gauss-Seidel approach since the couplings only occur in the reaction/source terms. Because we immediately used the updated information $c_j^h(\mathbf{x}, t^{n+1})$ for $j < i$ when we solved the ith equation in Equations (4), the Gauss-Seidel approach usually converges faster than the Jacobi approach described shortly. The second advantage is that the Gauss-Seidel approach only needs about one-half of the storage required by the Jacobi approach. The drawback of Gauss-Seidel approach is that it is strongly sequential. In other words, we can only solve the ith equation in Equations (4) after we solved the first $i - 1^{\text{th}}$ equations in Equations (4). In the Jacobi approach, we choose the estimated values for $c_i(\mathbf{x}, t^{n+1})$ ($i = 1, 2, \ldots, N$) in the coefficients of Equations (4) to be the known values $c_j^h(\mathbf{x}^*, t^n)$ at the foot of the characteristics at the previous time level t^n for all $j = 1, 2, \ldots, N$. This way, Equations (4) are naturally decoupled, linearized, and reduced to N individual linear variable-coefficient advective-diffusive-reactive transport equations in the form of (10) which may be solved in parallel. This is important when the Equations (4) involve a large number of species. In practical applications, we should choose the appropriate iterations based on the nature of the problems considered and on our computational resources.

In the above, a simple noniterative sequential linearization was chosen because of the relatively mild nonlinearities and very good numerical results that were subsequently produced. We can also use iterative schemes and higher-order predictor rules to solve system (1)–(5). Our numerical methods used in this work are insensitive to the linearization chosen and are applicable to any linearization procedure.

REFERENCES

1. Bear, J. (1979) Hydraulics of Groundwater, McGraw-Hill.
2. Celia, M.A., Russell, T.F., Herrera, I, and Ewing, R.E. (1990) "An Eulerian-Lagrangian localized adjoint method for the advection-diffusion equation", Advances in Water Resources, 13(4), 187–206.
3. Ewing, R.E., Yuan, Y., and Li, G. (1989) "Time stepping along characteristics for a mixed finite element approximation for compressible flow of contamination by nuclear waste in porous media", SIAM J. Numer. Anal., 26(6), 1513–1524.
4. Ewing, R.E. and Wang, H. (to appear) "An Eulerian-Lagrangian localized adjoint method with exponential-along-characteristic test functions for variable-coefficient advective-diffusive-reactive equations", Proceedings of Fifth Korean Advanced Institute for Science and Technology International Workshop in Analysis, Taejon, Korea.
5. Ewing, R.E. and Wang, H. (1993) "Eulerian-Lagrangian localized adjoint methods for linear advection or advection-reaction equations and their convergence analysis", Computational Mechanics, 12 1/2, 97–121.
6. Ewing, R.E., Wang, H., and Celia, M.A (1994) "A 3-dimensional finite element simulation to transport of nuclear-waste contamination in porous media", Proceedings of the Eighth International Conference of the International Association for Computer Methods and Advances in Geomechanics, Morgantown, West Virginia.

COMPARISON OF SEVERAL FORMULATIONS FOR THE STORAGE AND CONDUCTIVE TERMS OF NON-LINEAR FLOW EQUATIONS

GERMAN GALARZA*, JESUS CARRERA* and LINDA ABRIOLA**
* ETSECCPB. Universitat Politècnica de Catalunya
C/ Gran Capitán s/n, Edif. D2. 08034 Barcelona
** University of Michigan, Ann Arbor
48109 Michigan USA

ABSTRACT

Different numerical formulations for solving non-linear flow equations are compared in terms of both accuracy and computational cost. The generic term "formulation" refers here both to the treatment of the storage term and the computation of conductance terms. The storage term is computed (1) using the specific capacity parameter (capacitive scheme) and (2) expressing mass variations as the difference between water contents at sequential times (conservative schemes). Regarding conductance terms, all one needs is to obtain element-wise conductivities because the Finite Element Method (FEM) is used and only linear basis functions are considered here. We study four different alternatives to compute such conductivities. The resulting system of equations is solved, in all cases, by means of a full Newton-Raphson method. The formulations are evaluated by modelling six synthetic problems, which consider a wide variety of characteristics, including both unsaturated and unconfined flow. Comparison of the results shows that capacitive schemes, for the storage term, combined with the hydraulic conductivity corresponding to the mean element's pressure, for the conductance term, are the best alternatives, not only in terms of accuracy but also in terms of efficiency.

INTRODUCTION

Choosing a formulation for solving non-linear flow problems is associated to three critical issues. The first one relates to the evaluation of mass variations when the unsaturated flow equation is posed in h-based form. Classical treatment is based on taking derivatives in the storage term of the Richards equation. This drives the modeler to compute mass variations in time by means of the specific capacity parameter, which is obtined from the retention curve. Neuman (1973), and Carrera and Galarza (1993) adopt this approach, using either the finite element or finite differences techniques. An alternative, mixed formulation (they use moisture contents and pressure heads as explicit variables in the equations), consists of replace the storage term by the global mass balance. Milly (1985) and Celia et al. (1990) obtained good results with this formulation, which they attributed to its perfect mass balance. The second issue refers to the computation of cell

249

A. Peters et al. (eds.), Computational Methods in Water Resources X, 249–256.
© 1994 Kluwer Academic Publishers. Printed in the Netherlands.

or element conductivities, when this parameter depends upon the state variable. This problem is present in either unsaturated, unconfined or multiphase flow; its dificulty is increased when non-homogeneous medium is considered. Haverkamp and Vauclin (1979), compare nine different approaches for computing the conductivity in one-dimensional unsaturated flow problems. They concluded that geometric mean is the best alternative for the problems solved. Russo (1992), presents a procedure for defining block conductivity (associated to a certain domain) in an heterogeneous and unsaturated medium. The third important issue when seeking a general formulation is the time integration of the equation. In this paper, only the first two of this issues are studied. The last one, best time formulation, was discussed in a previous work [see Galarza et al., (1992)] for the same problems considered here.

MATHEMATICAL PROBLEM

Both unsaturated and unconfined flow are considered in this work. The former is controlled by

$$\frac{\partial \theta}{\partial t} = \nabla(K\nabla(\psi + z)) + q_r \tag{1}$$

where θ is moisture content, $K[L/T]$ is the hydraulic conductivity three-dimensional tensor, $\psi[L]$ is pressure head, $q_r[T^{-1}]$ is a sink-source term and $t[T]$ is time. Depending upon the type of medium, there are particular relationships between ψ, K and θ. van Genuchten (1980) and Brooks and Corey (1964) are two works presenting empirical mathematical expressions for those relationships. Solution of (1) requires specifying initial and boundary conditions. Cases where the vertical flow is negligible in unconfined aquifers may be modelled using the Dupuit assumptions, which leads to the following two-dimensional equation.

$$S\frac{\partial h}{\partial t} = \nabla(T\nabla h) + q_r \tag{2}$$

here, $h[L]$ is piezometric head, $T[L^2/T]$ is the transmisivity two-dimensional tensor and S is the storage coefficient. Transmissivity is the only physical paramater which depends on the state variable (piezometric head).

NUMERICAL FORMULATION

Storage term

The left hand side term in (1) may be expanded in two ways:

$$\frac{\partial \theta}{\partial t} = [S_w S_s \frac{\partial \psi}{\partial t} + \phi \frac{\partial S_w}{\partial t}] \quad (3) \qquad \frac{\partial \theta}{\partial t} = [S_w S_s + \phi \frac{\partial S_w}{\partial \psi}]\frac{\partial \psi}{\partial t} \quad (4)$$

As the specific storage, S_s $[L^{-1}]$ (S_w is saturation degree and ϕ is porosity), is very small, the first term in the right hand side of (3) and (4) may be ommitted in many cases. In fact, mass variation is mainly governed by the second term when the medium is partially saturated. As a consequence, the evaluation of such term is critical in the quality of results. The numerical method used in this work, Finite Element in space and Finite Differences in time, drive us to evaluate the mass variation in the discretized domain as a matrix product of the storage matrix times the temporal derivative vector. Since a lumped scheme is used for the storage matrix, the generic i-th component of the product, evaluated at time $k + \epsilon$, for the two alternatives, (3) and (4), is expressed, respectively, as

$$\left[\frac{\partial \theta}{\partial t}\right]_i^{k+\epsilon} = \frac{\psi_i^{k+1} - \psi_i^k}{t^{k+1} - t^k} \sum_{e=1}^{l} \left(\sum_{j=1}^{n} \int_{\Omega_e} \xi_i^e \xi_j^e \left[\phi \frac{S_w^{k+1} - S_w^k}{\psi_i^{k+1} - \psi_i^k} + S_s^{k+\epsilon} S_w^{k+\epsilon}\right] d\Omega\right) \quad (5)$$

$$\left[\frac{\partial \theta}{\partial t}\right]_i^{k+\epsilon} = \frac{\psi_i^{k+1} - \psi_i^k}{t^{k+1} - t^k} \sum_{e=1}^{l} \left(\sum_{j=1}^{n} \int_{\Omega_e} \xi_i^e \xi_j^e \left[\phi \left(\frac{\partial S_w}{\partial \psi}\right)^{k+\epsilon} + S_s^{k+\epsilon} S_w^{k+\epsilon}\right] d\Omega\right) \quad (6)$$

These expressions represent the storage change in the volume associated to the $i - th$ node per unit time. Here, l is the number of elements to which node i belongs, n is the total number of nodes of element e, ξ_i and ξ_j are the basis functions for nodes i and j and Ω is the problem domain. Variability of the physical parameters S_w and $\partial S_w / \partial t$ within the element hinders an exact computation of the integral. However, it is possible to approximate a mean value for those parameters in the element, which allows us an analytical treatment of the integral since we are using linear basis functions. We consider two possibilities to approximate the non-linear parameters: first, taking the value corresponding to the mean pressure within the element; second, computing the parameter with the pressure at the node i. Combining every one of these in (5) and (6), the final expressions of the storage term are:

$$\left[\frac{\partial \theta}{\partial t}\right]_i^{k+\epsilon} = \frac{\psi_i^{k+1} - \psi_i^k}{t^{k+1} - t^k} \sum_{e=1}^{l} \left(\left[\phi \frac{S_w(\psi_e^{k+1}) - S_w(\psi_e^k)}{\psi_i^{k+1} - \psi_i^k} + S_s S_w(\psi_e^{k+\epsilon})\right] \sum_{j=1}^{n} \int_{\Omega_e} \xi_i^e \xi_j^e d\Omega\right)$$

$$(7)$$

$$\left[\frac{\partial \theta}{\partial t}\right]_i^{k+\epsilon} = \frac{\psi^{k+1} - \psi^k}{t^{k+1} - t^k} \sum_{e=1}^{l} \left(\left[\phi \frac{\partial S_w}{\partial \psi}(\psi_e^{k+\epsilon}) + S_s S_w(\psi_e^{k+\epsilon})\right] \sum_{j=1}^{n} \int_{\Omega_e} \xi_i^e \xi_j^e d\Omega\right) \quad (8)$$

$$\left[\frac{\partial \theta}{\partial t}\right]_i^{k+\epsilon} = \frac{\psi_i^{k+1} - \psi_i^k}{t^{k+1} - t^k} \sum_{e=1}^{l} \left(\left[\phi \frac{S_w(\psi_i^{k+1}) - S_w(\psi_i^k)}{\psi_i^{k+1} - \psi_i^k} + S_s S_w(\psi_i^{k+\epsilon})\right] \sum_{j=1}^{n} \int_{\Omega_e} \xi_i^e \xi_j^e d\Omega\right)$$

$$(9)$$

$$\left[\frac{\partial \theta}{\partial t}\right]_i^{k+\epsilon} = \frac{\psi_i^{k+1} - \psi_i^k}{t^{k+1} - t^k} \sum_{e=1}^{l} \left(\left[\phi \frac{\partial S_w}{\partial \psi}(\psi_i^{k+\epsilon}) + S_s S_w(\psi_i^{k+\epsilon})\right] \sum_{j=1}^{n} \int_{\Omega_e} \xi_i^e \xi_j^e d\Omega\right) \quad (10)$$

$$\psi_e = \frac{1}{n} \sum_{j=1}^{n} \psi_j \quad ; j \in e \quad (11) \qquad \psi^{k+\epsilon} = \epsilon \psi^{k+1} + (1-\epsilon) \psi^k \quad 0.5 \le \epsilon \le 1 \quad (12)$$

Conductive term

Development of the FEM over the conductive term results in the product of the nodal pressure vector ($\boldsymbol{\psi}$) times the conductance matrix ($\boldsymbol{A}$), whose generic i, j component is:

$$A_{i,j}^{k+\epsilon} = \sum_{e=1}^{mt} \Big[\sum_{m=1}^{3} \sum_{m=1}^{3} \int_{\Omega_e} \Big(\frac{\partial \xi_i}{\partial x_m} \frac{\partial \xi_j}{\partial x_n} K_{m,n}(\psi^{k+\epsilon}) \Big) d\Omega \Big] \tag{13}$$

where $K_{m,n}$ is the component in directions x_m, x_n of the conductivity tensor and mt is the total number of elements conecting nodes i and j. The difficulty of evaluating (13) is caused by the dependence of the physical parameter, conductivity or transmissivity, upon the state variable. In this paper we have considered four different possibilities for evaluating a mean value (K^e) within the element, which allows us to bring it out of the integral. Those are:

$$K_e = K(\psi_e) \quad (14) \qquad K_e = \frac{1}{n} \sum_{i=1}^{n} K(\psi_i) \quad (15)$$

$$K_e = \Big[\sum_{i=1}^{n} 1/K(\psi_i) \Big]^{-1} \quad (16) \qquad K_e = \Big[K(\psi_1) K(\psi_2) ... K(\psi_n) \Big]^{1/n} \quad (17)$$

Formulations tested

Regardless of the scheme we use, application of FEM to the flow equations results in a non linear system, generalized in the expression (19).

$$F(\boldsymbol{\psi}^{k+\epsilon}) = \boldsymbol{A}(\boldsymbol{\psi}^{k+\epsilon})\boldsymbol{\psi}^{k+\epsilon} + \frac{\psi^{k+1} - \psi^k}{t^{k+1} - t^k} \boldsymbol{D}(\boldsymbol{\psi}^{k+\epsilon}) - \boldsymbol{b}(\boldsymbol{\psi}^{k+\epsilon}) = 0 \tag{18}$$

where matrix $\boldsymbol{A}$, whose order is n by n, can be computed by use the equation (13) combined with (15), (16), (17) or (18). $\boldsymbol{D}$ is, in this case, a diagonal n-component matrix whose i-th component is the righ hand side in expression (7) to (10), without the termporal derivative factor. On the other hand, the sink-source term vector $\boldsymbol{b}$ is computed by means of equation (14) together with (15), (16), (17) or (18). Obviously, additional terms may be added to $\boldsymbol{b}$ when other sink-source terms are present. Combining the alternatives for the three terms result in a large number of numerical formulations. Table 1 presents those which are considered to test in this paper.

COMPARISON

Quality of the formulations will be tested by modelling six problems, whose particular details are given by Galarza (1993). We consider accuracy and computation time as qualification parameters. The comparison procedure is as follows. Each problem is solved two times with every formulation. First, we use a refined time-space discretization, so as to ensure that no significant difference exists between these solutions and one obtined with more refined grids. At this level, also,

Table 1: Definition of the formulations tested (F_i), according to the treatment given to the storage and conductive terms.

PARAMETER	STORAGE CHANGE $\phi\frac{\partial S_w}{\partial \psi}(\psi_e)\frac{\Delta\psi}{\Delta t}$	STORAGE CHANGE $\phi\frac{\partial S_w}{\partial \psi}(\psi_i)\frac{\Delta\psi}{\Delta t}$	STORAGE CHANGE $\phi\frac{\Delta S_w(\psi_e)}{\Delta t}$	STORAGE CHANGE $\phi\frac{\Delta S_w(\psi_i)}{\Delta t}$
CONDUCTANCE of mean pressure	F_1: Eqs. 8,13,14	F_2: Eqs. 10,13,14	F_3: Eqs. 7,13,14	F_4: Eqs. 9,13,14
CONDUCTANCE Geometric mean		F_5: Eqs. 10,13,17		
CONDUCTANCE Arithmetic mean		F_6: Eqs. 10,13,15		
CONDUCTANCE Harmonic mean		F_7: Eqs. 10,13,16		

results obtained with all formulations are virtually identical. Those are considered as the accurate solutions. A second group of runs are made using a coarse time-space grids. Then, ranks of the formulations are stablished by comparing pressure heads and computation time of those two series of runs. In all cases, the time weighting parameter (ϵ) is equal to 0.5 (it corresponds to the Crank-Nicholson scheme)

Comparison Criteria

The formulations are ranked according to three parameters associated to the accuracy and three to the computation time. Thus, it is choosen a series of nob_j observations belonging to the space-time domain of the problem j. We compute the value of pressure or piezometric head obtained with both the coarse (ψ^c) and refined (ψ^r, which corresponds to the exact solution in problem 4) grids. This is used to define the three comparison criteria for the formulation i; sum of relative errors (SRE), sum of relative rankings (SRR) and maximum error (ME), as follows:

$$SRE_i = \sum_{j=1}^{6}\left[\sum_{l=1}^{nob_j}(\psi_{j,l}^r - \psi_{i,j,l})^2 / \sum_{k=1}^{7}\sum_{l=1}^{nob_j}(\psi_{k,j,l}^c - \psi_{j,l}^r)^2\right] \quad (19) \qquad SRR_i = \sum_{j=1}^{6}RR_{i,j}$$

$$(20)$$

$$ME_i = max_j\left[\sum_{j=1}^{nob_j}(\psi_{j,l}^r - \psi_{i,j,l}^c)^2 / \sum_{k=1}^{7}\sum_{l=1}^{nob_j}(\psi_{k,j,l}^c - \psi_{j,l}^r)^2\right] \qquad (21)$$

where $RR_{i,j}$ is the ranking obtained by formulation i when solving problem j. On the other hand, efficiency of the numerical formulations can be tested with analogous expressions; sume of relative times (SRT) sume of relative rankings in time (SRRT) and maximum relative time (MRT). In this case, the pressure

Table 2: Quadratic errors (QE) obtained in the problems by the formulations tested.

PROBLEM	1	2	3	4	5	6
FORMULATION						
$F-1$	$0.327\ 10^{-3}$	$0.737\ 10^{-4}$	$0.324\ 10^{-2}$	$0.975\ 10^{-3}$	49.90	1.480
$F-2$	$0.427\ 10^{-3}$	$0.801\ 10^{-4}$	$0.314\ 10^{-2}$	$0.125\ 10^{-3}$	57.50	1.570
$F-3$	$0.154\ 10^{-2}$	$0.835\ 10^{-4}$	$0.395\ 10^{-2}$	$0.915\ 10^{-3}$	36.10	25.30
$F-4$	$0.427\ 10^{-3}$	$0.801\ 10^{-4}$	$0.406\ 10^{-2}$	$0.102\ 10^{-2}$	56.80	35.00
$F-5$	$0.284\ 10^{-3}$	$0.643\ 10^{-4}$	$0.315\ 10^{-2}$	$0.958\ 10^{-3}$	104.0	1.740
$F-6$	$0.678\ 10^{-3}$	$0.801\ 10^{-4}$	$0.316\ 10^{-2}$	$0.959\ 10^{-3}$	50.30	2.400
$F-7$	$0.794\ 10^{-3}$	$0.544\ 10^{-4}$	$0.314\ 10^{-2}$	$0.957\ 10^{-3}$	421.0	1.170

Note: $\epsilon = 1.0$ for the storage term in the formulation $F-3$

difference $(\psi^c - \psi^r)$ is replaced in (19), (20) and (21) by the CPU time and no square exponents are taken.

Accuracy Evaluation

Table 2 displays the absolute quadratic errors QE defined in (22) for the 7 formulations in the 6 problems. A careful analysis of this table points out that the quality of each formulation is problem-dependent. Notice that both the best and worst formulation change for every problem. On the other hand, it is seen that results of unconfined problems (1 and 2) are more uniform than those of unsaturated cases. This is probably due to the relationships between physical parameters and state variables. In the unconfined problems, this relationship is close to linear while unsaturated functions (retention and permeability curves) are far from linear. That implies critical numerical conditions, and the consequent problems for poor formulations, which is more relevant in problems 5 and 6. Obviously, homogeneity of results is also dependent on the discretization degree. We tried, however, that discretizations in all problems were similar poor. Thus, the prescribed time steps size in the second set of runs (coarse discretization) were twenty times bigger than those corresponding to the accurate runs.

Table 3: Ranking of the formulations according to precision criterion.

Ranking	Formulation	QRE	Formulation	SRR	Formulation	ME
1	$F-2$	0.50042	$F-2$	17	$F-2$	0.15517
2	$F-1$	0.60262	$F-1$	20	$F-5$	0.16213
3	$F-5$	0.64169	$F-5$	20	$F-6$	0.16229
4	$F-6$	0.70127	$F-7$	20	$F-1$	0.16500
5	$F-7$	1.13625	$F-6$	27	$F-3$	0.36848
6	$F-4$	1.17646	$F-3$	29	$F-4$	0.50976
7	$F-3$	1.24130	$F-4$	35	$F-7$	0.54281

Table 3 displays the rankings of the formulations according to the criteria. As it can be seen, formulations 1, 2, 5 and 6 are ranked at the top. In addition to this, notice that there is not a significant change in the ranking when changing the comparison criterion. This indicates that results of the comparison have a good degree of generality.

Efficiency Evaluation

The CPU time spent by each algorithm for solving each problem using coarse discretizations is presented in table 4. The table contains a great variety in the absolute magnitude of the times, not only from one problem to another but also from one formulation to another for a given problem. This is a consequence of the poor time disretizations, which compel the program to reduce time increments depending on the quality of the formulation.

Table 4: Absolute CPU time (seg) spent by formulations for solving the problems

PROBLEM	1	2	3	4	5	6
FORMULATION						
$F-1$	28.00	6.000	13.00	71.00	13.00	1453.
$F-2$	33.00	8.000	15.00	450.0	16.00	1666.
$F-3$	32.00	8.000	14.00	76.00	13.00	1464.
$F-4$	31.00	8.000	16.00	434.0	21.00	1818.
$F-5$	51.00	10.00	24.00	138.0	34.00	3937.
$F-6$	42.00	8.000	22.00	136.0	22.00	3697.
$F-7$	42.00	8.000	22.00	133.0	23.00	3648.

A careful analysis of the results points out that the formulations in which element transmissivities are computed as an average of nodal values (5, 6 and 7) take significantly longer to run. This can be explained because those formulations compute the relative transmissivity (or permeability) function more than one time in each element, while other formulations (1,2,3,4) do such computation once per element.

Table 5: Ranking of formulations according to efficiency criteria.

Ranking	Formulation	SRT	Formulation	SRRT	Formulation	MTR
1	$F-1$	0.54152	$F-1$	6	$F-1$	0.10811
2	$F-3$	0.60471	$F-3$	17	$F-3$	0.14286
3	$F-2$	0.90914	$F-2$	22	$F-7$	0.20630
4	$F-6$	0.93820	$F-4$	25	$F-6$	0.20907
5	$F-7$	0.94038	$F-7$	27	$F-5$	0.23944
6	$F-4$	0.94204	$F-6$	31	$F-4$	0.30181
7	$F-5$	1.12401	$F-5$	40	$F-2$	0.31293

Rankings based on efficiency criteria are presented in table 5. We can observe that the comparison criteria are now more homogeneous than in table 3. In any case, formulation 1 appears to be the most efficient, because it obtains the best place independently on the criterium. Judging both accuracy and efficiency rankings, we can conclude that, for the range of the problems considered, that formulation (F_1) is the most robust.

SUMMARY

Two tests carried out on a varied type of problems, solved by means of the Finite Element Method, show that capacitive schemes perform best in terms of both accuracy and efficiency for expanding the storage term of the unsaturated flow equation. Computation of the conductive element's integral was also tested, considering various approximations for defining the mean conductivity or transmissivity value within the element. Results indicate that the conductivity corresponding to the mean pressure head is a very good approximation, especially when convergence rate is used for qualifying the schemes. Comparisons in terms of absolute CPU time and solution error points out that the quality of a particular numerical formulation is strongly dependent on the problem used for testing.

AKNOWLEDGMENTS. This work has been performed with funding from ENRESA (Spanish nuclear waste disposal company).

REFERENCES

Brooks, R.H. y Corey, A.T. (1964) "Hydraulic properties of porous media", Colorado State Univ., Hydrology papers no. 3, Fort Collins, Colorado, 27.

Carrera, J. Y Galarza, G. (1993) "Transporte de solutos: 2. Métodos de solución", La zona no saturada y la contaminación de las aguas subteráneas. Teoría medición y modelos. pp: 83:107 CIMNE, Barcelona.

Celia, M.A., Bouloutas, E.T. y Zarba R. (1990) "A general mass-conservative numerical solution for the unsaturated flow equation". Water resources res. vol. 26, No. 7, pp: 1483-1496.

Galarza, G., Carrera, J., Abriola, L., y Medina A. (1992) "Comparison of several formulations for the time discretization of non-linear flow equations", Proc. of the Ninth Int. Conf. in Computational Methodos in Water Resources. Computational Methods in Water Resources, IX, 1, 193-202.

Galarza, G. (1993) "Calibración Automática de Parámetros en Problemas no Lineales de Flujo y Transporte". Ph.D Dissertation. Politecnical University of Catalonia.

Haverkamp, R. y Vauclin, M. (1979) "A note on estimating finite difference interblock hydraulic conductivity values for transient unsaturated flow problems", Water Resour. Res. Vol. 15 No 1. pp: 181-187.

Knight, J.H. y Philip J.R. (1974) "Exact solutions in nonlinear diffusion", J. Eng. Math., 8, 219-227.

Milly, P.C.D. (1985) "A mass-conservative procedure for time stepping in models of unsaturated flow", Adv. Water resour., 8,32-36.

Neuman, S.P. (1973) "Saturated-unsaturated seepage by finite elements". ASCE, J. hydraul. Div. 99(HY12): 2233-2250.

Russo, D. (1992) "Upscaling of hydraulic conductivity in partially saturated heterogeneous porous formation". Water Resour. Res. Vol. 28 No. 2, 397-409.

Van Genuchten, M.Th. (1980) "A closed-form equation for predicting the hydraulic conductivity of unsaturated soils". Soil Sci. Am., J., 44, 892-989.

ON THE CHOICE OF MODEL FOR GROUNDWATER TRANSPORT

T. GIMSE[1] and C. TEGNANDER[2]
[1]Dept. of Informatics, Univ. of Oslo, P.O.Box 1080, 0316 Oslo, Norway
[2]Dept. of Mathematics, Univ. of Oslo, P.O.Box 1053, 0316 Oslo, Norway

In general, the equations modelling transport processes in porous media have one hyperbolic part (convection, Darcy flow), and one parabolic part (diffusion, capillary effects). A large class of numerical methods for the transport equations is based on splitting the problem in two (operator splitting), and solving the hyperbolic and the parabolic parts more or less independently. In that case, the solution of the hyperbolic part is usually the most critical, since hyperbolic equations are known to possess discontinuous solutions. In this paper we discuss some different approaches to the hyperbolic part of the transport equations, indicating that simplifying assumptions and formal manipulations may affect the solutions severely. A number of numerical tests are provided.

1. INTRODUCTION.

Mathematical and numerical modelling of contaminant transport in porous media is an area which has received an increasingly amount of attention lately. Basic to the mathematical modelling is to consider the conservation of mass and momentum (and if temperature effects are modelled; energy). In addition, different kinds of constituency relations may be introduced. Before any numerical computations are performed one also often make simplifying assumptions based on the knowledge of the physics involved. One important simplifying assumption that is widely acknowledged, is to assume that the air present in the unsaturated zone has infinite mobility. In other words, one assumes that the air is at atmospheric pressure, and that it moves away without interfering with water and/or contaminant. This assumption is essential for deriving Richards' equation. On the other hand, inspired by the two-phase litterature, one may include the air as a separate phase, affecting the flow. This gives the fractional flow approach. In this paper we will study the transition from that approach to the Richards' equation. We will also focus on the problems of manipulating the equations prior to numerical treatment. It is well-known that

257

A. Peters et al. (eds.), Computational Methods in Water Resources X, 257–264.

for hyperbolic equations this may alter the solution severely, and Celia and others [Celia *et al.*], have pointed out problems when applying difference schemes to different forms of the equations. The problem, however, is more fundamental, and one should be very careful on the choice of equations. Particularly if the hyperbolic part is dominating and treated separately, which is often the case using operator splitting techniques, e.g., [Espedal and Ewing] and [Douglas and Russell].

2. FLOW EQUATIONS.

We consider Darcy's law and the continuity equation [Hillel, p.202], and we will derive Richards' equation [Richards] and the fractional flow equation [Vauclin] from these by using different assumptions. We will see that Richards' equation is a special case of the fractional flow equation.

Notation. We denote t = time; z = vert. or horiz. coord.; g = grav. acceleration; P_j = pressure within the phase j, where $j = w$ for water and $j = a$ for air; $P_c = P_w - P_a$ is the capillary pressure; θ_j = volumetric content of the phase j; h_j = pressure head of the phase j, depending of θ_j; h_c = capillary pressure head; ρ_j = spec. mass of the phase j; v_j = velocity in Darcy's sense (spec. discharge); μ_j = dynamic viscosity of the phase j; and λ_j = mobility of the phase j with $\lambda_j = \frac{k_{rel,j}}{\mu_j}$ where $k_{rel,j}$ = rel. perm. of the phase j.

Remark. θ_w =volume of water/total volume, but it can also be defined as the product $\theta_w = nS_w$, where n is the porosity and S_w is the water saturation.

2.1. Richards' equation.

For the derivation of Richards' equation, we suppose that the air is at atmospheric pressure everywhere, that is, $P_a = 0$. Thus, we will just consider the water-phase ($j = w$) and we omit the indices on the variables.

Darcy's law gives:

$$(2.1) \qquad\qquad v = -K(h)\nabla H$$

where K is the hydraulic conductivity, depending on h, and $H = h + gz$ is the hydraulic head including both the suction and the gravitational component. ∇ is the gradient operator.

Denote $\partial\theta/\partial t$ by θ_t. Then the continuity equation is written:

$$(2.2) \qquad\qquad -\nabla v = \theta_t$$

We apply the continuity equation to Darcy's law to get:

$$(2.3) \qquad\qquad \nabla\big(K(h)\nabla(h + gz)\big) = \theta_t$$

Since relations between θ and h are provided (see [Hillel, p.205]), we may rewrite (supposing everything is smooth!) (2.3) to have:

$$(2.4) \qquad C(h)h_t - \nabla(K(h)\nabla h) - \frac{\partial K(h)}{\partial z}g = 0$$

where $C(h) = d\theta/dh$.

Or we can write (2.4) as a θ-based equation with $D(\theta) = K(\theta)/C(\theta)$:

$$(2.5) \qquad \theta_t - \nabla(D(\theta)\nabla\theta) - \frac{\partial K(\theta)}{\partial z}g = 0$$

Remark. Since $P_a = 0$ we have the relation $h = h_c = \int \frac{dP_c}{\rho_w}$. Then Darcy's law (2.1) can be written:

$$v = -\frac{K(h)}{\rho_w}(\nabla P_c + \rho_w g)$$

where ρ_w is constant. Define the mobility by

$$(2.6) \qquad \frac{K(h)}{\rho_w} = \lambda_w.$$

We will use this notation when we consider the case $P_a \neq 0$.

2.2. Fractional flow equation.

Suppose now that $P_a \neq 0$, i.e., $P_w = P_c + P_a$. Thus, Darcy's law and the continuity equation give rise to a system of two equations, one for each phase, and we can write Darcy's law for the phase velocity (of the phase j) on the following form (see [Vauclin]):

$$(2.7) \qquad v_i = -\lambda_i(\nabla P_i + \rho_i g).$$

In the rest of this paper we assume that the ρ_i-s are constant (incompressible flow).

Then we can write the continuity equation on the form

$$(2.8) \qquad (\theta_i)_t + \nabla v_i = 0$$

Substract v_a from v_w by using (2.7) twice:

$$(2.9) \qquad \frac{v_w}{\lambda_w} - \frac{v_a}{\lambda_a} = -(\nabla P_c + \Delta\rho g)$$

where $\Delta\rho = \rho_w - \rho_a$ is the difference between the specific masses in the two phases of water and air. Define the total velocity $v = v_w + v_a$, and consider the flow function

$F_j = v_i/v$ with $F_w + F_a = 1$. Replace v_w on the left hand side in (2.9) with vF_w and denote $\theta_w := \theta$. This gives:

$$F_w = f_w - \frac{E_w}{v}\nabla\theta - \frac{G_w}{v}$$

where $f_w = \frac{\lambda_w}{\lambda_a + \lambda_w}$, $E_w = -\lambda_w(1 - f_w)\frac{dP_c}{d\theta}$, $G_w = \Delta\rho g\lambda_w(1 - f_w)$.

Finally replace v_w in (2.7) with the expression for F_w multiplied with v. This gives us the fractional flow equation:

$$\theta_t + \nabla(vf_w - E_w\nabla\theta + G_w) = 0$$

In the vertical direction:

$$(2.10) \qquad \theta_t + \frac{\partial}{\partial z}(vf_w - E_w\theta_z + G_w) = 0$$

3. MODEL COMPARISON.

In this section we will study the behaviour of the fractional flow model as it approaches the Richards model. We do this by using explicit expressions for the mobilities and the fractional flow functions, and then let the air mobility tend to infinity. Thus, we start by considering (2.10), letting $\lambda_a \to \infty$. Then $f_w \to 0$, $E_w \to -\lambda_w\frac{dP_c}{d\theta}$, and $G_w \to \Delta\rho g\lambda_w$. Using (2.6) and identifying $D(\theta)$ we see that (2.10) $\to$ (2.5) as $\lambda_a \to \infty$ and the air density is ignored (one dimension).

We are interested in seeing exactly how this limit behaves, and for this purpose we ignore the diffusive term and consider the hyperbolic part of the equations. Thus, we are interested in

$$\theta_t + \big(vf_w + \Delta\rho g\lambda_w(1 - f_w)\big)_z = 0$$

which we may rewrite as

$$\theta_t + \big(f_w(v + \gamma\lambda_a)\big)_z = 0.$$

To this end let $\lambda_w = \theta^2$ and $\lambda_a = \lambda(1 - \theta)^2$, i.e. the flow function is

$$(3.1) \qquad \frac{\theta^2}{\theta^2 + \lambda(1 - \theta)^2}(v + \gamma\lambda(1 - \theta)^2)$$

We assume that v and γ are constants. To study the difference between the models, we consider a Riemann problem, that is, an initial value problem defined by two constant states separated by a discontinuity; $\theta = \theta_L$ for $x < x_0$ and $\theta = \theta_R$ for $x > x_0$. We pick θ_L and θ_R such that the resulting solution is a shock, and compute the shock speed by the Rankine-Hugoniot condition, e.g. [LeVeque]. Thus, for this particular case (3.1) we would have:

$$(3.2) \qquad s = \frac{\frac{\theta_L^2}{\theta_L^2 + \lambda(1 - \theta_L)^2}(v + \gamma\lambda(1 - \theta_L)^2) - \frac{\theta_R^2}{\theta_R^2 + \lambda(1 - \theta_R)^2}(v + \gamma\lambda(1 - \theta_R)^2)}{\theta_L - \theta_R}.$$

We have examined this expression for a large number of values for θ_L and θ_R, and the behaviour depends on both the relative values of v and γ, as well as the parameter λ. The general feature, however, is that the shock speed is closely related to the value of λ. It is easy to see that as $\lambda \to \infty$, (3.2) approaches:

$$(3.3) \qquad\qquad s = \gamma(\theta_L + \theta_R),$$

in agreement with the corresponding shock speed from (2.5). Below, in Figs. 1-3 , we have solved the Riemann problem with $\theta_L = 0.8$ and $\theta_R = 0.2$ at $x_0 = 0.5$, for $t = 5$, where $\gamma = 0.01$ and $v = 0.025$. We have used the Lax-Friedrichs scheme, with 10 000 grid points in the interval $[0, 1]$. Considering only $\lambda \geq 1$, the shock speeds are in the range 0.03—0.01.

We observe that for $\lambda = 1$ (Fig.1) the shock is replaced by a Buckley-Leverett type composite wave. For $\lambda = 10$ (Fig.2) the shock moves much faster (with a speed about 0.028) than it does for large values of λ, where the speed converges to 0.01 (In (Fig.3) we have plotted the Riemann-problem with $\lambda = 10000$) as we saw in (3.3).

This simple example shows that Richards' equation cannot always be used instead of fractional flow, so one has to be careful when choosing models for transport in porous media. For realistic data [Rouse] the viscosity ratio μ_w/μ_a between water and air may be of order 10^2, and since $\lambda_j = \frac{k_{rel,j}}{\mu_j}$ the same applies to λ. Thus, we may still be in the range where the two approaches differs significantly.

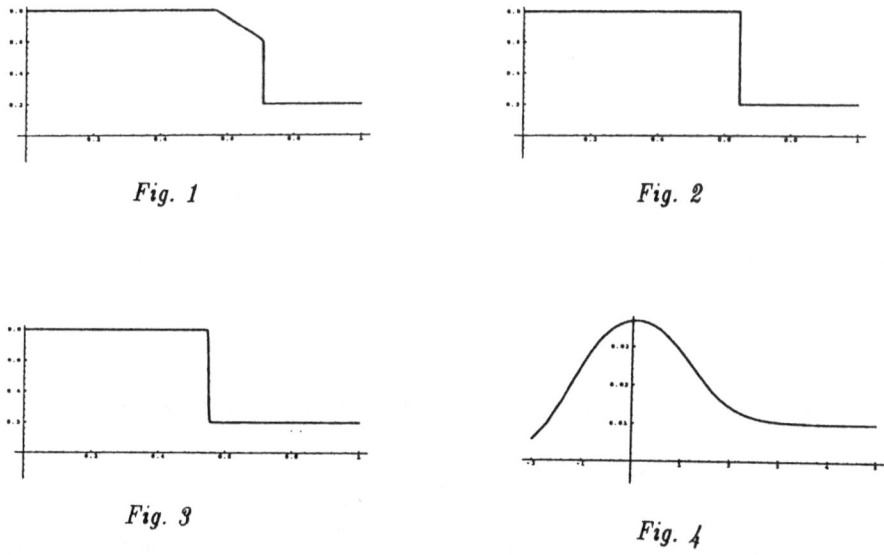

Fig. 1   Fig. 2

Fig. 3

Fig. 4

Finally we remark that for $0 \leq \lambda \leq 1$, the two phases interchange roles, and the shock speed is thus increasing from 0 (for $\lambda = 0$) to its maximum that lies near $\lambda = 1$. We will not discuss the value of the maximum speed nor for which λ it obtains this value since this will depend of the value of the constants γ and v. In (Fig.4) we have plotted the Rankine Hugoniot condition s as a function of λ when we keep the constants $v = 1/40$ and $\gamma = 1/100$. The x-axis expresses 10^x instead of x to underline what happens for $x \leq 1$. Observe that γ can be described as the gravitational part of the equation and v is the hydraulic gradient. If the ratio between them changes, the function $s(\lambda)$ behaves differently [Gimse and Tegnander].

4. DIFFERENT FORMULATIONS.

Finally we will briefly comment on the difference between the two formulations (2.4) and (2.5). We will refer to these as the $h-$based and the $\theta-$based versions of Richards' equation, respectively. Obviously, when the solution is smooth, these two formulations are formally equivalent. On the other hand, for numerical purposes one has to be careful, which is already noted by several authors [Celia *et al.*]. Our point of view is slightly different.

Over the years, a variety of operator splitting techniques and other methods based on the solution of hyperbolic equations, have been applied to transport modelling [Espedal and Ewing], [Douglas and Russell] and [LeVeque]. It is well known that formal manipulations on hyperbolic equations is a risky business. We will demonstrate this in general by showing that the two formulations obey different Rankine-Hugoniot conditions. We consider a moving shock with left and right states θ_L and θ_R respectively, and similar states $h_L = h_L(\theta_L)$, $h_R = h_R(\theta_R)$ in the h-based version.

Ignoring diffusion, (2.4) and (2.5) in one dimension can be written as respectively

$$(4.1) \qquad\qquad C(h)h_t - \frac{\partial K(h)}{\partial z}g = 0$$

where $C(h) = d\theta(h)/dh$, and

$$(4.2) \qquad\qquad \theta_t - \frac{\partial K(\theta)}{\partial z}g = 0.$$

We choose $K(\theta) = \theta^m$ as an empirical formula [Hillel, p.200], and let for simplicity $g = 1$. Then (4.2) can be written as

$$(4.3) \qquad\qquad \theta_t + (\theta^m)_x = 0$$

The Rankine-Hugoniot condition expresses then the shock-speed s_θ as

$$(4.4) \qquad\qquad s_\theta = \sum_{i+j=m-1} \theta_L^i \theta_R^j$$

For the functional dependency $\theta(h)$ several models are proposed [Hillel]. We choose a simple relation, and let $\theta(h) = (\frac{h}{a})^{-1/b}$.

Thus $C(h) = -\frac{1}{ba}(\frac{h}{a})^{-\frac{1+b}{b}}$, and (4.1) can be expressed as

$$(4.5) \qquad h_t + ab(\frac{h}{a})^{\frac{1+b}{b}}((\frac{h}{a})^{-m/b})_z = 0$$

or we can write the whole equation on conservative form effectuing derivation and reintegration of (4.5):

$$(4.5) \qquad h_t + \frac{ma^{\frac{1}{b(m-1)}}b}{1-m+b}(h^{\frac{1+b-m}{b}})_z = 0$$

This equation has the following Rankine-Hugoniot condition:

$$s_h = \frac{ma^{\frac{1}{b(m+1)}}b}{1+b-m}\frac{h_L^{\frac{1+b-m}{b}} - h_R^{\frac{1+b-m}{b}}}{h_L - h_R}$$

Since $h = a\theta^{-b}$ we can write the speed s_h as a function of θ_L and θ_R which gives us:

$$(4.7) \qquad s_h = \frac{mb}{1+b-m}\theta_L\theta_R \frac{\sum_{i+j=b-m}\theta_L^i\theta_R^j}{\sum_{i+j=b+1-m}\theta_L^i\theta_R^j}$$

The expressions for s_θ (4.4) and s_h (4.7) are irreducible. Thus the Rankine-Hugoniot conditions can only be equal if $b = -1$, i.e. the relation between θ and h is linear, which corresponds to $C =$ constant.

We have showed that the two formulations of Richards' equation are equivalent when the solution has discontinuities. Numerically one problem is immediately posed: How to implement the h-based version of the Richards' equation when it is not on conservative form. Celia [Celia et al.] has earlier discussed this problem in some detail. We will end this section by computing two different approximations using the Lax-Friedrichs' scheme.

We now use Lax-Friedrichs' difference scheme as we did in the proceding section to solve a Riemann problem of (4.3). Then we will solve the same problem of (4.4) by using two different approximations of $C(\theta)$ in the corresponding Lax-Friedrichs' scheme.

For simplicity let $K(h) = h^2$ and $\theta(h) = h^{-2}$. Then $C(h) = -2h^{-3}$ and (4.1) is on the form

$$h_t + \frac{1}{2h^{-3}}(h^2)_z = 0$$

Let $h_L = 0.8, h_R = 0.2$ be the left and right states to the Riemann problem in $x_0 = 0.5$ with $t = 0.2$. We still consider 10 000 grid points at the intervall $[0, 1]$.

Let $u_i^n = h(n\Delta t, i\Delta x)$. In the first case we consider a scheme where $\theta(h)$ is treated linearly, i.e. C being almost a constant:

$$(4.8) \qquad u_i^{n+1} = 0.5(u_{i+1}^n + u_{i-1}^n) - 0.5\frac{\Delta x}{\Delta t}(\frac{K(u_{i+1}^n)}{C(u_{i+1}^n)} - \frac{K(u_{i-1}^n)}{C(u_{i-1}^n)})$$

As to this second scheme we treat C more generally by taking an average of two values outside the derivation:

$$(4.9) \quad u_i^{n+1} = 0.5(u_{i+1}^n + u_{i-1}^n) - 0.5\frac{\Delta x}{\Delta t}\left(\frac{2}{C(u_{i+1}^n) + C(u_{i-1}^n)}\right)(K(u_{i+1}^n) - K(u_{i-1}^n))$$

In both cases (4.8) and (4.9) we see that the approximated solutions are not described by a shock, (Fig.5) and (Fig.6) respectively, and they also behave very differently. This example indicates the importance and difficulties of choosing a good numerical approximation to the non-conservative h-based form.

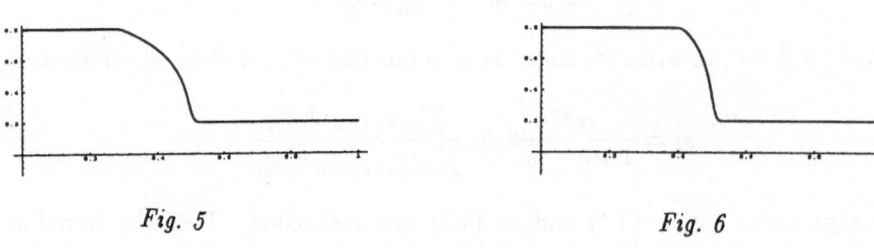

Fig. 5   Fig. 6

Acknowledgement.

The authors are supported by the Norwegian Research Council.

REFERENCES

Celia, Bouloutas, Zarba, Water Resources Research 26 (1990), no. 7, 1483–1496.
Douglas and Russell, SIAM J. Numer. Anal. 19 (1982), 871–885.
Espedal and Ewing, Comp. Meth. in Appl. Mech. Eng. 64 (1987), 113–135.
Gimse and Tegnander, *Work in progress*.
Hillel, Fundamentals of Soils Physics (1980).
LeVeque, Numerical Methods for Conservation Laws (1990).
Richards, Physics 1 (1931), 318–333.
Rouse, Elementary Mechanics of Fluids (1946).
Vauclin, In: Unsaturated Flow in Hydrologic Modeling, Theory and practice (1989), 53–91.

SOLUTION OF FLOW AND TRANSPORT PROBLEMS USING A COLLOCATION DISCRETIZATION OF A DOMAIN DECOMPOSITION ALGORITHM

Joseph Guarnaccia*, Ismael Herrera** and George Pinder*
*University of Vermont, 109 Votey Bldg., Burlington, VT 05405
**Instituto de Geofisica, UNAM, Apdo Postal 22-582, 14000 Mexico, D. F.

INTRODUCTION

We consider herein the two-dimensional simulation of fluid flow and contaminant transport in porous media using the collocation finite element method in conjunction with the domain decomposition conjugate gradient (DD-CG) algorithm proposed by Herrera et al. (1994). The motivation for using domain decomposition is that it allows us to take advantage of parallel processing and to treat subdomains differently from a computational point of view.

As detailed in Herrera et al. (1994), the subdomains arising from the domain decomposition are linked together using a conjugate gradient iterative algorithm to enforce continuity of the normal flux across the interfaces separating subdomains. An important issue to be discussed below is the procedure used to transpose the theory, derived for the continuous variable case, into an algorithm applicable to the discrete system arising from the application of the collocation finite element method.

In this paper we will discuss the collocation discretization and the use of the Hermite cubic basis to represent the dependent variables. We then detail the procedure used to calculate the variables on the interfaces required for the conjugate gradient algorithm. Finally, we present some convergence results for a model flow and transport problem.

MODEL PROBLEM

Consider for this discussion the following flow and transport model. The first equation to be solved is the incompressible groundwater flow equation, written as:

$$Lh(\mathbf{x}) = -\nabla \cdot \left[\underset{\sim}{k}(\mathbf{x}) \cdot \nabla h(\mathbf{x}) \right] = Q(\mathbf{x}), \quad [\mathbf{x}] \in \Omega \tag{1}$$

where $\underset{\sim}{k}(\mathbf{x})$ is the hydraulic conductivity tensor, which for this discussion is assumed to be diagonal, $h(\mathbf{x})$ is the fluid pressure head, and $Q(\mathbf{x})$ represents a point source or sink function. If, given proper boundary conditions, equation (1) is solved for $h(\mathbf{x})$, we can employ Darcy's law to evaluate the fluid velocity vector, $\mathbf{v}(\mathbf{x})$. The fluid velocity in turn is used to solve the second equation describing contaminant transport:

$$\varepsilon \frac{\partial c(\mathbf{x},t)}{\partial t} + \mathbf{v}(\mathbf{x}) \cdot \nabla c(\mathbf{x},t) - \nabla \cdot \left[\underset{\sim}{D}(\mathbf{x}) \cdot \nabla c(\mathbf{x},t) \right] = Q(\mathbf{x}) \left[c_Q - c(\mathbf{x},t) \right] \tag{2}$$

where ε is the porosity, $c(\mathbf{x},t)$ is the contaminant concentration, $\underset{\sim}{D}(\mathbf{x})$ is the dispersion-diffusion tensor which has components defined as:

$$D_{ij}(\mathbf{x}) = a_T |\mathbf{v}(\mathbf{x})| \delta_{ij} + (a_L - a_T) \frac{v_i(\mathbf{x}) \, v_j(\mathbf{x})}{|\mathbf{v}(\mathbf{x})|} + D_m \delta_{ij} \tag{3}$$

where i and j symbolize spatial direction, a_L and a_T are the longitudinal and transverse dispersivities respectively, D_m is the coefficient of molecular diffusion, $v_i(\mathbf{x})$ and $v_j(\mathbf{x})$ are the components of the velocity vector, and $|\mathbf{v}(\mathbf{x})|$ is the velocity magnitude. In addition, $Q(\mathbf{x})$ in (1) represents point sources and sinks of contaminant and c_Q is the concentration associated with Q.

A. Peters et al. (eds.), Computational Methods in Water Resources X, 265–272.

COLLOCATION FINITE ELEMENTS

The domain is discretized into a finite number of rectangular elements with nodes located at element boundary intersections (figure 1). Since the dependent variables, h(x) and c(x,t), require continuous second order derivatives, the C^1 continuous Hermite cubic functions are chosen as the basis. The bicubic Hermite approximation of a general function, $u(x,y,t)$, is:

$$u(x,y,t) \approx \widehat{u}(x,y,t) = \sum_{i=1}^{I} \sum_{j=1}^{J} \left[\begin{array}{c} U_{ij}(t)\, \phi_{ij}^{00}(x,y) + \partial U/\partial x_{ij}(t)\, \phi_{ij}^{10}(x,y) \\ + \partial U/\partial y_{ij}(t)\, \phi_{ij}^{01}(x,y) + \partial^2 U/\partial x \partial y_{ij}(t)\, \phi_{ij}^{11}(x,y) \end{array} \right] \quad (4)$$

where $\widehat{u}(x,y,t)$ is the discretized approximation, the grid (figure 1) has I nodes in the x-direction, J nodes in the y-direction, (i,j) is the nodal index at which is defined four time dependent (if applicable) undetermined coefficients (the functional value, the two first order derivatives and the cross derivative), and four space dependent Hermite bicubic basis functions (ϕ^{00}, ϕ^{10}, ϕ^{01} and ϕ^{11}, see Lapidus and Pinder, 1982, page 83 for functional form). These approximations are then substituted into their respective balance equations (1) and (2) yielding two equations in 2[4(IJ)] nodal unknowns (2 dependent variables, 4 undetermined coefficients per node, and IJ nodes).

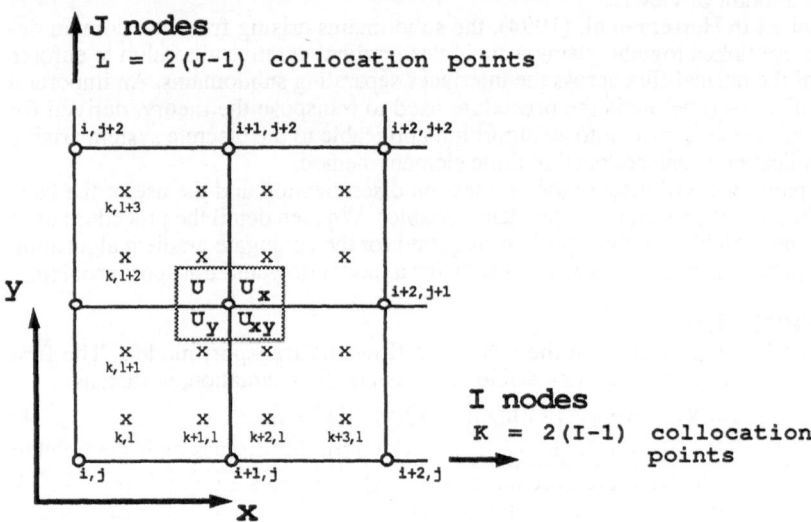

Figure 1 - Depiction of a collocation finite element mesh showing nodal locations (open circles with index i,j), approximate location of collocation points (x's with index k,l), and the degrees of freedom associated with each node (U's).

The nodal unknowns are determined by employing the collocation method: a method of weighted residuals where the weighting function is the displaced Dirac delta function (see Lapidas and Pinder [1982] for a detailed discussion). As a result, the residual errors incurred by using approximated dependent variables in the governing equations are driven to zero at specified points in the domain, called collocation points. If the Gauss quadrature points are chosen as the collocation points, the method is called orthogonal collocation. Using the general function $\widehat{u}(x,y,t)$ as a surrogate for each approximated dependent variable, a system of linear algebraic equations is generated by imposing the interpolation constraints:

$$\widehat{u}\ (x_k,y_l,t\ ) = u\ (x_k,y_l,t\ ), \ k = 1,\ 2..., \ 2(I\text{-}1); \ l = 1,\ 2,..., \ 2(J\text{-}1), \tag{5}$$

where (x_k,y_l) are the locations of the collocation points (see figure 1). This results in four collocation points per element, and yields $2[4(I\text{-}1)(J\text{-}1)]$ equations (2 balance equations each written at $4(I\text{-}1)$ $(J\text{-}1)$ collocation points). Note that equation generation requires no formal integrations, and therefore, collocation is computationally analogous to the finite difference method in that equations are written at points in the domain.

To augment the collocation equations such that the number of equations equals the number of unknowns requires that a sufficient number of boundary conditions be specified. Recall that we have four undetermined coefficients at each node. In order that the number of unknowns equal the number of equations, boundary conditions must be imposed such that two of the four undetermined coefficients be specified at midside nodes and three be specified at corner nodes. Therefore, for this discussion the following rule holds: a Dirichlet condition requires specification of both the function and the derivative tangent to the boundary, and a Neumann condition requires specification of both the normal and the cross derivatives. For example, along x-oriented boundary nodes, Dirichlet data includes U_{ij} and $\partial U/\partial x_{ij}$, while Neumann data includes $\partial U/\partial y_{ij}$ and $\partial^2 U/\partial x \partial y_{ij}$.

All other variables requiring spatial representation are defined nodally and interpolated into the element using bilinear Lagrange polynomials. This interpolation is shown for a general function $g(x,y)$ as:

$$g\ (x,y) \approx \widehat{g}\ (x,y) = \sum_{i=1}^{I} \sum_{j=1}^{J} G_{ij}\ \Gamma_{ij}(x,y) \tag{6}$$

where $\widehat{g}\ (x,y)$ is the approximated function, G_{ij} is the coefficient defined at node (i,j) and represents the value of the function at the node, and $\Gamma_{ij}(x,y)$ is the space dependent bilinear Lagrange basis defined at node (i,j) (see Lapidus and Pinder [1982] page 83 for functional form). The combination of using a continuous permeability field and a Hermite interpolated head variable has the advantage of yielding a continuous velocity field to be used in the transport equation.

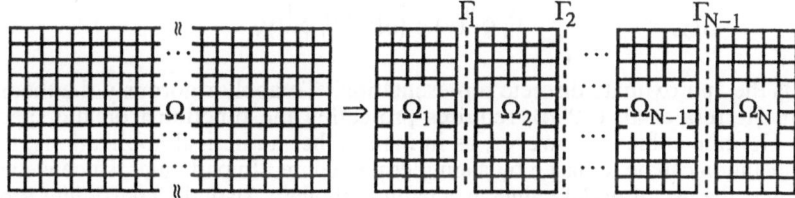

Figure 2: Definition sketch of the domain decomposition algorithm. The domain is partitioned into N subdomains such that, $\Omega = \Omega_1 \cup \Omega_2 ... \cup \Omega_{N-1} \cup \Omega_N$, separated by interfaces Γ_i, $i = 1,2,...,N-1$. Note that the partitioning cuts Ω in one dimension only, and that horizontal partitioning is equally feasible.

APPLICATION OF THE DD-CG ALGORITHM

In this section we present a procedural summary of the implementation of the DD-CG algorithm proposed by Herrera et al. (1994) using the collocation discretization presented above. Consider the domain decomposition shown in figure (2). For this discussion we require that the partitioning cuts Ω in one direction only so that the interfaces Γ_i have no nodes in common. This is not a necessary requirement, however, it allows each interface problem to be computed independently.

Let us present the solution of the flow equation (1) for illustrative purposes and present later the specific features of the transport equation (2) which require special attention. As described in Herrera et al. (1994), we seek the solution to (1) as the linear combination of an inhomogeneous and a homogeneous problem. Using an initial guess for the solution of equation (1) as Dirichlet data for interface nodes, the solution to the inhomogeneous problem (equation 2.4 in Herrera et al., 1994) is obtained in each independent subdomain using the collocation method and a direct solver for each system of equations. This calculation results in a solution which has a continuous functional value but which in general has a jump in the mass flux across the interfaces.

The next step is to solve the homogeneous problem (equation 3.1 in Herrera et al., 1994) requiring that its solution has a jump in mass flux across the interfaces which is equal and opposite to that obtained for the inhomogeneous problem. This is referred to as the jump condition in Herrera et al. (1994, equation 2.6). From this we define the jump in mass flux across an interface Γ_i:

$$f(\mathbf{x}) = - \underline{k}(\mathbf{x}) \, [\![\partial h / \partial n]\!] \tag{7}$$

where the double bracket denotes the jump operator: $[\![\bullet]\!]_\Gamma \equiv (\bullet)_{\Gamma^+} - (\bullet)_{\Gamma^-}$, and the derivative is with respect to the positive normal (n) direction. This "boundary value problem with prescribed jumps on Γ" will be solved using the CG procedure proposed in Herrera et al. (1994).

During the CG procedure, equation (7) is evaluated using the jump in the Neumann data after the solution of each set of subdomain boundary value problems. $f(\mathbf{x})$ then becomes the Dirichlet data on each Γ_i for the solution of each subdomain homogeneous problem and is also used for the evaluation of the weights α and β (defined in Herrera et al. [1994]). Because the dependent variables are represented by the space spanned by Hermite cubics, the Dirichlet data and thus $f(\mathbf{x})$ must also be represented by the same space. This Dirichlet data can be thought of as the one-dimensional Hermite representation of $f(\mathbf{x})$ along the interface Γ_i. For example, if the interface is along the y-axis, the discrete approximation of $f(y)$ is:

$$f(y) \approx \hat{f}(y) = \sum_{j=1}^{J} \hat{F}_j \, \phi_j^{\,0}(y) + \left(\partial \hat{F} / \partial y \right)_j \phi_j^{\,1}(y) \tag{8}$$

where $\hat{f}(y)$ is the approximate discrete representation. To apply rigorously the theory presented in Herrera et al. (1994), requires projecting the function f on the space spanned by Hermite cubics. However, as in many other numerical applications, the projection operation can be replaced by a simpler procedure. In particular, since we are using Hermites, the use of nodal values is a natural choice. Also, the coefficient $\mathbf{k}$ on Γ can be approximated in different manners. Other options can be considered, and in the numerical experiments presented here, three different procedures for calculating the Hermite coefficients in (8) have been applied.

Method 1: The Hermite coefficients are calculated using nodal values of $\underline{k}(\mathbf{x})$ and the coefficients of the Hermite representation for $h(\mathbf{x})$. That is, if for example, the x-direction is normal to Γ and the y-direction is tangential, then the coefficients in (8) are:

$$\hat{F}_j = - (k_{xx} \, [\![\partial h / \partial x]\!])_j \tag{9a}$$

$$\left(\partial \hat{F} / \partial y \right)_j = - \{ (\partial k_{xx} / \partial y \, [\![\partial h / \partial x]\!])_j + (k_{xx} \, [\![\partial^2 h / \partial x \partial y]\!])_j \} \tag{9b}$$

where in equation (9b) we have used the chain rule. Note that since $\underline{k}(\mathbf{x})$ is represented by linear Lagrange along Γ, the term $(\partial k_{xx}/\partial y)$ at the node can be discontinuous. If in the heterogeneous case $(\partial k_{xx}/\partial y)$ at node j is discontinuous then the value is approximated by the average of the values defined on either side of the node. We also note that even though $h(x)$ is represented by Hermites, the product $k_{xx} [\![\partial h/\partial x]\!]$, is not. Therefore, this method of calculating the coefficients for $f(x)$ is not optimal from a theoretical point of view.

Method 2: Here we apply a more rigorous approach where we project the function $f(x)$ on the space spanned by Hermite polynomials as discussed in Herrera et al.(1994). The coefficients in equation (8) are calculated by solving the following system of equations:

$$\left\{ \begin{aligned} \langle \hat{f}, \phi_j^0 \rangle &= \langle f, \phi_j^0 \rangle \\ \langle \hat{f}, \phi_j^1 \rangle &= \langle f, \phi_j^1 \rangle \end{aligned} \right\} ; \qquad j = 1, 2,..., J \tag{10}$$

where $\hat{f}$ is defined by equation (8), the operation described by $\langle \bullet, \bullet \rangle$ is for example, $\langle g, h \rangle = \int_{\Gamma} g(x) h(x)\, dx$, and the integrations required in (10) are evaluated using two point Gauss quadrature. In addition, evaluation of f on the right hand side of (10) is achieved by evaluating equation (7) at the Gauss points. Note that nodal values of f and its derivative need not be evaluated. The system of equations defined by (10) is of order $2*(J-1)$, and the matrix is symmetric with a half bandwidth of four.

Method 3: Here we approximate the permeability on Γ by a constant. Actually in applying the CG procedure the value of the constant is irrelevant and $f(x)$ becomes:

$$f(x) = - [\![\partial h/\partial n]\!] \tag{11}$$

which has the advantage that $f(x)$ is in the space spanned by Hermite cubics. That is, the coefficients for $f(x)$ are calculated directly from the coefficients of the Hermite representation for $h(x)$, where for example if the x-direction is normal to Γ and the y-direction is tangential, then the coefficients in equation (8) are:

$$\hat{F}_j = [\![\partial h/\partial x]\!]_j \text{ and } (\partial \hat{F}/\partial y)_j = [\![\partial^2 h/\partial x \partial y]\!]_j \tag{12}$$

Finally, with respect to the evaluation of the scalar weights α and β (in Herrera et al., 1994), given the Hermite representation for $f(x)$, the α weight is for example:

$$\alpha^k = \sum_{i=1}^{N-1} \langle \hat{R}^k, \hat{R}^k \rangle_{\Gamma_i} / \langle \hat{P}^k, \widehat{AP}^k \rangle_{\Gamma_i} \tag{13}$$

where the integrations are computed using two point Gauss quadrature, and β is defined in an analogous way. Note that the integrations in equations (10) and (13) require four point Gauss quadrature to be exact, however, computational experiments show that a two point scheme provides the best convergence results. The reason for this result is currently being investigated.

RESULTS FOR THE FLOW PROBLEM

Using a series model flow problems, we will compare the convergence attributes of the DD-CG algorithm using the three methods of evaluating equation (8).

Let us now consider an example from the oil industry, the repeated five-spot well field pattern. Borrowing a problem from Russell et al. (1986), a square domain, 304.8m on edge, is discretized into 48 by 48 6.35m elements. As shown in figure 3, a source well is located in the lower left corner of the mesh, and a sink well is located in the upper right corner; they each pump $18.6 m^3 d^{-1} m^{-1}$. No flow boundary conditions prevail. The permeability field is in general isotropic and heterogeneous. It is generated in the following manner. Each dimension is divided into seven equal sections of seven nodes each. A patch is defined by the inner product of each dimension section. As shown in figure (3a), to each of the resulting 49 patches is assigned a constant isotropic k value defined by: $k = 10^s k_0$, where s is a random number, $0 < s < 1$, and for these examples $k_0 = 8.6 \times 10^{-7}$ m/d. Thus, the maximum variation in k between patches is an order of magnitude.

For the results shown, the grid is partitioned into four 48 by 12 element subdomains. Results were obtained for both vertical (constant in x) and horizontal (constant in y) partitioning. The location of the partitions is shown in figure 3. For the homogeneous case and for four different realizations of the k field, we note the number of iterations, m, for which the convergence condition, $\| R^m \|_2 \le 0.001 \| R^0 \|_2$, is satisfied. Table 1 summarizes the results.

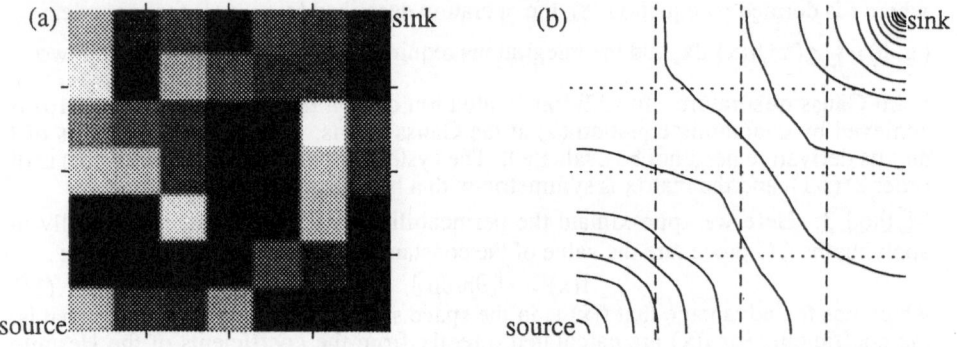

Figure 3: a) plot of the domain showing permeability patch distribution for one realization (light patches reflect higher permeabilities), and tic marks on the boundary representing domain partition locations by connecting marks on opposite faces. b) solution of the head field for the permeability distribution in (a), vertical partition locations are shown by dashed lines and horizontal by dotted lines.

Table 1- Iterations required to solve of the flow problem using four subdomains.

k field	method (1) horizontal partitions	vertical partitions	method (2) horizontal partitions	vertical partitions	method (3) horizontal partitions	vertical partitions
homogeneous	20	20	20	20	20	20
realization 1	82	82	65	69	52	39
realization 2	85	79	69	66	58	54
realization 3	92	97	65	77	58	53
realization 4	88	84	69	71	60	54

Upon review of Table 1 the following conclusions can be drawn. Use of the projection operation [method(2)] substantially improves convergence relative to method (1) and appears to be a cost effective procedure. In most cases that we have treated method (3) provided the fastest convergence. However, in some cases method (3) failed to converge, while the convergence of method (2) is always granted. Thus, a reasonable strategy is to use method (3) and the to use method (2) only when necessary.

SOLUTION OF CONTAMINANT TRANSPORT BY DD-CG

If we discretize equation (2) in time with an implicit backward difference, it can be rewritten in the form:

$$[\varepsilon/\Delta t + Q(\mathbf{x})]\, c^{n+1}(\mathbf{x}) + \mathbf{v}(\mathbf{x})\, \nabla c^{n+1}(\mathbf{x}) - \nabla \cdot \left[\underset{\approx}{D}(\mathbf{x})\, \nabla c^{n+1}(\mathbf{x})\right] = g(\mathbf{x}) \qquad (11)$$

where $\Delta t = (t^{n+1} - t^n)$, n is the time level, $c^{n+1}(\mathbf{x}) = c(\mathbf{x}, t^{n+1})$ and $g(\mathbf{x}) = (\varepsilon/\Delta t)\, c^n(\mathbf{x}) + c_Q\, Q(\mathbf{x})$ is the forcing function. Due to the presence of the first order term this equation is non-symmetric, and the method proposed by Herrera et al. (1994) is not directly applicable. The modifications required are being developed at present. Thus, in this paper we apply a procedure that was quite successful in our numerical experiments. It consists of applying the CG algorithm as discussed in Herrera et al. (1994), realizing that the algorithm will not provide optimal convergence properties, nor will convergence be guaranteed. Also, the definition of the jump condition on Γ_i is modified to be:

$$f(\mathbf{x}) = [\![\partial c/\partial n]\!] \qquad (12)$$

which is in the space spanned by Hermite cubics. Note that the advective part of the flux does not appear in (12) because by definition, $[\![c]\!] = 0$.

Results

We test the application of the DD-CG procedure on the transport equation (11) using the results from the flow problem to define the velocity field. Additional problem definition includes: $\varepsilon = 0.1$, $a_L = 10\, a_T$ (a_L is a variable for analysis below), $D_m = 0$, and c_Q at the source $= 1.0$. We note that with respect to finite element discretizations there is a constraint on time step size, determined from the grid Courant number (Co), generally Co ≤ 1 (Huyakorn and Pinder, 1983). Since mass is transported a relatively short distance per time step, we choose for these examples an extreme domain decomposition which cuts the domain into one element strips. This results in 48 subdomains of dimension 48 by 1 elements. An example of the solutions for the homogeneous case and a heterogeneous case are presented in figure 4.

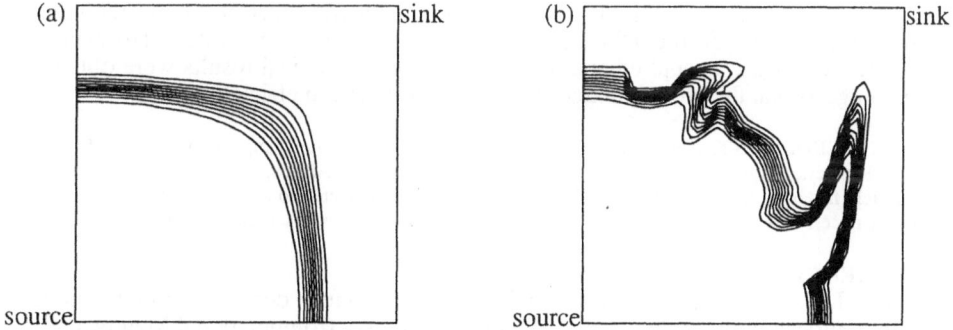

Figure 4: Contour plots for the solution of contaminant transport at 250 days ($\Delta t = 1$ day, $a_L = 0.635$m) for (a) the homogeneous k-field and (b) the heterogeneous k-field shown in figure 3a.

Figure 5 presents two computational experiments. In figure 5a, using the homogeneous flow problem as an example, we show the average number of iterations to convergence (using the same convergence criterion as for the flow problem) per Δt as a function of grid Peclet number, Pe=(grid spacing)/a_L. Observe that as dispersion decreases, the number of iterations also decreases. This is because the coupling between subdomains becomes weaker as dispersion is reduced.

In figure 5b, we compare results that were obtained applying this method to the transport problems shown in figure 4. It can be seen that the number of iterations required for the heterogeneous problem is slightly larger than for the homogeneous problem, but the method is still highly efficient. Also note in that figure, that the number of iterations increases in a more or less linear fashion with Δt.

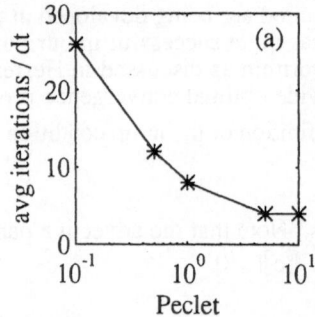 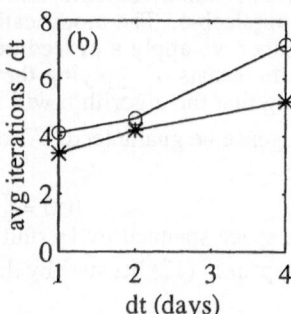

Figure 5 - (a) the relationship between Pe and iteration count for the homogeneous problem ($\Delta t=1$d). (b) the relationship between time step size and iteration count for the solution to the problems in figure 4 (Pe = 10): the homogeneous problem (*'s) and the heterogeneous problem (o's).

SUMMARY

In this paper we applied the DD-CG algorithm proposed by Herrera et al. (1994) using a collocation discretization of the equations describing groundwater flow and contaminant transport. For the solution of the flow problem we applied the theory directly and found that approximating the permeability on the interface by a constant reduces the number of iterations required for convergence, in most cases. However, in order to grant convergence it is necessary to use a more precise representation of the interface flux.

The non-symmetric character of the contaminant transport problem precludes a direct application of the DD-CG algorithm, in its present state. However, a modification of it was applied and quite satisfactory numerical results were obtained. This suggests that the proposed algorithm is both robust and efficient.

ACKNOWLEDGMENT. The authors wish to acknowledge the financial support of the U.S. Environmental Protection Agency (EPA) assistance I.D. No. CR-820499. Although the research described in this article has been funded by the U.S. EPA, it has not been subjected to agency review and therefore does not necessarily reflect the views of the Agency and no official endorsement should be inferred.

REFERENCES

Herrera, I., J. Guarnaccia and G. F. Pinder (1994), "Domain decomposition methods for collocation procedures", in A. Peters et al. (eds.), Computational Methods in Water Resources, Kluwer Academic Publishers, Dordrecht, this volume.

Huyakorn, P. S., and G. F. Pinder (1983), Computational Methods in Subsurface Flow, Academic Press, New York.

Lapidus, L., and G. F. Pinder (1982), Numerical Solution of Partial Differential Equations, John Wiley, New York.

Russell, T., M. Wheeler and C. Chiang (1986), "Large-scale simulation of miscible displacement by mixed and characteristic finite element methods", W. Fitzgibbon (ed.), SIAM, Philadelphia, pp 85-107.

DOMAIN DECOMPOSITION METHOD FOR COLLOCATION PROCEDURES

I. Herrera

Institute of Geophysics, National University of Mexico,
Apartado Postal 22-582, Mexico D.F., 14000 Mexico

J. Guarnaccia & G.F. Pinder
University of Vermont,
109 Votey Building
Burlington, Vermont 05405, USA

ABSTRACT

Combining collocation procedures with domain decomposition methods presents complications that must be overcome in order to profit from the advantages of parallel computing. Recently, Herrera supplied formulations which effectively combine these methods. In this paper such formulations an their implementations are presented and discussed. In a companion paper, also presented in this conference, the application of this method to nonlinear flow and transport of a multiphase system is discussed.

1. INTRODUCTION

Domain decomposition techniques have received much attention in recent years. They constitute a natural route to parallelism. Using them it is possible to transform large discrete systems into smaller ones. Also, domains of irregular shape can be decomposed into regular subdomains in which tensor-product discretizations can be applied. In addition, domain decomposition techniques are quite suitable for carrying out grid refinements in regions where they are required, as where the coefficient variability is high.

For elliptic problems, domain decomposition methodologies are well developed (see, for example [1-5]). In many instances time dependent problems of parabolic type can be treated in the same manner, because for usual time discretizations, they give rise to an elliptic problem at each time-step. In this case, the grid Green function has a rapid exponential decay and this property can be exploited to minimize the amount of work that is required for applying the domain decomposition method [6-8]. Also, domain decomposition methods are frequently applied by means of a preconditioned conjugate gradient iteration. When this is done, the type of preconditioner to be used is case dependent (see for example [1]).

In this paper and a companion one [9], we address the problem of

A. Peters et al. (eds.), Computational Methods in Water Resources X, 273–280.
© 1994 Kluwer Academic Publishers. Printed in the Netherlands.

combining collocation (see, for example [10]), with domain decomposition methods. In the present paper, the procedure we propose and its foundations, are explained and in the companion one [9], the numerical implementation is illustrated and discussed in connection with flow and transport problems. Actually, the method here presented has considerable generality and its applicability is no restricted, by any means, to cases where the discretization procedure used is collocation.

As a matter of fact, the method is to a large extent independent of the discretization procedure applied to the basic equations, because it is based on general properties of elliptic partial differential equations which are symmetric and positive definite. Given a decomposition of a region into subdomains, the problem is transformed into one which exclusively involves the internal boundaries (Γ) separating the subdomains from each other. Then, a positive definite transformation is associated with this latter problem. Up to this point the discussion refers to the continuous problem. Then, a discretization is introduced in an abstract manner, by means of a finite dimensional subspace of functions defined on Γ. It is shown that the restriction of the transformation previously introduced to such subspace, also possesses the positive-definiteness property and it is suitable for applying the Conjugate Gradient Method. The use of preconditioners, in conjunction with our method, is also possible and it is being the subject of ongoing research [11].

Here the method is explained for elliptic equations but it can also be applied to a large class of parabolic equations when a step by step procedure, for integration of time is used, either by means of a θ-scheme fixed in space or an Eulerian-Lagrangian approach (see for example [12,13]). In such case, an elliptic problem has to be solved at each time-step, and domain decomposition methods converge rapidly, because of the radius of influence of fundamental solutions in such situations are usually small, the radius decreasing with decreasing time-step (see, for example [6]).

2.- DOMAIN DECOMPOSITION FORMULATION

In this Section and the following one, we outline the general ideas of the method, leaving aside some technical details. Later, in Section 4, the method is presented in a more systematic and rigorous manner. In addition a more thorough analysis is being prepared [11].

Consider the boundary value problem (BVP), defined in Ω (Fig. 1a), which consists in satisfying

$$\mathcal{L}u \equiv -\nabla \cdot (K \cdot \nabla u) + Ru = f_\Omega, \qquad \text{in } \Omega \qquad (2.1)$$

together with Dirichlet boundary conditions:

$$u(\underline{x}) = u_\partial(\underline{x}), \text{ in } \partial\Omega \qquad (2.2)$$

It will be assumed that for every $\underline{x}$ of Ω, the matrix K is positive definite and $R \geq 0$. In addition, suitable smoothness conditions for the coefficients in Eq. (2.1) are assumed, so that the general existence theorems for partial differential equations (see for example [14]) grant the existence of a unique solution of the BVP.

The region Ω will be divided in two subregions, Ω_I and Ω_{II}, with internal boundary Γ (Fig. 1a). Let $\partial\Omega_I$ and $\partial\Omega_{II}$, be the boundaries of Ω_I

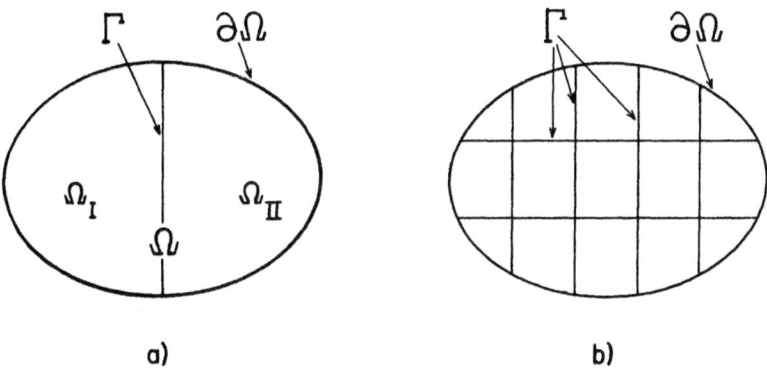

Figure 1.- Illustration of domain decompositions.

and Ω_{II}, respectively and define

$$\partial_I\Omega=\partial\Omega\cap\partial\Omega_I \quad \& \quad \partial_{II}\Omega=\partial\Omega\cap\partial\Omega_{II} \qquad (2.3)$$

Observe that $\partial\Omega_I=\Gamma\cup\partial_I\Omega$ and $\partial_{II}\Omega=\Gamma\cup\partial\Omega_{II}$. In what follows a procedure is presented for constructing the solution (u) of the BVP, by exclusively solving boundary value problems in Ω_I and Ω_{II}, separately. Such procedure will be the "domain decomposition" method. It must be mentioned that although in order to make the explanations that follow simpler, the region Ω is divided into two subregions only, the results to be presented remain valid even if the region Ω is divided into many subdomains, as long as Γ is interpreted as the union of all the internal boundaries separating the subdomains from each other (Fig. 1b).

To this end define the function v_I, in Ω_I, as the unique solution of the boundary value problem

$$\mathcal{L}v_I = f_\Omega, \quad \text{in } \Omega_I \qquad (2.4)$$

subject to the boundary conditions

$$v_I(\underline{x}) = u_\partial(\underline{x}), \text{ in } \partial_I\Omega \text{ and } v_I(\underline{x}) = V(\underline{x}), \text{ in } \Gamma, \qquad (2.5)$$

where $V(\underline{x})$ is a suitably chosen function, defined on Γ. In particular, $V(\underline{x})$ can be chosen to be identically zero on Γ, and for simplicity we assume this in what follows. The function v_{II} is defined similarly, replacing I by II, above. Having v_I and v_{II} at hand, the function $v(\underline{x})$ is defined in Ω by

$$v(\underline{x}) = \begin{cases} v_I(\underline{x}), & \underline{x}\in\Omega_I \\ v_{II}(\underline{x}), & \underline{x}\in\Omega_{II} \end{cases} \qquad (2.6)$$

Observe that the function $v(\underline{x})$, so defined, is continuous in Ω. However, the normal derivative of v may be discontinuous across Γ. In addition, observe that the construction of $v(\underline{x})$ only requires solving boundary value problems formulated in Ω_I and Ω_{II}, separately.

Proposition 1

Let the function $u(\underline{x})$, be the solution of the BVP in Ω. Then, the function $w(\underline{x})$, defined in Ω by $w \equiv u-v$, is the unique solution of the boundary value problem with prescribed jumps (BVPJ) on Γ, defined by the following conditions:

"*The differential equation* $\mathcal{L}w=0$ *is satisfied in* Ω_I *and* Ω_{II}, *separately,* $w(\underline{x})=0$ *for every* $\underline{x} \varepsilon \partial\Omega$, w *is continuous across* Γ *and its first order partial derivatives have jump discontinuities across* Γ, *which satisfy the jump condition*

$$\left[K \cdot \nabla w\right] \cdot \underline{n} = -\left[K \cdot \nabla v\right] \cdot \underline{n}, \qquad on \; \Gamma" \qquad (2.7)$$

Here, the square-bracket stands for the value of the "jump" across Γ (value on the "positive" side minus value on the "negative" one) and $\underline{n}$ is taken pointing towards the positive side.

Proof.- Because $\mathcal{L}w=\mathcal{L}u-\mathcal{L}v=0$ in Ω_I and Ω_{II}, and $w = u-v=o$, in $\partial\Omega$ Also,

$$\left[K \cdot \nabla w\right] \cdot \underline{n} = \left[K \cdot \nabla u\right] \cdot \underline{n} - \left[K \cdot \nabla v\right] \cdot \underline{n} = -\left[K \cdot \nabla v\right] \cdot \underline{n} \; on \; \Gamma \qquad (2.8)$$

since the function u, has first order continuous derivatives.

3. REDUCTION TO A PROBLEM FORMULATED ON Γ

For the sake of clarity, we introduce the following notation. Given any function $s(\underline{x})$ defined in Ω, we write $S:\Gamma \longrightarrow R^1$, for the restriction of $s(\underline{x})$ to Γ, which is defined for every $\underline{x} \in \Gamma$ by $S(\underline{x}) \equiv s(\underline{x})$. In particular, $W:\Gamma \longrightarrow R^1$ will be the restriction to Γ of the function $w(\underline{x})$ introduced in Proposition 1. It will be shown that when $W(\underline{x})$ is known on Γ, the construction of $w(\underline{x})$ in Ω only requires solving two boundary value problems: one in Ω_I and the other one in Ω_{II}.

Proposition 2

Let $w_I(\underline{x})$ be defined for every $\underline{x} \in \Omega_I$, as the unique solution of the boundary value problem

$$\mathcal{L}w_I=0, \qquad in \; \Omega_I \qquad (3.1)$$

subject to the boundary conditions in $\partial\Omega_I$, given by

$$w_I(\underline{x}) = 0, \quad in \; \partial_I\Omega \; and \; w_I(\underline{x}) = W(\underline{x}) \; in \; \Gamma, \qquad (3.2)$$

In addition, define $w_{II}(\underline{x})$ for every $\underline{x} \in \Omega_{II}$, replacing I by II in the definition of w_I, given above. Then

$$w(\underline{x})=w_I(\underline{x}), \; for \; every \; \underline{x} \in \Omega_I \; and \; w(\underline{x})=w_{II}(\underline{x}), \; for \; every \; \underline{x} \in \Omega_{II} \qquad (3.3)$$

Proof.- According to Proposition 1, the restrictions of $w(\underline{x})$ to Ω_I and Ω_{II}, are solutions of the first and second of these boundary value problems, respectively. Thus, Eqs. (3.3) follow from the uniqueness of solution properties of these problems.

The mapping A

In view of Proposition 2, the key to obtain a domain decomposition method is to develop a procedure for constructing the function $W(\underline{x})$, which is restriction of $w(\underline{x})$ to Γ. In what follows, we develop such procedure using exclusively solutions of boundary value problems formulated in Ω_I and in Ω_{II}, separately. In view of Propositions 1 and 2, this will complete the outline of the domain decomposition method we propose in this paper. The following auxiliary mapping, will be useful in the sequel.

Given a function $S \in L^2(\Gamma) \equiv H^0(\Gamma)$, construct the function s_I, defined in Ω_I, as the unique solution of the boundary value problem

$$\mathcal{L}s_I = 0, \qquad \text{in } \Omega_I \qquad (3.4)$$

subject to the boundary conditions in $\partial \Omega_I$, given by

$$s_I(\underline{x}) = 0, \quad \text{in } \partial_I\Omega \text{ and } s_I(\underline{x}) = S(\underline{x}) \text{ in } \Gamma, \qquad (3.5)$$

In addition, define $s_{II}(\underline{x})$ for every $\underline{x} \in \Omega_{II}$, replacing I by II in the definition of s_I. The function $s(\underline{x})$ is then defined in Ω, by

$$s(\underline{x}) = \begin{cases} s_I(\underline{x}), & \underline{x} \in \Omega_I \\ s_{II}(\underline{x}), & \underline{x} \in \Omega_{II} \end{cases} \qquad (3.6)$$

The function $AS: \Gamma \longrightarrow R^1$, is defined by:

$$AS(\underline{x}) = -\left[K \cdot \nabla s \right] \cdot \underline{n}, \qquad \underline{x} \varepsilon \Gamma \qquad (3.7)$$

Formulation of the problem on Γ

Define the function J on Γ by:

$$J \equiv \left[K \cdot \nabla v \right] \cdot \underline{n}, \qquad \text{on } \Gamma \qquad (3.8)$$

Where $v(\underline{x})$ is given by Eq. (2.6). Then, using the mapping A and in view of Eq. (2.7), the problem can be stated as follows:

"Find $W(\underline{x})$ on Γ, such that

$$AW = J, \qquad \text{on } \Gamma" \qquad (3.9)$$

In the next Section, it will be shown that after discretization, the solution of Eq. (3.9), can be constructed applying the Conjugate Gradient (C-G) procedure. This is because the mapping A is positive definite, as it is shown next.

Proposition 3

In the <u>linear</u> <u>subspace</u> of $L^2(\Gamma)$ for which $AS \in L^2(\Gamma) \equiv H^0(\Gamma)$, the mapping A, is symmetric and positive definite.

Proof. - In view of the fact that $\mathcal{L}s = 0$ in Ω, except on Γ where a jump discontinuity of the first order derivatives occurs. Application of the generalized divergence theorem [15], yields:

$$- \int_\Omega s \nabla \cdot (K \cdot \nabla s) d\underline{x} = - \int_{\partial\Omega} s(K \cdot \nabla s) \cdot \underline{n} \, d\underline{x} +$$

$$\int_\Gamma s \left[K \cdot \nabla s \right] \cdot \underline{n} \, d\underline{x} + \int_\Omega (K \cdot \nabla s) \cdot \nabla s \, d\underline{x} \qquad (3.10)$$

Observe that

$$\int_\Omega s \nabla \cdot (K \cdot \nabla s) d\underline{x} = \int_\Omega Rs^2 \, d\underline{x}; \quad \int_{\partial\Omega} s(K \cdot \nabla s) \cdot \underline{n} \, d\underline{x} = 0 \qquad (3.11a)$$

and

$$-\int_\Gamma s \left[\mathbf{K} \cdot \nabla s \right] \cdot \underline{n} \, d\underline{x} = \int_\Gamma (AS)S \, d\underline{x} \qquad (3.11b)$$

by virtue of the conditions imposed on $s(\underline{x})$ and the definition of the mapping A. Therefore

$$(AS*S) = \int_\Gamma (AS)S \, d\underline{x} = \int_\Omega \left\{ (\mathbf{K} \cdot \nabla s) \cdot \nabla s + Rs^2 \right\} d\underline{x} \geq 0 \qquad (3.12)$$

Where, the inner product operation in $L^2(\Gamma)$ is denoted by (*). Finally, observe that the equality in relation (3.12) holds when the function $s(\underline{x})$ is identically zero, exclusively, in which case so is $S(\underline{x})$.

Remark 1.- Observe that the linear subspace of $H^0(\Gamma)$, we are referring to in Proposition 3, is $H^1(\Gamma)$. Indeed, the image of $H^1(\Gamma)$ under the mapping A, is $H^0(\Gamma)$ [14]. This is a linear subspace of $H^0(\Gamma)$, since $H^1(\Gamma) \subset H^0(\Gamma)$. However, $H^1(\Gamma)$ is not closed with respect to the metric of $H^0(\Gamma) = L^2(\Gamma)$.

4. DISCRETIZATION AND CONJUGATE GRADIENT FORMULATION

The fact that the transformation A introduced in the previous Section is positive definite, permits applying the conjugate gradient method (C-G) to the problem of obtaining the function $W(\underline{x})$, on Γ. However, it is necessary to discretize the problem before C-G can be applied. The procedure to be presented is independent of the actual set of approximating functions used to carry out the discretization. What is essential is that, after discretization, the problem is no longer formulated on the whole space $L^2(\Gamma)$, but instead, it is formulated in a finite dimensional subspace of $L^2(\Gamma)$.

Thus, let $\mathfrak{G} \subset H^1(\Gamma) \subset H^0(\Gamma) = L^2(\Gamma)$ be such finite dimensional subspace which is necessarily closed in $L^2(\Gamma)$, since it is finite dimensional. Elements belonging to $\mathfrak{G}$ will wear a hat. In particular, for any $S \in L^2(\Gamma)$, we write $\hat{S}$ for its projection on $\mathfrak{G}$, with respect to the $L^2(\Gamma)$ inner product. In addition, for every $\hat{S} \in \mathfrak{G}$, $\widehat{A\hat{S}} \in \mathfrak{G}$ is defined as the projection of $A\hat{S} \in H^0(\Gamma) = L^2(\Gamma)$, on $\mathfrak{G}$. Thus, in this manner a well defined mapping $\hat{A}$ of $\mathfrak{G}$ into itself, is obtained ($\hat{A}: \mathfrak{G} \longrightarrow \mathfrak{G}$).

Remark 2.- Observe that we could associate a square matrix with the mapping A, since it is a linear transformation of a finite dimensional space into itself. However, such matrix will not be used in what follows, because in applications it is too costly to construct.

The Conjugate Gradient formulation

Projecting Eq (3.9) on $\mathfrak{G}$, the following "discrete" version of the problem is obtained:

"*Find* $\hat{W} \in \mathfrak{G}$ *on* Γ, *such that*

$$\hat{A}\hat{W} = \hat{J}, \qquad\qquad on \; \Gamma" \qquad (4.1)$$

Here the function J is defined on Γ, by Eq (3.8). This problem is suitable to be solved by means of the C-G method. This is due to the fact that $\hat{A}$ is finite dimensional and positive definite.

Proposition 4

The mapping $\hat{A}: \math6 \longrightarrow \math6$, is (symmetric and) positive definite.

Proof.- This is a well known result of Linear Algebra. It can be easily derived from the fact that

$$(\hat{\hat{A}\hat{S}},\hat{W}) = (\hat{AS},\hat{W}) \qquad (4.2)$$

whenever $\hat{S}$ and $\hat{W}$ belong to $\math6$.

The C-G algorithm

Because of the special character of the application of the Conjugate Gradient method, the corresponding algorithm is given in detail. Recall the notations that have been adopted and in particular, that when a low case letter is used for a function defined in Ω, the corresponding capital letter is reserved to denote the restriction of such function to Γ. The C-G algorithm follows:

Given a tolerance $\varepsilon>0$, define $w^0 \equiv 0$ in Ω.

$R^0=J$

$\hat{P}^0=\hat{R}^0$

$k=0$

(@)   Do

"*Construct p^k in Ω, satisfying $\math{L}p^k=0$ in Ω_1 and Ω_2 separately, homogeneous boundary conditions in $\partial\Omega$ and*

$p^k=\hat{P}^k \qquad$ *in Γ*"

(Recall in what follows that $A\hat{P}^k(\underline{x}) = -\left[K\cdot\nabla p^k\right]\cdot\underline{n}$, on Γ)

$\alpha^k = \hat{R}^k*\hat{R}^k/\hat{P}^k*A\hat{P}^k$

$w^{k+1}=w^k+\alpha^k p^k$

$R^{k+1}=\hat{R}^k-\alpha^k A\hat{P}^k$

If $\|\hat{R}^{k+1}\|_\infty \leq \varepsilon$, set $u=v+w^{k+1}$ in Ω, and stop.

$\beta^k = \hat{R}^{k+1}*A\hat{P}^k/\hat{P}^k*A\hat{P}^k$

$\hat{P}^{k+1}=\hat{R}^{k+1}-\beta^k\hat{P}^k$

$k=k+1$ go to (@)

5. EXTENSIONS

Firstly, it must be observed that in the preceding discussion nothing has been assumed about the dimension of the space in which the problem is defined. Thus, the domain decomposition method developed in this paper is applicable independently of the number of dimensions. Secondly, in addition to elliptic problems, it is applicable to time dependent problems, because many of them at each time step adopt the form of Eq (2.1), after the time discretization has been applied.

There are several options for treating differential operators which are not self-adjoint, such as those occurring in transport-diffusion problems. The most rigorous one, is to apply an extension of the theory and this is being developed at present. Another one, more direct and simple was successfully applied in the companion paper [9]. Another very interesting possible extension that should be studied, is combining the method here presented with Eulerian-Lagrangian procedures, such as those developed in [12, 13].

REFERENCES

1.- D.E. Keyes and W.D. Gropp., "A comparison of Domain Decomposition Techniques for Elliptic Partial Differential Equations and Their Parallel Implementation", SIAM, J. SCI. STAT. Comput., $\underline{8}$(2), pp 166-202, 1987.

2.- P.E. Bjorstad and O.B. Widlund, "On some trends in elliptic problem solvers", In G. Birkhoff & A. Schoenstadt (eds.),"Elliptic Problem Solvers II", Academic Press, New York, 1984.

3.- T.F. Chan., "Domain Decomposition Algorithms and Computational Fluid Dynamics", In J. Dongarra, I. Duff, P. Gaffney & S. McKee (eds.),"Vector and Parallel Computing", pp 65-82, Ellis Horwood, Chichester, U.K., 1989.

4.- Q.V. Dihn., R. Glowinski and J. Periaux., "Solving Elliptic Problems by Domain Decomposition Methods with Applications", In G. Birkhoff & A. Scoenstadt,"Elliptic Problem Solvers II", Academic Press, New York, 1984. Academic Press, Inc, pp 395-425, 1984

5.- M. Dryja and O.B. Windlund,"Some domain decomposition algorithms", In *Iterative methods for large linear systems*, Academic Press, San Diego, CA, 1989.

6.- M. Israeli, L. Vozovoi and A. Averbuch., "Domain decomposition methods for solving parabolic PDEs on multiprocessors"., Applied Numerical Mathematics, $\underline{12}$, pp 193-212, 1993.

7.- Y.A. Kuznetsov,"New algorithms for approximate realization of implicit difference schemes"., Sov. J. Numer . Anal. Math. Modeling, $\underline{3}$ (2), pp. 99-114, 1988.

8.- Y.A. Kuznetsov,"Domain descomposition methods for unsteady convection-diffusion problems", in: R. Glowinski and A. Lichnewsky, eds., Computing Methods in Applied Sciences and Engineering, SIAM, Philadelphia, PA, pp. 211-227, 1990.

9.- J. Guarnaccia, I. Herrera and G.F. Pinder,"Solution of flow and transport problems by a combination of collocation and domain decomposition procedures", In this volume, 1994.

10.- L.R. Bentley, G.F. Pinder and I. Herrera,"Solution of the Advective-Dispersive Transport Equation Using a Least Squares Collocation, Eulerian-Lagrangian Method". Numer. Methods for Partial Differential Equations, 5(3), pp. 227-240, 1989.

11.- I. Herrera, J. Guarnaccia and G.F. Pinder,"Combining Domain Decomposition and Collocation", to be published.

12.- I. Herrera, R.E. Ewing, M.A. Celia and T.F. Russell,"An Eulerian-Lagrangian Localized Adjoint Method: The Theoretical Framework".- Numerical Methods for Partial Differential Equations $\underline{9}$(4), pp 431-457, 1993.

13.- M.A. Celia, T.F. Russell, I. Herrera and R.E. Ewing,"An Eulerian-Lagrangian Localized Adjoint Method for the Advection-Diffusion Equation. Advances in Water Resources, $\underline{13}$(4), pp. 187-206, 1990.

14.- J.L. Lions and E. Magenes,"Non-homogeneous boundary value problems and applications", Springer-Verlag, New York & Heidelberg, pp357, 1972.

15.- M.B Allen, I. Herrera and G.F. Pinder,"Numerical modeling in Science and Engineering", John Wiley, New York, pp 418, 1988.

A Comparative Study of Particle Tracking Techniques for Numerically Solving the Convection-Dispersion Equation

K. Huang and M. Th. van Genuchten
U. S. Salinity Laboratory, USDA, ARS, 4500 Glenwood Dr., Riverside, CA 92501, USA

Abstract

Characteristics-based particle tracking techniques combined with standard finite element or finite difference methods have been widely used for solving convection-dispersion type solute transport equations. The popularly-used single-step backward tracking technique (SRPT) and a mixed technique (HYBRID), obtained by combining SRPT with continuous forward tracking, were evaluated by means of a large number of numerical tests. Solutions obtained with SRPT and HYBRID were both found to be free of oscillations. Although HYBRID generally worked better than SRPT, both schemes suffered from numerical dispersion. Numerical dispersion always increased with increasing grid Peclet number, and usually with decreasing cell Courant number. The numerical problems were especially evident when sharp concentration peaks (local maxima) or valleys (local minima) existed. Numerical dispersion in these cases was found to result mainly from interpolation errors in the particle tracking techniques. We developed a modified SRPT method which continuously tracks the local maximum or minimum concentrations for use in the interpolation of the convective components. The modified SRPT was far more effective in eliminating numerical dispersion than current particle tracking methods.

Introduction

Convection-dispersion type equations (CDE's) are generally solved numerically using standard finite difference or finite element methods. Such methods are relatively accurate for dispersion-dominated transport problems, but often face numerical difficulties when convective transport dominates dispersion and the concentration fronts are steep. The numerical difficulties are manifested by artificial dispersion of sharp concentration fronts and/or numerical oscillations at or near the sharp fronts. Although some of the numerical dispersion and oscillation problems can be avoided by grid refinement, such an approach may greatly increase the computational effort.

Several alternative methods have been developed to avoid or limit numerical problems in the standard solution schemes. One of the more popular methods for

A. Peters et al. (eds.), Computational Methods in Water Resources X, 281–290.
© *1994 Kluwer Academic Publishers. Printed in the Netherlands.*

this purpose is the Eulerian-Lagrangian approach which solves separately for the convective and dispersive components of the transport equation. The convective transport problem is solved on a moving coordinate system by tracking a moving particle along a characteristic line, while the dispersion problem is solved on a fixed Eulerian grid system (*Douglas and Russell*, 1982; *Neuman*, 1984). Several characteristics-based particle tracking techniques currently exist for calculating the convective component. Among these, the single-step reverse particle tracking (SRPT) is probably the more popular one because of its simplicity, whereas a mixed technique (HYBRID), obtained by combining SRPT with continuous forward tracking, is relatively more robust for handling the convection problem for any value of the grid Peclet number. Although SRPT effectively eliminates numerical oscillations, this scheme may still produce serious numerical dispersion for problems involving large Peclet numbers. Several approaches have been used to reduce numerical dispersion. For example, *Casulli* (1987) suggested reducing the spatial step, *Huang et al.* (1992) implemented a more precise method for tracking the concentration front, while *Goblet and Cordier* (1993) used a spectral element method to achieve a more accurate interpolation when the SRPT is implemented. Although some improvements were made, none of these schemes were completely successful in eliminating numerical dispersion. Whereas the HYBRID scheme at present seems to produce the most accurate results, its numerical problems (notably numerical dispersion) have still not been adequately addressed.

In this paper we evaluate the numerical behavior of the SRPT and HYBRID particle tracking techniques by means of several numerical tests. The SRPT scheme will be implemented using a modified numerical method which tracks the local extremes of a concentration profile.

The Eulerian-Lagrangian Approach

For simplicity, we consider here one-dimensional transport in a porous medium with uniform water flow and constant transport parameters as follows

$$R\frac{\partial c}{\partial t} = D\frac{\partial^2 c}{\partial x^2} - v\frac{\partial c}{\partial x} - \mu c + \gamma \tag{1}$$

where c is the solute concentration, t is time, x is distance, R is the retardation factor, v is the pore-water velocity, D is the dispersion coefficient, and μ and γ are rate constants for first-order decay and zero-order production, respectively. Equation (1) will be solved subject to the initial condition

$$c(x,0) = C_i(x) \tag{2}$$

The inlet boundary condition is taken to be

$$c(0,t) = C_0(t) \tag{3}$$

while at the lower boundary, $x = L$, a zero-gradient is assumed:

$$\frac{\partial c}{\partial x}(L,t) = 0 \tag{4}$$

where $C_i(x)$ and $C_0(t)$ are prescribed functions of x and t, respectively.

Equation (1) may be rewritten in Lagrangian form as

$$R\frac{Dc}{Dt} = D\frac{\partial^2 c}{\partial x^2} - \mu c + \gamma \tag{5}$$

where Dc/Dt is the Lagrangian derivative

$$\frac{Dc}{Dt} = \frac{\partial c}{\partial t} + v^* \cdot \frac{\partial c}{\partial x} \tag{6}$$

in which $v^* = v/R$, and where c now describes the concentration of a fluid particle moving along a characteristic path defined by the equation

$$\frac{dx}{dt} = v^* \tag{7}$$

According to Neuman's operator splitting approach, the transport problem is decoupled into two parts, one covering pure convection, and the other dispersion and all remaining transport processes. The split convection problem is required to satisfy the homogeneous Lagrangian derivative

$$\frac{D\bar{c}}{Dt} = 0 \tag{8}$$

along the characteristic line. The "convective component", $\bar{c}$, of the concentration c, is solved independently by the method of characteristics, and subsequently substituted into the discretized Lagrangian derivative of equation (6) as follows

$$R\frac{c - \bar{c}}{\Delta t} = D\frac{\partial^2 c}{\partial x^2} - \mu c + \gamma \tag{9}$$

where Δt is the time increment. The residual transport problem, Eq. (10), can be solved for c using standard fixed-grid methods such as finite elements or finite differences.

Two different characteristics-based particle tracking methods are considered in this study. One method is the SRPT single-step reverse particle tracking (*Neuman*, 1981; *Galeati et al.*, 1992) in which the characteristic paths are traced backward. The other approach is the HYBRID scheme proposed by *Neuman* (1984) in which the convective components of steep concentration fronts are tracked forward with the help of moving particles clustered around each front. The convection problems away from the fronts are solved using backward particle tracking. When a front dissipates in time, its forward tracking stops automatically and the corresponding cloud of particles is eliminated.

Numerical Experiments

This section presents results of several numerical tests comparing the relative accuracy of the SRPT and HYBRID particle tracking methods. For the HYBRID method we initially (at $t=0$) introduced 60 particles (2 per element) in areas with large concentration gradients. The analytical solutions of (1) subject to (2), (3) and (4) were used to verify the numerical results of both particle tracking methods. Assuming $v=10$ and $\Delta x = 1.0$ (any consistent set of units may be used), a conservative tracer ($R = 1$), and no production or decay ($\mu=\gamma=0$) for all examples, concentration distributions were calculated for a variety of dispersion coefficients, D, and time steps, Δt, such that the values of the grid Peclet number, $Pe = v\Delta x/D$, and the grid Courant number, $Cu = v\Delta t/\Delta x$, differed from case to case. Comparisons with the analytical solutions provide stringent tests on the performance of the numerical methods, especially their ability to correctly track sharp fronts.

We first simulated concentration distributions for $C_0 = 1$ and $C_i = 0$ assuming a relatively small Peclet number of 10. Results indicated a fairly good match between the exact solution and the SRPT and HYBRID particle tracking methods. However, as shown in Figure 1, both methods suffered from numerical dispersion when the grid Peclet number, Pe, was increased to 100. The simulations were obtained with fractional Courant numbers of 1.5 ($t = 2$) and 0.5 ($t = 4$). When integer Courant numbers (e.g., $Cu = 2$) were used, the numerical results (dots in the figure) essentially duplicated the exact solution.

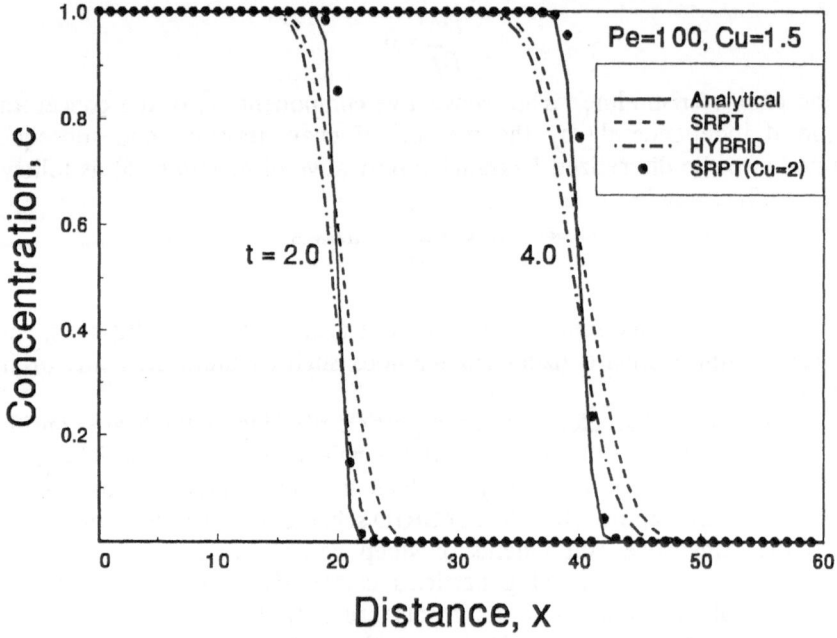

Fig. 1. Simulated single-front concentration distributions
obtained with SRPT and HYBRID.

The above comparisons pertain to single-front concentration distributions. Simulations were also carried out for steep, multi-peak initial distributions assuming solute-free input ($C_0 = 0$). Figure 2 shows the simulated distributions using a fractional Courant number of 1.25 and a relatively large Peclet number of 100. Calculated concentrations obtained with both SRPT and HYBRID clearly suffered from numerical dispersion, especially near the concentration peaks and valleys. Notice that the numerical dispersion in Figure 2, increased with simulation time, or equivalently, with the number of time steps during which particle tracking was implemented. The HYBRID method in most cases was found to be more accurate than SRPT, especially for relatively high Peclet numbers.

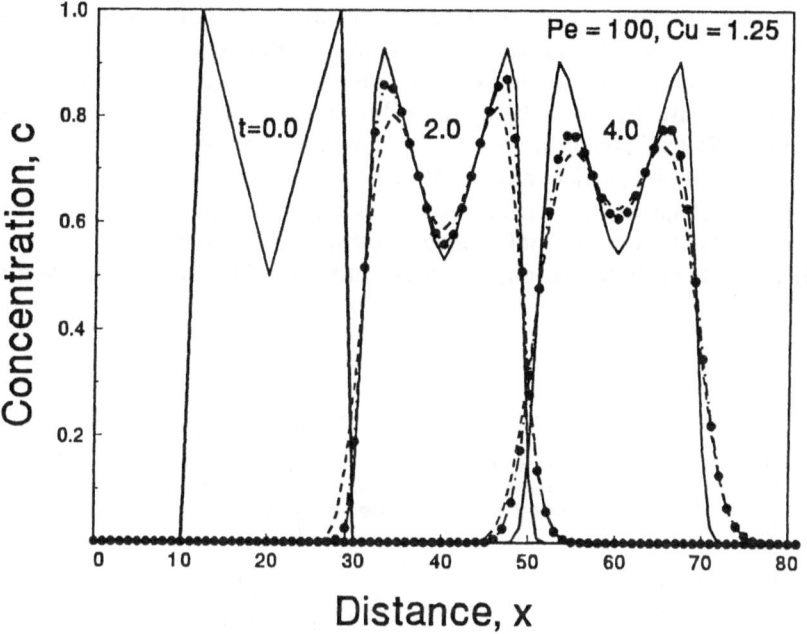

Fig. 2. Simulated multi-peak concentration distributions obtained with SRPT (dashed lines) and HYBRID (dot-dashed lines).

The grid Peclet number, Pe, is generally viewed as a major factor determining the extent of numerical dispersion (i.e., increasing Pe leads to more numerical dispersion). This finding is not necessarily true for transport problems having moderately steep concentration fronts. As an example, the SRPT was used to simulate the transport of a multi-peak concentration profile with relatively small gradients and assuming a value of 1000 for Pe. Such a high value for Pe often leads to serious numerical instabilities in numerical transport studies. The simulation was performed using fractional Courant numbers ranging from 0.044 to 1.31, values which usually produce serious numerical dispersion when particle tracking techniques are implemented. The simulated results (not shown here) were very accurate, even near the peaks. These results suggest that the performance of a numerical scheme is affected not only by the Peclet number, but also by the spatial gradient. Actually, we found that the spatial gradient is often more important than the value of *Pe* because of potentially large interpolation errors near steep fronts.

Implementation

The numerical experiments above indicate that SRPT and HYBRID perform relatively poorly for transport problems involving steep concentration gradients. In general, particle tracking techniques become more dissipative when the Peclet number increases or the Courant number decreases. Our calculations indicate that the numerical problems, mainly numerical dispersion, in both schemes are likely a result of interpolation errors near sharp fronts, especially near concentration extremes. One should expect some errors of this type when simple linear interpolation schemes are used in SRPT to solve for the convective components near steep fronts. Figure 4 shows schematically how a steep concentration peak, *ABC*, is tracked using SRPT. According to the Lagrangian point of view, i.e., equation (9), the concentration of a particle moving along its characteristic path remains constant. Therefore, the shape of the concentration distribution at time $t_{k+1} = t_k + \Delta t$ should be the same as that at t_k because only convective transport is considered. However, the peak *ABC* has been flattened and smeared into *A'B'C'* by the SRPT scheme.

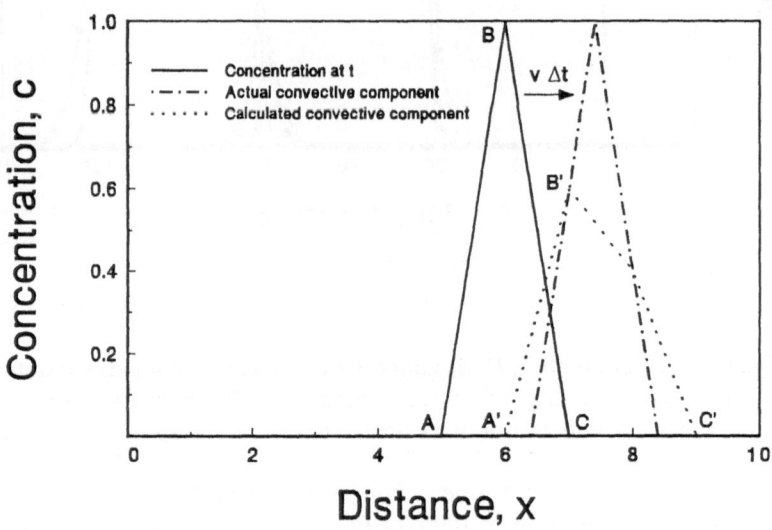

Fig. 3. Schematic illustration of the convective component
as calculated with the SRPT technique.

Higher-order interpolation formula, such as cubic Hermitian equations, could possibly improve the accuracy. However, higher-order equations often involve spatial derivatives which may be difficult to determine accurately. Alternatively, small spatial steps should also reduce the interpolation errors. The additional forward-moving particles introduced with the HYBRID scheme then serve primarily to add extra interpolation points in areas with steep fronts, thus making HYBRID more accurate than SRPT. Although the moving particles used in HYBRID help to reduce artificial damping of the numerical results as compared to SRPT, they are not useful if they can not trace the exact paths of the local extremes. To reduce the interpolation errors near the peaks, which are key points dominating the shape of a concentration distribution, we propose a modified SRPT scheme as described below.

Assume that the concentration c^k at time t_k is known. Equation (6) subject to auxiliary conditions (2) through (4) will be used to solve for c^{k+1} at time step $t_{k+1} = t_k + \Delta t$ using the following consecutive steps:

1. *Determine the Local Maximum/Minimum Concentrations.* At the beginning of the simulation, determine the local maximum and minimum concentrations, C^p at x^p ($p=1,...,P$), from the initial concentration distribution and the inlet boundary condition at $x=0$ (P represents the number of the selected concentration extremes). The inlet boundary must also be viewed as an extreme if the inlet concentration differs significantly from the first nodal concentration.

2. *Continuous Forward Tracking.* Once determined, the local minimum and maximum concentration points are considered to be moving particles and tracked continuously forward along the characteristic path. Assume x_p^k is the position of particle p at t_k. Position x_p^{k+1} of this particle at $t_{k+1}=t_k+\Delta t$ can then be calculated from equation (8) as follows

$$x_p^{k+1} = x_p^k + \int_{t_k}^{t_{k+1}} v \cdot dt \qquad (p=1,2,...,P) \qquad (10)$$

3. *Single-Step Reverse Tracking (Modified Method of Characteristics).* Consider a fictitious particle that moves during time step Δt from location x_n' at t to a new location x_n which is the fixed Eulerian coordinate of the finite element node n. Based on equation (8), the initial particle location x_n' can be tracked backward as (*Neuman*, 1984)

$$x_n' = x_n - \int_{t_k}^{t_{k+1}} v \cdot dt \qquad (11)$$

Once the backward position x_n' has been determined, the corresponding concentration, i.e., the convective component $\bar{c}_n$ for node n, can be computed by interpolating between nodal values using the finite element formulation

$$\bar{c}_n = \sum_{i=1}^{N_p} c_i(t)\varphi_i(x_n') \qquad (N_p = N+P) \qquad (12)$$

where the $\varphi_i(x)$ are the usual finite element basis functions. The $c_i(t)$ in equation

(13) represent concentrations of all nodes (N) and moving particles (P). The concentrations of the moving particles are used in (13) to increase the number of interpolation points in areas having high concentration gradients, thus improving the accuracy of the interpolations in these areas. Concentrations of the moving particles, c_p, are generally more accurate than those of the nodes, c_n, since they are calculated independently from the characteristics approach as will be discussed later.

4. *Finite Element Approximation*. Solutions for c^{k+1} are subsequently obtained by applying the Galerkin finite element method with linear basis functions to (6). Since the Galerkin method is relatively standard, we do not further review here its application to the solution of equation (6). However, we emphasize that the Lagrangian derivative should be approximated by

$$\frac{Dc_n}{Dt} \approx \frac{c_n^{k+1} - \overline{c}_n}{\Delta t} \tag{13}$$

This discretization is based on the view that node n is a fictitious particle reaching x_n at t_{k+1}.

5. *Dispersion Correction for Moving Particles*. Finally, the concentration of moving particle p at time t_{k+1} is corrected by the dispersive component which is estimated from the finite element interpolation as follows

$$c_p^{k+1} = c_p^k + \sum_{i=1}^{N} (c_i^{k+1} - \overline{c}_i)\varphi_i(x_p) \tag{14}$$

Note that the term $(c_i^{k+1} - \overline{c}_i)$ is the dispersive component of the concentration at node i. Repeating steps 2 to 5 yields the complete solution at the next time step.

Examples

The modified SRPT scheme was found to produce encouraging results which agreed well with the analytical solution for several test cases. Figure 4 shows simulated results for a steep-peak initial distribution assuming $Pe = 1000$ and $Cu = 0.9$. The proposed modified scheme yielded accurate solutions, whereas SRPT and HYBRID produced considerable numerical dispersion.

Figure 5 shows that the modified SRPT scheme generated also more accurate results than either SRPT or HYBRID for a multi-peak initial distribution. In particular, the sharp concentration peaks and valleys were more accurately simulated with the modified scheme. Accurate simulation of local concentration minima or maxima is sometimes important when predicting contaminant transport in the subsurface. As opposed to the modified SRPT scheme, SRPT and HYBRID both generated serious numerical dispersion. For this simulation we assumed a Peclet number of 100, and a Courant number of 1.25.

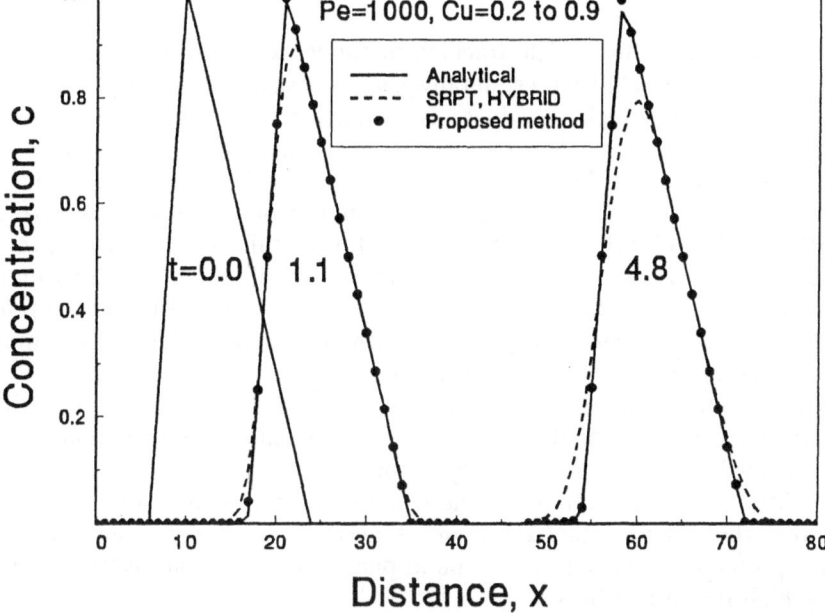

Fig. 4. Calculated concentration distributions for a single-peak using SRPT, HYBRID, and the proposed modified SRPT method.

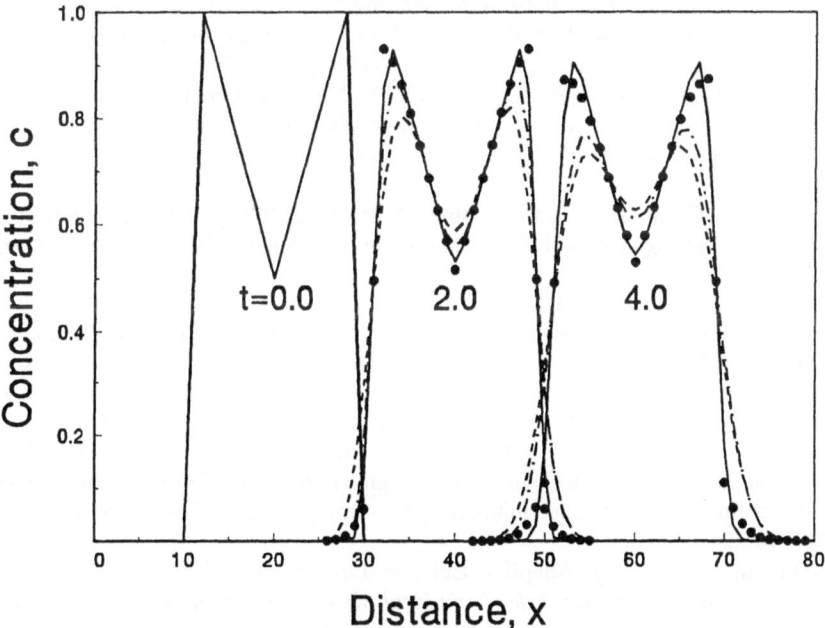

Fig. 5. Simulated multi-peak concentration distributions obtained with SRPT (dashed lines), HYBRID (dot-dashed lines), and the proposed modified SRPT method (dots) using Pe = 100, Cu = 1.25.

Conclusions

The SRPT and HYBRID particle tracking methods were both found to be free of numerical oscillations. HYBRID generally gave somewhat better results than SRPT, especially for relatively high Pe values. Both particle tracking techniques suffered from numerical dispersion when Pe was relatively large and sharp concentration fronts existed. Numerical dispersion produced with SRPT and HYBRID increased when smaller time steps (relatively small fractional Courant numbers) were used. Still, SRPT and HYBRID both gave satisfactory results for transport problems involving moderate gradients in the concentration distributions, even for relatively high Pe values (e.g., Pe as high as 1000). This fact indicates that the Peclet number is not the only factor determining the performance of a numerical scheme; the results are also affected by the concentration gradient.

Our results indicate that numerical dispersion is caused primarily by inaccurate determination of the convective components because of interpolation errors, especially in areas having large concentration gradients. The proposed modified SRPT virtually eliminated these interpolation errors. Preliminary tests showed that the modified SRPT scheme is accurate and very robust for solving transport problems involving sharp multi-peak concentration distributions and relatively high Peclet numbers.

References

1.   Casulli, V. (1987) "Eulerian-Lagrangian methods for hyperbolic and convection dominated parabolic problems", in C. Taylor et al. (eds.), *Computational Methods for Nonlinear Problems*, Pineridge, Swansea, Wales, pp. 239-269, Chap. 8.

2.   Douglas, J., Jr., and Russell, T.F. (1982) "Numerical methods for convection-dominated diffusion problems based on combining the method of characteristics with finite element or finite difference procedures", *SIAM J. Numerical Analysis*. 19, 871-885.

3.   Galeati, G., Gambolati, G., and Neuman S. P. (1992) "Coupled and partially coupled Eulerian-Lagrangian model of freshwater-seawater mixing", *Water Resources Research* 28(1), 149-165.

4.   Goblet, P., and Cordier, E. (1993) "Solution of flow and mass transport equations by means of spectral elements", *Water Resources Research* 29(9), 3135-3144.

5.   Huang, K., Zhang, R., and van Genuchten, M. Th. (1992) "A simple particle tracking technique for solving the convection-dispersion equation", in T. F. Russel et al. (eds.), *Numerical Methods in Water Resources, Proc., Ninth International Conference on Computational Methods in Water Resources*, Elsevier, New York, pp. 86-97.

6.   Neuman, S. P. (1981) "An Eulerian-Lagrangian numerical scheme for the dispersion-convection equation using conjugate space-time grids", *J. of Computational Physics* 41, 270-294.

7.   Neuman, S. P. (1984) "Adaptive Eulerian-Lagrangian finite element method for advection-dispersion", *Int. J. for Numerical Methods in Engineering* 20, 321-337.

A LAGRANGIAN-EULERIAN FINITE ELEMENT METHOD WITH ADAPTIVE GRIDDING FOR ADVECTION-DISPERSION PROBLEMS

Y. IJIRI* and K. KARASAKI
Earth Sciences Division
Lawrence Berkeley Laboratory
University of California
Berkeley, California 94720
USA

In the present paper, a Lagrangian-Eulerian finite element method with adaptive gridding for solving advection-dispersion equations is described. The code creates new grid points in the vicinity of sharp fronts at every time step in order to reduce numerical dispersion. The code yields quite accurate solutions for a wide range of mesh Peclet numbers and for mesh Courant numbers well in excess of 1.

INTRODUCTION

The advection-dispersion equation used for simulating subsurface transport of solutes has been solved by a number of numerical methods including Eulerian, Lagrangian and Lagrangian-Eulerian methods. In recent years, many attempts have been made to eliminate numerical oscillation and dispersion, which are especially troublesome for advection-dominated problems. The mixed Lagrangian-Eulerian methods have been gaining popularity for solving these problems.

In the mixed Lagrangian-Eulerian methods, the advection-dispersion equation is decomposed into two parts, one controlled by pure advection and the other by dispersion (Neuman, 1981,1984). The advected concentration profiles are calculated by Lagrangian approaches such as particle tracking methods, whereas the dispersed concentration profiles are numerically solved by conventional techniques such as the finite difference method or finite element method on fixed Eulerian grids.

Neuman (1984) proposed an adaptive scheme for calculating the advected profile. In his paper, continuous forward particle tracking is used for nodes in the vicinity of sharp fronts and single-step reverse particle tracking is used for nodes away from sharp fronts. However, his method still suffers from some numerical dispersion due to the interpolation scheme used in the tracking methods. Furthermore, the accuracy of the results is highly dependent on the number of particles introduced in the model.

The interpolation scheme of Neuman (1984) was improved by Cady and Neuman (1988). In their scheme, the fixed grid is covered by a cloud of front-tracking particles, the concentration at the grid is calculated from the concentrations of the cloud particles by triangulating these cloud particles according to the algorithm of Sloan and Houlsby (1984) and then the residual dispersion finite element equations are solved on this local grid using linear functions. They noted that their approach is based on a mesh refinement idea.

---

*Now at Taisei Corporation, Technology Research Center, 344-1 Nasemachi, Totsukaku, Yokohama 245, Japan.

A. Peters et al. (eds.), Computational Methods in Water Resources X, 291–298.

Another mesh refinement approach with a Lagrangian-Eulerian method has been developed by Yeh (1990). His approach successfully reduces numerical dispersion by zooming the sharp-front elements in which the gradients of concentration are steep, and activating hidden fine-mesh nodes in the elements.

Karasaki (1987,1988) developed a numerical code that employs a mixed Lagrangian-Eulerian scheme with adaptive gridding, which is called TRINET, for three-dimensional fracture networks of channels. This approach avoids numerical dispersion by creating new Eulerian grid points instead of interpolating the advected profile back to the fixed Eulerian grid. Another important feature of his approach is that he used tracking methods not for particles but for nodes. Therefore, the number of particle introduced in the model is not an issue.

In the present paper, the scheme used in TRINET is extended and applied to a two-dimensional porous medium model, TRIPOR. In addition to the ability to handle a two-dimensional continuous medium, the main advantage of TRIPOR is its capability to create new nodes anywhere in a two-dimensional domain while TRINET creates new nodes only along existing channels. The present paper describes the numerical approach and some applications for one- and two- dimensional problems in order to demonstrate the capability of the method. Comparing results of the preliminary studies against analytical solutions suggests that the present method gives accurate results for a wide range of Peclet numbers and for Courant numbers well in excess of 1.

NUMERICAL APPROACH

The flow field is first solved by a simple Galerkin finite element method in order to calculate the velocity profile for the entire domain. Since linear shape functions are currently used to calculate the flow field, the velocity is assumed to be uniform within a given element. The code then solves the solute transport problem expressed by the advection-dispersion equation written as

$$\frac{\partial c}{\partial t} = \nabla (D \cdot \nabla c - vc) + q \tag{1}$$

where c is solute concentration, D is dispersion coefficient, v is pore water velocity, q is a source term and ∇ is the gradient operator. Since the equation is decomposed into two parts as described earlier, advection and dispersion problems are calculated separately. The advection part is solved by tracking methods and the dispersion part is solved by a Galerkin finite element method with fixed Eulerian grids. Triangular elements are used for finite element discretization in TRIPOR.

Tracking methods for advection

The tracking methods used in this approach are done in the same manner as described in Neuman (1984) except that they are applied not to particles but to nodes. Therefore, no particles are introduced in the model. First, backward tracking is applied to obtain an advected concentration for all nodes (Figure 1(a)). The advected concentration of node j at

time $t+\Delta t$, $\bar{c}_j^{t+\Delta t}$, is given by that of the point x_j^*, which is tracked backward along the streamline from node j.

$$x_j^* = x_j - \int_t^{t+\Delta t} v \, dt \qquad\qquad j=1,2,\cdots,N \tag{2}$$

$$\bar{c}_j^{t+\Delta t} = \bar{c}(x_j, t+\Delta t) = c(x_j^*, t) \tag{3}$$

where x_j is the location of node j, $c\,(x_j^*, t\,)$ is the concentration of x_j^* at time t, Δt is time step size and N is the number of nodes.

If the tracked point, x_j^*, does not correspond to a fixed node, $\bar{c}_j^{\,t+\Delta t}$ is calculated by a finite element interpolation scheme written as

$$\bar{c}_j^{\,t+\Delta t} = \sum_{m=1}^{3} c\,(x_m, t\,)\phi_m(x_j^*) \tag{4}$$

where $\phi_m(x_j^*)$ is the basis function evaluated at x_j^* and x_m denotes the vertex of the triangular element surrounding x_j^*.

Second, forward tracking is used for nodes where concentration gradient is greater than a user-defined tolerance, or for nodes defined by the user as moving nodes. The node j is tracked forward along the streamline to the point x_j', and the concentration of the point at time $t+\Delta t$, $\overset{\circ}{c}_j^{\,t+\Delta t}$, is given by that of node j at time t.

$$x_j' = x_j + \int_t^{t+\Delta t} v\,dt \qquad\qquad j = 1, 2, \cdots, N \tag{5}$$

$$\overset{\circ}{c}_j^{\,t+\Delta t} = \overset{\circ}{c}\,(x_j', t+\Delta t) = c\,(x_j, t\,) \tag{6}$$

If the tracked point, x_j', does not correspond to a fixed node, a new node is created at the point as shown in Figure 1(b). In this manner, sharp fronts are kept at exact positions during the simulation. If the sharp front has passed through the area, the created nodes in the area are not necessary and are eliminated.

Complete mixing is assumed for calculating the concentration of the node where flow converges from more than one position in TRIPOR. Complete mixing is also assumed in TRINET at each intersections of channels if a node is tracked beyond one channel for a given Δt.

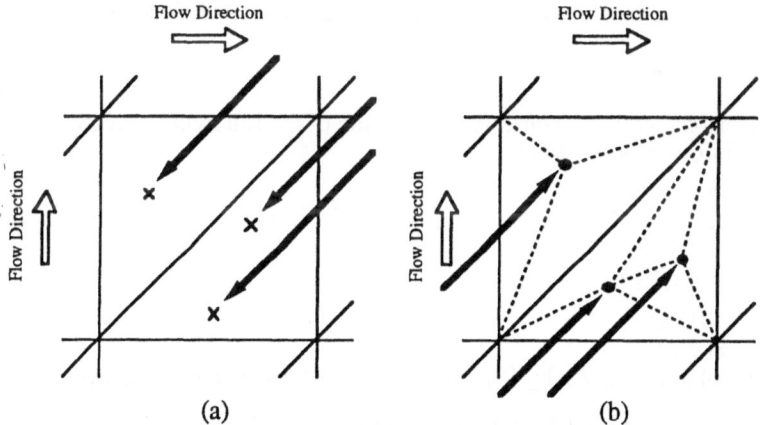

Figure 1. Tracking methods in TRIPOR: (a) backward tracking method, (b) forward tracking method: Cross denotes the point tracked backward. Solid circle and dotted line denote the new node and element, respectively, created by forward tracking method.

Finite element method for dispersion

As a Galerkin finite element method is used for discretization in space, the dispersion part of equation (1) can be written in matrix form as

$$[R]\left\{\frac{\partial c}{\partial t}\right\} + [P]\{c\} = \{F\} \tag{7}$$

$[R]$, $[P]$ and $\{F\}$ are given by

$$R_{ij} = \sum_{e=1}^{N} \int_{V^e} \phi_i^e \, \phi_j^e \, dV^e \tag{8}$$

$$P_{ij} = \sum_{e=1}^{N} \int_{V^e} (\nabla \phi_i^e) \cdot D \cdot (\nabla \phi_j^e) \, dV^e \tag{9}$$

$$F_i = \sum_{e=1}^{M} \int_{\Gamma^e} \phi_i^e \, n \cdot [D \cdot (\nabla c)] \, d\Gamma^e \tag{10}$$

where ϕ_i^e is the basis function of element e associated with node i, V^e is the region of the element, N is the set of elements connected to side i-j, Γ^e is the boundary of element e, M is the set of boundaries connected to node i and n is the unit vector normal to Γ^e and pointing outward. As a finite difference approximation is used for the time derivative, equation (7) can be written as

$$\left(\frac{R_{ij}}{\Delta t} + \theta P_{ij}\right) c_j^{t+\Delta t} = \left[\frac{R_{ij}}{\Delta t} - (1-\theta) P_{ij}\right] c_j^t + F_i \tag{11}$$

where $c_j^{t+\Delta t}$ and c_j^t are the concentrations of node j at time $t+\Delta t$ and t, respectively and θ is the weighting function for time.

APPLICATIONS

In the following preliminary studies, TRINET is used to model a porous medium by setting appropriate value of apertures for channels in a lattice configuration.

One-dimensional problem

The one-dimensional problem concerns the solution of the following advection-dispersion equation in a uniform velocity field.

$$\frac{\partial c}{\partial t} = D \frac{\partial^2 c}{\partial x^2} - v \frac{\partial c}{\partial x} \qquad\qquad 0 \leq x < \infty \tag{12}$$

subject to

$$c(x, 0) = 0 \qquad\qquad 0 \leq x < \infty$$
$$c(0, t) = 1 \qquad\qquad t > 0$$
$$c(x, t) \rightarrow 0 \qquad\qquad t > 0, x \rightarrow \infty$$

The analytical solution is given by Carslaw and Jaeger (1946) and can be written as

$$c(x, t) = \frac{1}{2} \operatorname{erfc}\left(\frac{x-vt}{\sqrt{4Dt}}\right) + \frac{1}{2} \exp\left(\frac{vx}{D}\right) \operatorname{erfc}\left(\frac{x+vt}{\sqrt{4Dt}}\right) \tag{13}$$

First, we solve the problem using a fine grid system [$\Delta x=0.01(0\leq x\leq 1)$]. Figure 2 and 3 show the results of our methods at $t=50$ for case A ($D=10^{-5}$, $v=10^{-2}$, $\Delta t=5$, Peclet number:$Pe=v\Delta x/D=10$, Courant number:$Cr=v\Delta t/\Delta x=5$) and case B ($D=10^{-6}$, $v=10^{-2}$,

Δt=5, Pe=100, Cr=5), respectively. In both cases, no nodes are defined as moving nodes. As can be seen in these figures, the results of both TRINET and TRIPOR agree very well with the analytical solution for a wide range of Peclet numbers and for Courant numbers well in excess of 1.

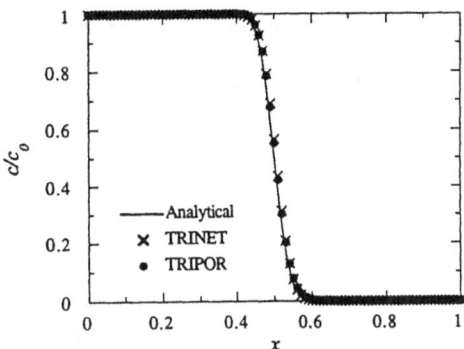

In order to demonstrate the capability of our methods, an irregular grid system is also used for solving this problem (case C). In this grid system, only the vicinity of the concentration boundary is discretized [Δx=0.01(0≤x≤0.2), no discretization for 0.2<x<1], and the nodes for 0≤x≤0.2 are defined as moving nodes. The parameters used here are the same as in case A except Pe and Cr. The results of TRIPOR at t=50 are shown in Figure 4. Although no nodes are initially set for 0.2<x<1, the code yields quite accurate results by creating new nodes in the vicinity of concentration front. This result suggests that hardly any attention has to be paid to discretizing the domain away from the concentration boundary.

Figure 2. Results obtained with TRINET and TRIPOR, and analytical solution for one-dimensional problem at t=50 for case A (Pe=10, Cr=5).

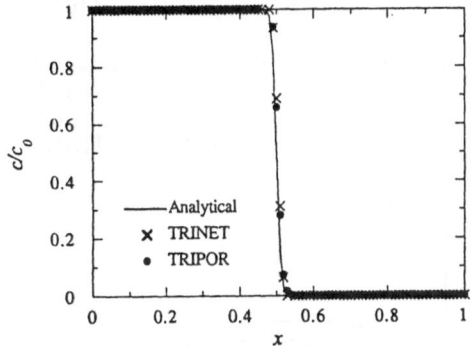

Figure 3. Results obtained with TRINET and TRIPOR, and analytical solution for one-dimensional problem at t=50 for case B (Pe=100, Cr=5).

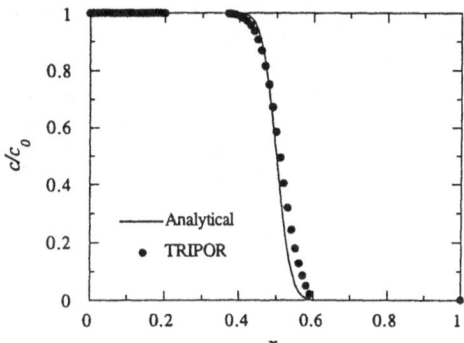

Figure 4. Results obtained with TRIPOR and analytical solution for one-dimensional problem at t=50 for case C [Δx=0.01 (0≤x≤0.2), no grid for 0.2<x<1].

Two-dimensional problem

The two-dimensional problem concerns the solution of the advection-dispersion equation in a uniform velocity field, which is written as

$$\frac{\partial c}{\partial t} = D_L \frac{\partial^2 c}{\partial c^2} + D_T \frac{\partial^2 c}{\partial c^2} - v \frac{\partial c}{\partial x} \tag{14}$$

subject to the following initial and boundary conditions

$$c\,(0, y, t) = 1 \qquad\qquad -a \le y \le a$$

$$c\,(0, y, t) = 0 \qquad\qquad y < -a,\ y > a$$

$$\lim_{y \to \pm\infty} \frac{\partial c}{\partial y} = 0, \quad \lim_{x \to \infty} \frac{\partial c}{\partial x} = 0$$

where D_L and D_T are longitudinal and transverse dispersion coefficient, respectively, and a is half length of a line source. The direction of flow is along the x-axis. The analytical solution of this problem is given by Javandel et al. (1984) and can be written as

$$c(x, y, t) = \frac{x}{4\sqrt{\pi D_L}} \exp\left(\frac{vx}{2D_L}\right) \int_0^t \exp\left[-\frac{v^2\tau}{4D_L} - \frac{x^2}{4D_L\tau}\right]\tau^{-3/2}$$

$$\cdot \left[\operatorname{erf}\left(\frac{a-y}{2\sqrt{D_T\tau}}\right) + \operatorname{erf}\left(\frac{a+y}{2\sqrt{D_T\tau}}\right)\right] d\tau \qquad (15)$$

A schematic view of this problem is shown in Figure 5. The parameters used in this problem are $D_L=10^{-2}$, $D_T=2.5\times10^{-3}$, $v=0.1$ and $a=0.5$. Two cases using different grid systems are analyzed. The geometric parameters for case D are $\Delta x=0.2$, $\Delta y=0.1$, $\Delta t=2(Pe=2, Cr=2)$ and those for case E are $\Delta x=0.5$, $\Delta y=0.1$, $\Delta t=5(Pe=5, Cr=1)$. The concentration distribution obtained with TRIPOR for case D, which is almost identical to that of analytical solution, is shown in Figure 6. The concentration profiles obtained with TRINET and TRIPOR at different coordinates are shown in Figure 7. Although smaller Peclet number (case D) yields better results, the results of all cases agree favorably well with the analytical solution.

Because the domain is discretized along the flow direction, new nodes are created along elements in TRINET and along the sides of elements in TRIPOR. Therefore, not much difference is seen between the results of the two codes. Further study is needed where the flow direction is variable in space to demonstrate the advantage of TRIPOR over TRINET for two-dimensional problems.

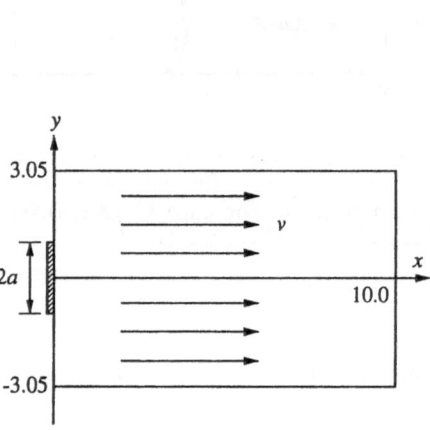

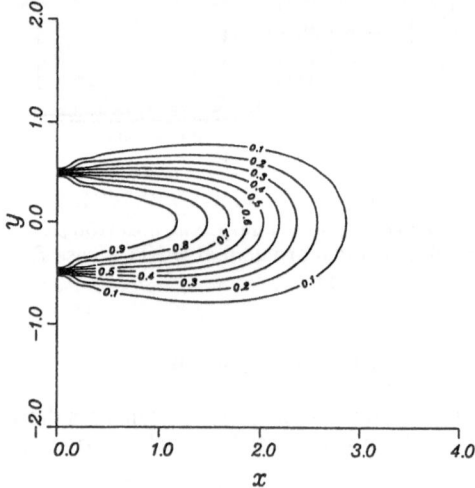

Figure 5. A schematic view of the two-dimensional model.

Figure 6. Concentration distribution obtained with TRIPOR at $t=20$ for case D.

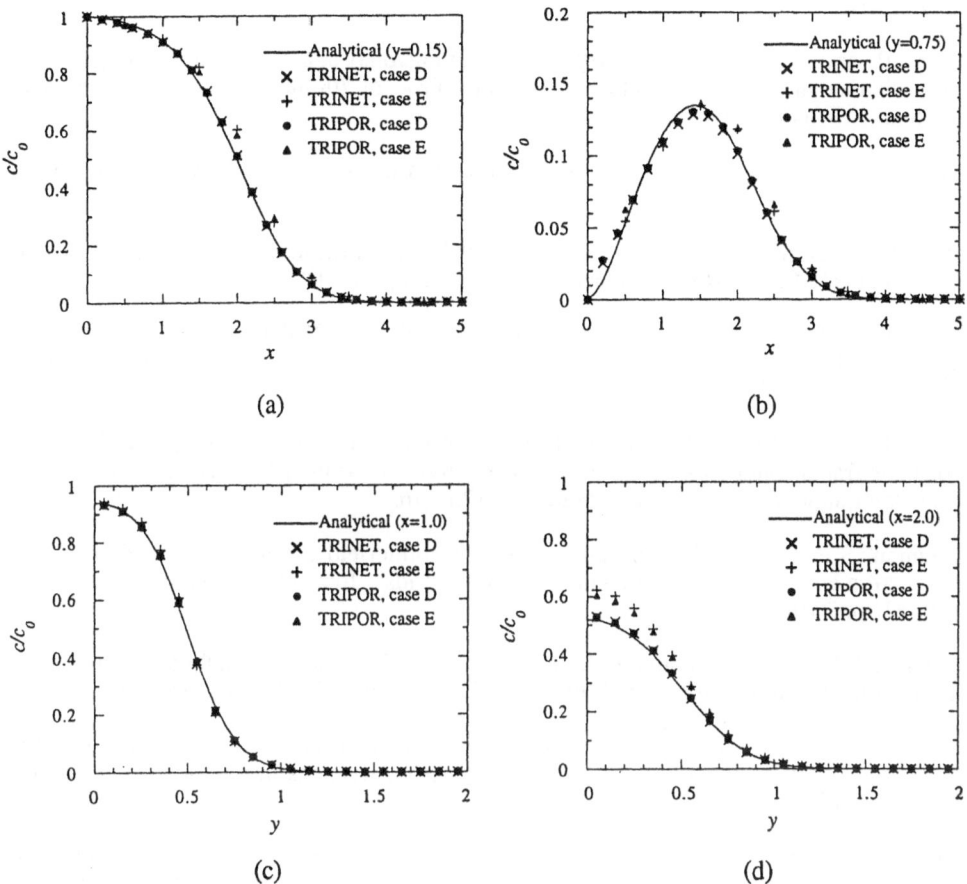

Figure 7. Concentration profiles obtained with TRINET and TRIPOR, and analytical solution for two-dimensional problem at $t=20$ for case D and E: (a) $y=0.15$, (b) $y=0.75$, (c) $x=1.0$, (d) $x=2.0$.

CONCLUSIONS

Our preliminary studies suggest that our method is capable of accurately solving advection-dispersion problems for a wide range of Peclet numbers and for Courant numbers well in excess of 1. Since the codes create new nodes in the vicinity of the concentration fronts at each time step, hardly any attention has to be paid to the discretization of space away from the concentration boundary.

The code, however, sometimes creates flat elements (high aspect ratio elements) when the tracked points are close to the sides of elements. These flat elements will affect the convergence of the matrices and the accuracy of the results will be decreased. Further studies are necessary in this respect.

ACKNOWLEDGMENT

This work is supported by TAISEI Corporation in Japan. Thanks are due to Christine Doughty for a review of the manuscript.

REFERENCES

Cady, R. and Neuman, S.P. (1988) "Three-dimensional adaptive Eulerian-Lagrangian finite element method for advection-dispersion", Proc. of 7th Int. Conf. on Computational Methods in Water Resources MIT, pp.183-193.

Carslaw, H.S. and Jaeger, J.C. (1946) Conduction of heat in solids, Clarendon Press, Oxford.

Javandel, I., Doughty, C. and Tsang, C.F. (1984) Groundwater Transport, Handbook of Mathematical Models, American Geophysical Union, pp.14-19.

Karasaki, K. (1987) "A new advection-dispersion code for calculating transport in fracture networks", Earth Sciences Division Annual Report 1986, Lawrence Berkeley Laboratory Report LBL-22090, pp.55-57.

Karasaki, K. (1988) "Modification to TRINET: A three-dimensional advection-dispersion code for fracture networks", Earth Sciences Division Annual Report 1987, Lawrence Berkeley Laboratory Report LBL-24200, pp.175-176.

Neuman, S.P. (1981) "A Eulerian-Lagrangian numerical scheme for the dispersion-convection equation using conjugate space-time grids", Jour. of Computational Physics, Vol. 41, pp.270-294.

Neuman, S.P. (1984) "Adaptive Eulerian-Lagrangian finite element method for advection-dispersion", Int. Jour. Numerical Methods in Engineering, Vol. 20, pp.321-337.

Sloan, S.W., and Houlsby, G.T. (1984) "An implementation of Watson's algorithm for computing 2-dimensional delaunay triangurations", Advances in engineering software, Vol. 6, pp.192-197.

Yeh, G.T. (1990) "A Lagrangian-Eulerian method with zoomable hidden fine-mesh approach for solving advection-dispersion equations", Water Resources Research, Vol. 26 No.6, pp.1133-1144.

LAGRANGE–GALERKIN APPROXIMATION FOR ADVECTION–DOMINATED NONLINEAR CONTAMINANT TRANSPORT IN POROUS MEDIA

P. KNABNER[1], J.W. BARRETT[2] and H. KAPPMEIER[1]

[1] University of Erlangen–Nürnberg
Institute of Applied Mathematics
Martensstr. 3, D 91058 Erlangen
Germany

[2] Imperial College
Department of Mathematics
London SW7 2BZ
United Kingdom

We describe an extension of the Lagrange–Galerkin approach for advection problems with nonlinear (non–)equlibrium adsorption, possibly with isotherms of the Freundlich type. Numerical examples indicate that this method allows for large time steps and still gives exact approximations of shocks and sharp fronts.

INTRODUCTION

A basic problem in soil science and subsurface hydrology is the numerical simulation of the transport of solutes by dispersion and advection, which in addition undergo various reactions like equilibrium and non–equilibrium adsorption. A typical model takes the form

$$\Theta \partial_t c + \rho \partial_t s - \nabla \cdot (\Theta \underline{D} \nabla c) + \underline{q} \cdot \nabla c = f,$$

$$\partial_t s = k(\varphi(c) - s),$$

where c denotes the dissolved and s the adsorbed concentration in the usual reference system, and $\Theta, \rho, \underline{D}, \underline{q}$ are the geohydrological quantities, assumed to be known, in standard

299

A. Peters et al. (eds.), Computational Methods in Water Resources X, 299–307.

notation. Furthermore $k < \infty$ is a rate parameter such that an equilibrium description is formally included by $k = \infty$. Various aspects of complication in this equation are:

- Possibly an advection–dominated flow, i.e. $\|\underline{q}\|L/\|\underline{D}\|$ large, where $L > 0$ is an appropriate length unit.
- Possibly a nonlinear and degenerate isotherm φ (i.e. unbounded derivative at $c = 0$)
- Non–equilibrium.

Up till now these aspects have only been considered seperately. In particular, for advection dominated linear transport problems, a variety of methods have been designed. Among these, the Lagrange–Galerkin Method (or Modified Method of Characteristics or Eulerian-Lagrangian Method or....) has the advantage to allow for large time steps without stability problems. From the vast amount of literature in mathematics and engineering, we refer to Douglas and Russel, 1982, and Neuman, 1984... Nevertheless there are only few papers, dealing with nonlinear problems. We will review these attempts and indicate our own approach.

NUMERICAL SCHEMES

We will restrict ourselves to a one–dimensional advection problem with equilibrium adsorption. The schemes to be developed can be extended to the non–equilibrium case. We will also indicate how to incorporate dispersion in a way analogous to the Lagrange–Galerkin method for linear problems. In principal our approach can be extended to multiple space dimensions, but to the expense of increased computational and algorithmic complexity. We postpone this point to a future paper.

Linear advection problems

To review the Lagrange–Galerkin method, we start with the linear initial–boundary problem

$$\partial_t u + b\partial_x u = 0 \quad \text{for} \quad t \in (0,T], x \in (0,L) \tag{1}$$

$$u(0,t) = h(t) \quad \text{for} \quad t \in (0,T], \tag{2}$$

$$u(x,0) = g(x) \quad \text{for} \quad x \in (0,L), \tag{3}$$

where $b = b(x,t)$ is a function of space and time.

The *method of characteristics* is based on the observation that if for fixed $\overline{x} \in (0,L), \overline{t} \in (0,T)$ we define the corresponding *characteristic* to be the (local) solution of

$$\frac{dx}{dt} = b(x,t), \qquad t \in (0,\bar{t})$$
$$x(\bar{t}) = \bar{x}$$

$$(4)$$

then i.e. u is constant along characteristics. We can apply this locally in time, i.e. for a given time discretisation

$$0 = t_0 < \cdots < t_{n-1} < t_n < \cdots < T \tag{5}$$

we compute an approximate solution at level t_n by *backtracing* along the characteristics till we reach level t_{n-1}. At each time level we assume the approximation to be given by linear finite elements (splines) for a given spatial discretization

$$0 = x_0 < \cdots < x_{j-1} < x_j < \cdots < L, \tag{6}$$

denoted by U^n. Therefore it is sufficient to backtrace only the nodal values, denoted by U_j^n for node x_j and time t_n. The advantage of such an approach is the possiblity at large time steps, i.e. of Courant numbers far beyond 1. Nevertheless we aim at a scheme which has no or little discretization error in time, such that the accuracy of the solution can be enhanced only by spatial refinement, probably in a local and adaptive manner, in particular in the presence of sharp fronts and shocks, as it will be the case for the nonlinear problems to be discussed. Therefore it is of paramount importance that the characteristics are approximated precisely. If we assume that the coefficient b is constant on each space–time cell $(x_{j-1}, x_j) \times (t_{n-1}, t_n)$ with a value $b_{j-1/2}^{n-1/2}$, then (4) can be resolved exactly on (t_{n-1}, t_n) for the nodal points, as

$$\frac{dx}{dt} = b(x,t) = b(x) \quad \text{for} \quad t \in (t_{n-1}, t_n),$$
$$x(t_n) = x_j, \qquad x(t_{n-1}) = \bar{x}_j$$

$$(7)$$

with the unknown backtraced node $\bar{x}_j$ to be determined, i.e.

$$\int_{\bar{x}_j}^{x} \frac{dx}{b(x)} = \int_{t_{n-1}}^{t_n} dt =: \Delta t^n$$

and thus $\bar{x}_j \in [x_{j-(m+1)}, x_{j-m}]$, where m and $\bar{x}_j$ are such, that

$$\sum_{k=1}^{m} \frac{h^{j+1-k}}{b_{j+1/2-k}^{n-1/2}} + \frac{x_{j-m} - \bar{x}_j}{b_{j-1/2-m}^{n-1/2}} = \Delta t^n \tag{8}$$

where $h^j := x_j - x_{j-1}$, or equivalently:

Let $m \geq 0$ be the first Integer, for which

$$\sum_{k=1}^{m+1} \frac{h^{j+1-k}}{b_{j+1/2-k}^{n-1/2}} > \Delta t^n, \tag{9}$$

then

$$\overline{x}_j := x_{j-m} + b_{j-m-1/2}^{n-1/2} \left[\sum_{k=1}^{m} \frac{h^{j+1-k}}{b_{j+1/2-k}^{n-1/2}} - \Delta t^n \right]$$

and finally

$$U_j^n := U^{n-1}(\overline{x}_j). \tag{10}$$

If the coefficient is a general (smooth) space dependent function, we can approximate it by piecewese constants by

$$b_{j-1/2} := \frac{1}{2}(b(x_j) + b(x_{j-1})) \tag{11}$$

and as we only introduce a spatial error still have convergence for fixed Δt^n and $h := \max h^n \to 0$:

$$U_j^n \to u(x_j, t^n) \tag{12}$$

The Lagrange–Galerkin method now consists in a combination of backtracing and the Galerkin approach, i.e. for an initial-boundary value problem with the differential equation

$$\partial_t u - \partial_x(D\partial_x u) + b\partial_x u = f$$

u^n is determined by

$$hU_j^n + \Delta t \int_0^L D\partial_x U^n \partial_x \chi_j \, dx = h(U^{n-1}(\overline{x}_j) + f(x_j, t^n)), \tag{13}$$

where χ_j are the hat basis functions, and for notational simplicity we restrict overselves from now on to equidistant discretizations. The trapezoidal rule has been used in (13) to approximate the mass matrix terms, but one may also think of other quadrature rules.

Linear advection with nonlinear equilibrium adsorption

We return to problem (1-3), but we substitute (1) by

$$\partial_t u + \partial_t \varphi(u) + q\partial_x u = 0 \tag{14}$$

where the isotherm φ is assumed to be monotonically increasing, possibly with $\varphi'(0) = \infty$, as it is the case for the Freundlich isotherm

$$\varphi(s) = As^p, \qquad A > 0, \qquad p \in (0, 1). \tag{15}$$

For the sake of simlicity we assume $q > 0$ to be constant. It has been pointed out by van Duijn and Knabner, 1992, that the presence of such a nonlinearity may have significant influence on the shape and travel speed of fronts, also in the presence of dispersion. Therefore a Lagrange–Galerkin method should incorporate the nonlinearity in the definition of the characteristics, in particular to achieve the precise travel speed also for fixed time discretization. The method should also be capable to approximate shocks.

The first approach only uses the linear advection part for backtracing according to (10), i.e. the pore velocity, and approximates $\partial_t \varphi(u)$ by the standard time difference quotient, i.e. the *pore velocity scheme* reads

$$U_j^n + \varphi\left(U_j^n\right) = U^{n-1}(x_j - q\Delta t) + \varphi\left(U_j^{n-1}\right). \tag{16}$$

This approach has been investigated e.g. by Yeh and Gwo, 1992, for linear φ, and order of convergence estimates for nonlinearities of the type (15) and including dispersion may be found in Dawson et. al., 1994 and Barrett and Knabner, 1995.

To incorporate φ in the definition of the characteristics, we can rewrite (14) as

$$\partial_t u + \frac{q}{1 + \varphi'(u)} \partial_x u = 0$$

and thus define the characteristic by

$$\frac{dx}{dt} = \frac{q}{1 + \varphi'(u(x(t), t))}, \quad t \in \left(t^{n-1}, t^n\right), \quad x(t^n) = x_j. \tag{17}$$

Here u is still unknown, such that despite of the previous considerations we are lead to a simple approximation of (17), e.g. one step of the implicit Euler scheme. This leads to the scalar nonlinear equation

$$\overline{x}_j + \frac{q\Delta t}{1 + \varphi'\left(U^{n-1}\left(\overline{x}_j\right)\right)} = x_j \tag{18}$$

and then

$$U_j^n = U^{n-1}(\overline{x}_j). \tag{19}$$

The explicit Euler method leads to a coupling of $\overline{x}_j$ and U_j^n. The approximation (18) has been used by Russell, 1983, for Burgers' equation, which by a change of variables can be cast in the form (14). A sufficient condition for the unique solvability of (18), e.g. by fixed point iteration, is given by

$$\left\| q\Delta t \frac{d}{dx} \left[\frac{1}{1 + \varphi'(U^{n-1}(x))} \right] \right\| < 1. \tag{20}$$

To be able to satisfy this condition for nonlinearities of the type (15) we are forced to *regularize*, i.e. to subsitute φ by φ_ε, which differs only on $[0, \varepsilon]$, e.g. defined by $\varphi_\varepsilon(s) = \varphi(\varepsilon)/\varepsilon\, s$, what we will do from now on. But even then, condition (20) requires small Δt in the presence of sharp fronts and cannot be fuffilled, if a shock develops.

For linear φ, condition (20) is void, and (18) is the exact evaluation of the characteristic, thus taking into account the retarded velocity caused by the adsorption. Therefore we may call this approach the *retarded velocity scheme*.

The new scheme

Our aim is to incorporate the 'relevant' part of the nonlinearity in the definition of the characteristic, but to avoid a condition like (20). Thus we write (14) in the form

$$\partial_t(\varphi(u) - \alpha(x)u) + (1 + \alpha(x)) \left[\partial_t u + \frac{q}{1 + \alpha(x)} \partial_x u\right] = 0 \qquad (21)$$

with a piecewise linear function α still to be determined and apply standard differencing to the first part and the approach of (7)–(12) to the second term (with $b(x) = q/(1 + \alpha(x))$). This leads to

$$U_j^n + \varphi(U_j^n) = (1 + \alpha_j)U^{n-1}(\overline{x}_j) - \alpha_j U_j^{n-1} + \varphi(U_j^{n-1}), \qquad (22)$$

where

$$\overline{x}_j = x_{j-m} + \frac{1}{1 + \alpha_{j-m-1/2}} \left[h \sum_{k=1}^{m} \left(1 + \alpha_{j+1/2-k}\right) - q\Delta t\right] \qquad (23)$$

and $m \geq 0$ is the first integer for which

$$h \sum_{k=1}^{m+1} \left(1 + \alpha_{j+1/2-k}\right) > q\Delta t. \qquad (24)$$

Note that for $\alpha(x) = 0$ we are back to the pore velocity scheme.
Instead of this we want to have

$$U_j^n = U^{n-1}(\overline{x}_j) \qquad (25)$$

which because of (22) is equivalent to

$$\alpha_j = \begin{cases} \dfrac{\varphi(U^{n-1}(\overline{x}_j)) - \varphi(U_j^{n-1})}{U^{n-1}(\overline{x}_j) - U_j^{n-1}}, & \text{if } U_j^{n-1} \neq U^{n-1}(\overline{x}_j) \\[2mm] \varphi'(U_j^{n-1}), & \text{otherwise.} \end{cases} \qquad (26)$$

Note that (23)–(26) is an implicit definition of $\alpha_j : \alpha_j$ depends on $\overline{x}_j$, but $\overline{x}_j$ depends on α_j (and possibly all $\alpha_1, \ldots, \alpha_{j-1}$). The one–dimensional situation simplifies this dependence insofar as, when the scheme is performed in the direction of the stream, i.e. for $j = 1, 2, \ldots$, the $\alpha_j's$ are determined successively such that (23), (26) infact is a nonlinear dependence of $\overline{x}_j$ and α_j, which can be resolved e.g. by fixed point iteration:

Given $\alpha_j^{(i)}, \overline{x}_j^{(i)}$ is determined by (23) and with $U_j^{n,(i)} := U^{n-1}(\overline{x}_j^{(i)})$ we determine $\alpha_j^{(i+1)}$ by (26) with $\overline{x}_j$ substituted by $\overline{x}_j^{(i)}$. The process can be started by $\alpha_j^{(0)} = 0$.

We may call this method the *fully implicit scheme.*

Our approach has some similarity to the one of Dahle et. al., 1994, but among other differences we can deal with time instead of space derivatives for the nonlinear part.

NUMERICAL RESULTS

We present computations for two examples, where the exact solution is known. In both cases $\varphi(u) = u^{1/2}$, regularized as indicated, and $q = 1$. Boundary and initial data are induced by the exact solutions. We use $\Delta x = 0.1, \Delta t = 0.93$, i.e. a Courant number for the related linear problem of 9.3 and $\varepsilon = 10^{-4}$. The first example is a shock, which in the case of non–equilibrium and/or dispersion would be smoothed to a travelling ware with the same speed (van Duijn and Knabner, 1992). Figure 1 shows that the pore velocity scheme is not capable in approximating the front speed and shape correctly. The same applies to the retarded velocity scheme (Figure 2).

Only the fully implicit scheme delivers the exact solution without temporal discretization error (Figure 3). The displayed time level is $t = 9.3$.

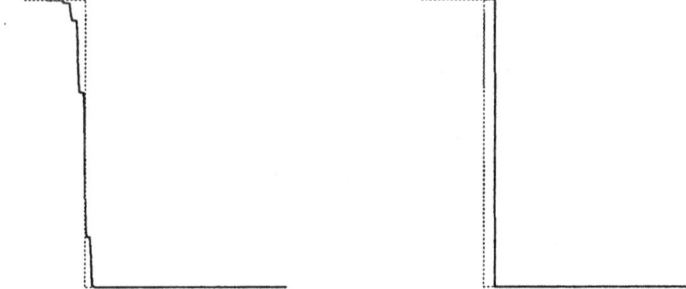

Figure 1: Shock approximation   Figure 2: Shock approximation
with pore velocity scheme   with retarded velocity scheme

The second example is the development of a pulse, taken from Grundy et. al., 1994.
It combines a shock with an expansion ware, and also by its changing monotonicity pro-

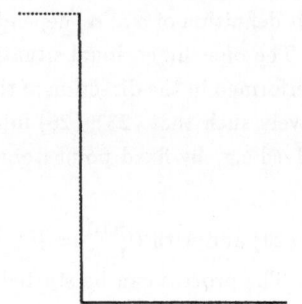

Figure 3: Shock approximation
with fully implicit scheme

duces harder requirements than the first example. Only the fully implicit scheme is able to approximate the shock very well and also the expansion wave reasonably well (Figure 5), the approximation by the pore velocity scheme induces an enormous amount of dispersion and also gives a wrong shock speed (Figure 4).

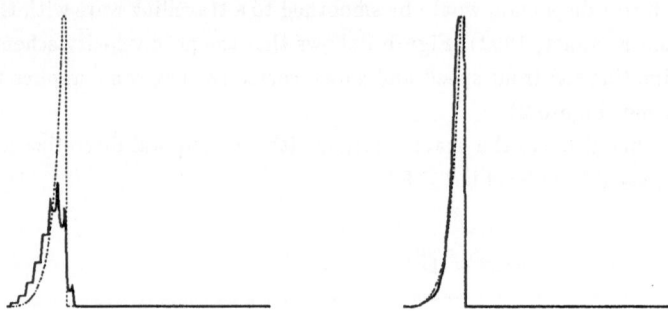

Figure 4: Pulse approximation   Figure 5: Pulse approximation
with pore velocity scheme   with fully implicit scheme

Results of the retarded velocity scheme cannot be displayed, as there were severe convergence problems in (18).

REFERENCES

Barrett, J. W. and Knabner P. (1995) "Lagrange Galerkin finite element approximation of transport of reactive solutes in porous media", in preparation.

Dahle, H. K., Ewing, R. E. and Russell, T. F., (1994) "Eulerian–Lagrangian localized adjoint methods for a nonlinear advection–diffusion equation", submitted to CMAME.

Dawson, C., Duijn, C. J. van and Wheeler, M.F. (1994) "Characteristic–Galerkin methods for contaminant transport with non–equilibrium adsorption kinetics", to appear in SIAM J. Numer. Anal.

Douglas, J. and Russell, T. F. (1982) "Numerical methods for convection–dominated diffusion problem based on combining the method of characteristics with finite element or finite difference procedures", SIAM J. Numer. Anal. 19, 871–885.

Duijn, C. J. van and Knabner P. (1992) "Travelling waves in the transport of reactive solutes in porous media: Adsorption and binary ion exchange, Part 1, 2", Transport in Porous Media 8, 167–194, 199–226.

Grundy, R., Duijn, C. J. van, Dawson, C. (1994) "Asymptotic profiles with finite mass in one–dimensional contaminant transport through porous media: The fast reaction case", in press in Q. J. Mech. Appl. Math.

Neuman, S. P. (1984) " Adaptive Eulerian–Lagrangian finite element method for advection–dispersion", Int. J. Num. Meth. Engrg. 20, 321–337.

Russell, T. F. (1983) "Galerkin time stepping along characteristics for Burgers' equation", in R. Stepleman et. al. (eds.), Scientific Computing, North–Holland, 183–192.

Yeh, G. T. and Gwo, J. P (1992) "A Lagrangian–Eulerian approach to modeling multicomponent reactive transport", in Computational Methods in Subsurface Hydrology, 419–427.

OPERATOR SPLIT FOR THREE DIMENSIONAL MASS TRANSPORT EQUATION

C. König
Ruhr–Universität Bochum
Theorie der Tragwerke und Simulationstechnik
Bochum
Germany

ABSTRACT: This paper presents a Finite Element Method for the analysis of three–dimensional flow and transport processes in porous media. By using accelerating algorithms the computation time for very large problems is considerably reduced. The resulting operation split method with extrapolation can be used together with a PCG solver. This method split the mass transport equation into two parts. In the first step dispersion/diffusion is calculated with the help of a Galerkin scheme while advection is solved by a least square method on the second step. With this combination an upwind–weighting in the flow direction is obtained, which can be interpreted as an artificial dispersion. The main advantage of this method is the achievement of a symmetrical set of equations and a well conditioned matrix. Consequently a PCG solver with a high acceleration in comparison to other common solvers can be used most efficiently for large three–dimensional mass transport equations. Various case studies are presented.

1.0 INTRODUCTION

In consequence of progressing pollution of ground water through agriculture, industry and waste repository the question often arises about the origin of the ground water as well as the spreading of the substances dissolved in it. For the calculation and prognosis of the spreading of substances in ground water, more and more three–dimensional simulation models are being applied. Since the Finite Element Method can describe the complex three–dimensional, inhomogeneous hydrogeological facts more precisely than other methods this technique is often used for the calculation of transport problems. In contrast to the analysis of ground water flow the following problems arise: stability and convergence criteria to prevent numerical dispersion and oscillation require fine discretisation in time and space. Especially in three–dimensional models the amount of nodes can easily rise up to a number of 100.000 unknowns, which have to be solved for a multitude of time steps. Due to convection the coefficient matrix becomes unsymmetric so that powerful available solution techniques for symmetric matrices can't be employed. The large amount of computation time used by the Gauss solver with its nearly N^3 basic operations (where N is the number of nodes) as well as the filling in of the entire matrix during the inversion process don't allow a conventional solution, due to storage limitation and processing performance of available workstations. An alternative, shown here, is the PCG technique. Its efficiency for ground water flow has been

A. Peters et al. (eds.), Computational Methods in Water Resources X, 309–316.

demonstrated [2].

In this paper the solution procedure is extended to mass transport problems, even in the case of a large number of unknowns. The resulting equations can be solved effectively on workstations.

2.0   MASS TRANSPORT EQUATION

Considering that the ground water flow is not influenced by the solute, the solute transport and the ground water flow are decoupled. The distribution of the substances in ground water is determined by convection, hydrodynamic dispersion and sorption as well as chemical and biological decay reaction. The hydrodynamic dispersion contains the physical phenomena of diffusion and hydromechanical dispersion. The partial differential mass transport equation is:

$$\frac{\partial c}{\partial t} + v \; \boldsymbol{grad} c + cq - div(\boldsymbol{D} \; \boldsymbol{grad} c) = qc^* + r, \quad (1)$$

where
$\quad c \quad$ solute concentration,
$\quad t \quad$ time,
$\quad v \quad$ velocity vector,
$\quad q \quad$ source,
$\quad \boldsymbol{D} = D_m \boldsymbol{I} + \boldsymbol{D}_d \quad$ diffusion/ dispersion tensor,
$\quad c^* \quad$ concentration source,
$\quad r \quad$ chemical and biological reaction.

For reasons of simplification the term r is disregarded in the further examination of the transport equation.

The initial and boundary conditions in the region Ω with boundary Γ are:

$$c(t_0) = \bar{c}_0 \qquad in \; \Omega \qquad for \; t_0 = 0,$$
$$c = \bar{c} \qquad on \; \Gamma_1,$$
$$\boldsymbol{D}(\boldsymbol{grad} c)\boldsymbol{n} = j_r \qquad on \; \Gamma_2,$$

where $\boldsymbol{n}$ is the outward unit vector normal to the surface.

3.0 OPERATOR SPLIT FOR THE TRANSIENT MASS TRANSPORT EQUATION

As a consequence of the problems mentioned above a technique has been developed which stabilizes the mass transport equation and leads to a symmetric system of equations. The mass transport equation can be split into two problems which can be solved in two steps [1]. This technique can only be applied to transient problems.
The transient mass transport equation can be written as:

$$\frac{\partial c}{\partial t} + A_1 c + A_2 c = f \tag{4}$$

where

$$A_1 = - \ div(\boldsymbol{D} \ \boldsymbol{grad}(\ )) + q(\ ), \tag{5}$$

$$A_2 = \boldsymbol{vgrad}(\ ), \tag{6}$$

$$f = qc^* . \tag{7}$$

The time discretisation results in:

$$\frac{\Delta c}{\Delta t} + \Theta(A_1 + A_2)\Delta c = - (A_1 + A_2)c^n + f, \tag{8}$$

where

$$c^{n+1} = c^n + \Delta c \tag{9}$$
$$0 \le \Theta \le 1 .$$

In the splitting technique the operators A1 and A2 are worked off step by step. As a first step the term Δc is substituted by $'e'$ and equation (8) becomes:

$$\frac{e}{\Delta t} + \Theta A_1 e = - (A_1 + A_2)c^n + f. \tag{10}$$

In the second step $'g'$ is determined using $'e'$ previously obtained from equation (10):

$$\frac{g}{\Delta t} + \Theta A_2 g = \frac{e}{\Delta t}. \tag{11}$$

The final solution for the time step n+1 is given by

$$c^{n+1} = c^n + g . \tag{12}$$

3.1 First step of the operator split

The Galerkin method applied for the first step can be written as:

$$\left[\int_\Omega \frac{1}{\Delta t}\phi_i\phi_j d\Omega + \Theta \int_\Omega D_{kl}\phi_{i,k}\phi_{j,l}d\Omega + \quad + \Theta \int_\Omega q\phi_i\phi_j d\Omega\right] e_j = \left[\int_\Omega D_{kl}\phi_{i,k}\phi_{j,l}d\Omega + \right.$$

$$\left. + \int_\Omega v_k\phi_i\phi_{j,k}d\Omega + \int_\Omega q\phi_i\phi_j d\Omega\right] c_j^n + qc_i^* \qquad (13)$$

The system of equations is symmetric because the convective term is on the right–hand side.

3.2 Second step of the operator split

By applying the least square method to equation (11) we obtain in the second step the discretised form [1]:

$$\left[\int_\Omega \frac{1}{\Delta t}\phi_i\phi_j d\Omega + \Theta \int_\Omega v_k\phi_i\phi_{j,k}d\Omega + \Theta \int_\Omega v\phi_{i,l}\phi_j d\Omega + \Theta^2 \int_\Omega \Delta t v_k v\phi_{i,l}\phi_{j,k}d\Omega\right] g_j =$$

$$\left[\int_\Omega \frac{1}{\Delta t}\phi_i\phi_j d\Omega + \Theta \int_\Omega v\phi_{i,l}\phi_j d\Omega\right] e_j. \qquad (14)$$

The result is again a symmetric equation system.

3.3 Valuation of the splitting technique

In [1] it has been shown that equation (14) can be interpreted as the discretized form of equation (11), weighted with a test function consisting of two parts (see Fig. 1). It represents an upwind weighting. The upwind weighting leads to a higher stability of the scheme and to a reduction of physically incorrect oscillations.

The Galerkin method in the first step, leading to equation (13), corresponds to a weighting function with the same test function as shown in Fig. 1 but with $v = 0$.

The operator splitting procedure is stable with $\Theta > 0.5$, as numerical studies have shown [1].

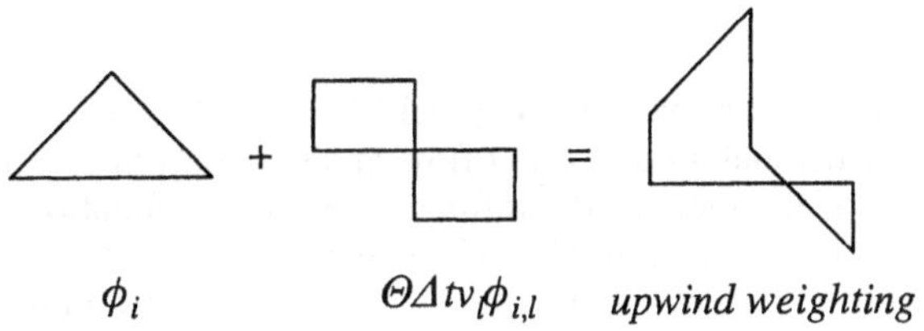

$$\phi_i \qquad\qquad \Theta \Delta t v_l \phi_{i,l} \qquad upwind\ weighting$$

Fig. 1 Test function of the operator split

In the equation (14) the term

$$\Theta^2 \Delta t v_k v_l \phi_{i,l} \phi_{j,k}$$

should be seen as an upwind weighting and can be interpreted as an artificial diffusion in the flow direction. It is responsible for the high stability of the procedure.

A further advantage of this technique is the symmetric system of equations in each operator step. As a result a PCG equation solver, highly effective for symmetric matrices, can be used. The operator splitting technique results in well conditioned matrices which accelerates the iterative solver.

4 APPLICATION

As part of a project in commission of the Federal Ministry of Research and Technology of Germany a three–dimensional ground water flow and transport model has been developed [4]. This numerical model simulates the influence of a contamination wave, propagating down the Rhine, upon the water quality in wells situated near the Rhine bank. Sorption and decay are not taken into consideration. The model comprises the water works Flehe (Düsseldorf), situated on the right hand side of the Rhine, the Rhine River and part of the left Rhine side.

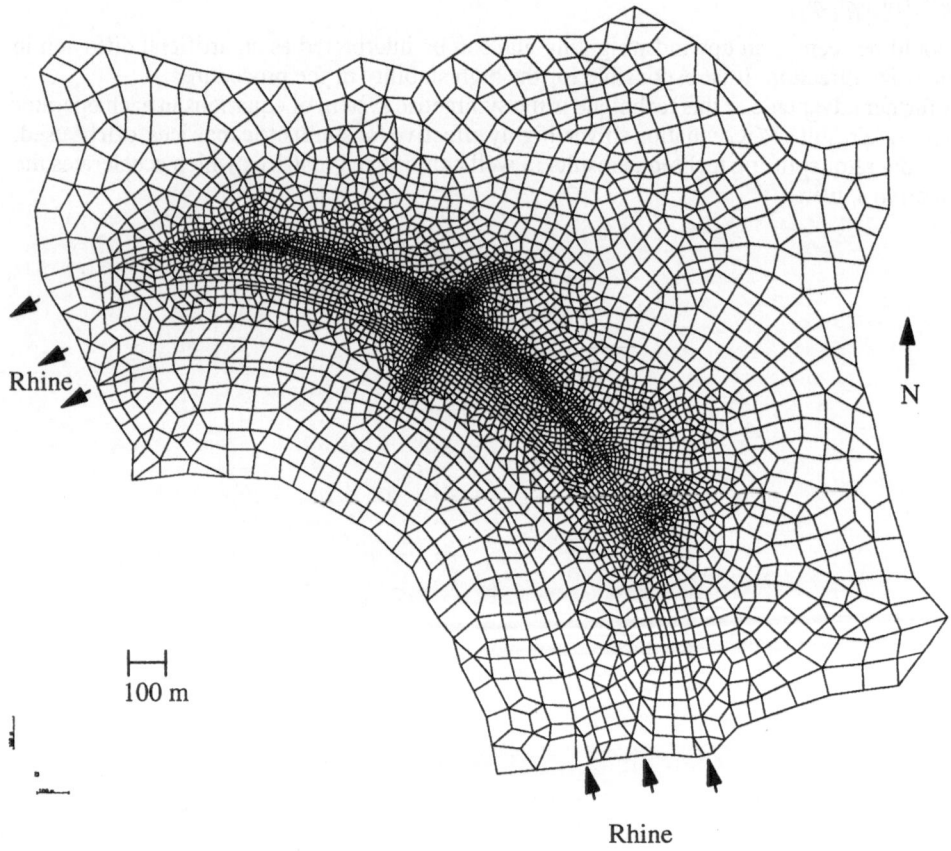

Fig 3 Horizontal mesh

The model supplies the water works with management measures to avert possible impairments of the water quality in the wells.

The three–dimensional groundwater flow and transport model is part of a regional horizontal model which comprises the whole area of the city of Düsseldorf (see Fig 3). In order to grasp the geological situation and to get an adequate discretisation in the depth of the aquifer, the model is subdivided into seven rows of layers, as shown in Fig 4. The top boundary represents the ground level and the bottom boundary the tertiary surface.

From an industrial plant situated upstream, chloride is continuously injected into the Rhine, producing a time depending chloride concentration. Chloride is particularly suitable as a tracer, since it doesn't react with other substances dissolved in the ground water. Therefore it can be used as a tracer in ground water transport problems. The development of chloride concentration has been used to calibrate the model. For the calibration some multi–level chloride measurements at several locations were available.

The flow and the transport are simulated with a Finite Element model having 39048 unknowns using 3,7 MB main memory. They are solved for 65 time steps using 16 seconds (IBM 6000/370) per time step. Only with the method presented here such a large problem could be reasonably solved on a workstation. Conventional methods would have far too much exceeded available storage and allowable computation time.

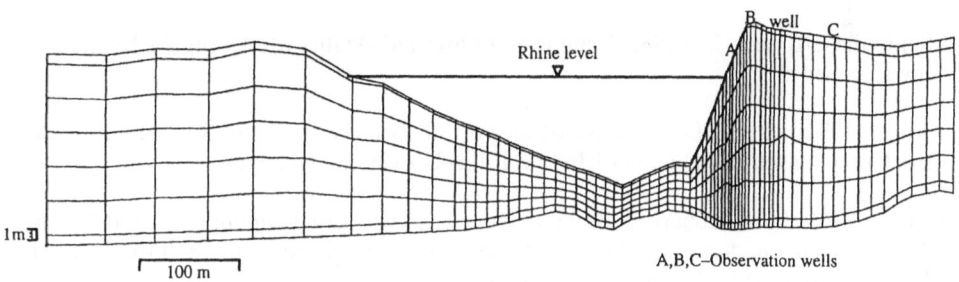

Figure 4. Vertical mesh

5 CONCLUSION

An operator splitting technique has been presented for the ground water mass transport equation. It leads to a symmetric equation system. This can be treated efficiently with a PCG solver. As a consequence very large 3–D problems, where the time and space discretisation fulfills the Peclet and Courant criteria can be attacked. Until now these problems had to be solved on a super–computer.

The second step of the operator split is discretized by the least square method. This leads to an upwind–type weighting function with a stabilizing characteristic, preventing numerical oscillations in the calculated concentrations. The power of the developed algorithm is demonstrated. It refers to a practical case where contamination could endanger the water quality. The method presented here allows today to simulate multi–component transport problems having up to 70.000 nodes . And this even on a workstation.

6 REFERENCES

[1] König, C. 1991. Numerische Berechnung des dreidimensionalen Stofftransports im Grundwasser. TWM Nr. 91–13, Institut für Konstruktiven Ingenieurbau, Ruhr–Universität Bochum.

[2] Schmid, G., Braess, D. 1988. Groundwater Flow and Quality Modelling, D. Reidel Publishing Company.

[3] AGM, 1991, Abschlußbericht: Stofftransport im Grundwasser, ein Simulationsmodell für das Genossenschaftsgebiet der LINEG, Ruhr–Universität Bochum.

[4] AGM, 1991, Abschlußbericht: Sicherheit der Trinkwassergewinnung aus Uferfiltrat bei Stoßbelastungen, ein dreidimensionales Grundwasser– und Transportmodell für die Stadtwerke Düsseldorf, Ruhr–Universität Bochum.

DELINEATING A CONTAMINANT PLUME BOUNDARY

W. A. McGRATH and G. F. PINDER
Dept. of Civil & Environmental Engineering
University of Vermont
Burlington, VT 05405
U.S.A.

A geostatistical method has been developed for contaminant plume boundary de-lineation. This guide for initial field investigations utilizes Bayesian updating in an effort to achieve maximum information gain, while utilizing all known information. The first step is to conditionally simulate hydraulic conductivity fields with vari-ograms, whose parameters are normally distributed random variables generated from a Monte-Carlo analysis. Flow and transport equations are solved to obtain multiple contaminant plumes. Two methods of plume analysis are considered for determining the optimum point for the drilling of a subsequent monitoring well. The first method discards plumes which do not agree with all field data, and analyzes the uncertainty of the remainder. The second analyzes the plume which best matches the field data. The result is a highly parallel process which converges upon an unknown contaminant plume boundary as data are collected.

INTRODUCTION

Currently, there is no standard protocol for the hydrogeological investigation of a contaminant plume boundary. Thus, investigations often proceed using a protocol that fails to take advantage of gained information. This paper presents a method for sequentially placing monitoring wells in locations that will optimally identify the boundary of a contaminant plume.

The approach presented herein applies to investigations where little or no informa-tion is known. It models an isotropic vertically integrated groundwater system, and assumes stationarity of hydrogeologic parameters. The method begins by analyses of the logarithm of hydraulic conductivity data for its statistical structure and fits a variogram. Due to the small data set and the inherent uncertainty of the variogram, the variogram parameters are considered normally distributed random variables. Val-ues are selected by a Monte-Carlo scheme. Conditional simulations of conductivity are generated with the variogram realizations, using the turning bands algorithm. Velocity fields are obtained by solving the flow equation using the conditionally sim-ulated conductivity fields coupled with boundary conditions of head which are also considered random normally distributed variables, and are selected by a Monte-Carlo scheme. Multiple concentration plumes are generated by solving the transport equa-

A. Peters et al. (eds.), Computational Methods in Water Resources X, 317–324.
© 1994 *Kluwer Academic Publishers. Printed in the Netherlands.*

tion with the velocity field realizations assuming that the residence time of the plume is known. The set of plumes is analyzed to determine the optimum location for the next monitoring well.

CONDITIONAL SIMULATIONS

It is assumed that the logarithm of hydraulic conductivity is normally distributed *Smith et al.*, (1979) denote the log-hydraulic conductivity at location x_j, $Y_j = \ln(K_j)$. A least-squares fit of measurements of Y_j data determines the 'best-fit' exponential variogram:

$$\gamma = \omega(1 - e^{h/\lambda}) \tag{1}$$

where the sill, ω, and correlation length, λ, are considered normally distributed random variables with mean and standard deviation equal to their best-fit variogram and user input values respectively, and h is the distance, or lag, between two points. Other variogram models can easily be incorporated.

Utilizing a Monte-Carlo scheme, uncorrelated sills and correlation lengths are obtained, thereby creating variogram realizations. These are used to generate I conditionally simulated hydraulic conductivity fields [*Delhomme*, 1979, *Smith et al.*, 1981] using the turning bands algorithm [*Tompson et al.*, 1989]:

$$Y_i = Y_i^* + [Y_{tb_i} - Y_{tb_i}^*] \qquad i = 1, 2, \ldots I, \tag{2}$$

where Y_i is the ith conditionally simulated hydraulic conductivity field, Y_i^* is the ith kriged field, Y_{tb_i} is the ith turning bands field adjusted to same variance as the kriged field, and $Y_{tb_i}^*$ is the ith kriged field using the data of the adjusted turning bands fields at the monitoring well locations.

FLOW SIMULATION

The steady-state flow equation is solved for each realization of the hydraulic conductivity field,

$$\nabla \cdot (K_i \nabla h_i) = 0 \qquad i = 1, 2, \ldots I. \tag{3}$$

Boundary conditions for head are generated by assuming the magnitude and direction of the hydraulic gradient are normally distributed random variables. The means and standard deviations of each are initially specified and updated as the process proceeds. By assuming a known value of head at a point, nodal boundary conditions for each simulation can be linearly estimated using gradient magnitude and direction values selected by a Monte-Carlo analysis. End of domain boundaries are considered Dirichlet, while no flow or Dirichlet boundary conditions are possible on the other two sides.

TRANSPORT SIMULATION

Darcy's law is used to generate I velocity fields from the set of I head and conductivity fields earlier generated:

$$\mathbf{v}_i = -\frac{1}{n} K_i \nabla h_i \qquad i = 1, 2, \dots I. \tag{4}$$

The solute transport equation is then solved to obtain multiple concentration plumes:

$$\frac{\partial c_i}{\partial t} + \mathbf{v}_i \cdot \nabla c_i - \nabla \cdot (\overline{\overline{D_i}} \cdot \nabla c_i) = 0 \qquad i = 1, 2, \dots I. \tag{5}$$

PLUME ANALYSIS

Analysis of the generated plumes is necessary to determine the optimum monitoring well location. Two plume analysis methods are considered, the *reduced plume population method*, and the *closest fit method*. The reduced plume population method reduces the plume set population by eliminating those plumes which do not closely match the field data, and then analyzing the resulting statistics. The closest fit method finds the plume which has minimum deviation from the data set, and uses this plume as the basis for the ongoing investigation.

Reduced plume population method
The reduced plume population method compares the calculated nodal concentration values with the observed values at the sample points, for each plume, P_i in the set of I plumes. If all deviations are not within a specified bound, η, plume P_i is discarded.

The mean, $\bar{c}_f$ and variance, $\sigma_{c_f}^2$ fields of the remaining plumes are calculated. The statistics of the flow parameters used to generate the variable head boundary conditions of the remaining plumes are calculated; i.e. mean and variance of the gradient magnitude, and gradient direction. These are used as the entering boundary condition statistics in the next simulation.

An action contour, or *fringe*, is defined as the contour of the minimum acceptable (action) concentration, c_{min} of the mean plume, $\bar{c}_f$. The fringe represents the expected value of the concentration plume boundary.

The decision for locating the subsequent monitoring well begins with generation of a decision variable field, δ_f, which encompasses nodal concentration uncertainty and distance from the fringe:

$$\delta_f = \text{minimum of} \begin{cases} \dfrac{\sigma_{c_f}}{|\log_{10}(\frac{\bar{c}_f}{c_{min}})|} \\ \alpha * \sigma_{c_f} \end{cases} \tag{6}$$

where α is a user defined constant (generally 3 to 10). The rationale behind this form is to force the decision variable to equal 0 away from the expected boundary (the fringe) and be a function of the uncertainty near it. Figure 1 shows an example of the decision variable field, δ_f with $\alpha = 4$.

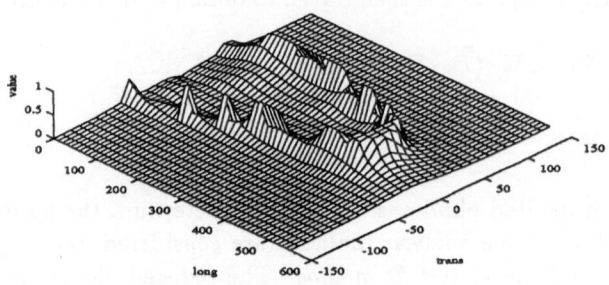

Figure 1: Decision Variable Field, δ_f

The decision variable field, δ_f is then adjusted to account for existing well locations (a new well should not be placed at or near an existing well), and nodal locations relative to the fringe. This is done by means of a variogram-like weighting function defined herein as the information worth function, γ_f. γ_f equals 0 at existing wells and asymptotically approaches 1 away from them. It is multiplied by the decision variable field, δ_f to obtain the adjusted decision variable field, δ_A:

$$\delta_A = \delta_f * \gamma_f \tag{7}$$

In essence, the decision variable field, δ_f is 'decreased' at and near the well locations. Examples of the information worth function, γ_f and the adjusted decision variable field, δ_A can be seen in figures 2, 3 and 4.

Each well has its own unique information worth function which equals 0 at the well location and radially approaches 1 away from it. The over-all information worth function γ_f is a minimum composite of all the well's functions. Mathematically it can be expressed as:

$$\gamma_o = \left(1 - e^{h_{jo}/\lambda_j}\right)_{min} \qquad j = 1, \text{ num. of existing wells.} \tag{8}$$

here, γ_o is the information worth at point x_0, h_{jo} is the distance from point x_0 to well j, and λ_j is the correlation radius associated with well j. The latter is a measure of the relative influence of well j, which is primarily a function of its distance from the fringe.

Large and small correlation radii equate with gradual and quick rises of the information worth function respectively. A well located far away from the fringe will have a large correlation radius, because it is not advantageous to locate a new well in its vicinity. A well near the fringe will have a small correlation radius which will promote the information worth function to quickly approach 1, so that new wells can be located in its vicinity, while insuring that it is an appropriate distance away.

Individual well's correlation radii, λ_j are proportional to a minimum correlation radius, λ_{min} by their respective distances from the fringe:

$$\lambda_j = \text{minimum of} \begin{cases} \lambda_{min}\frac{h_{(jf)min}}{h_{min}} \\ \lambda_{min} \end{cases} \tag{9}$$

Where, $h_{(jf)min}$ is the minimum distance from well j to the fringe, h_{min} is the distance from the well which will make the information worth function will equal 0.5 when λ_{min} is used. In other words, the information worth function associated with well, j with correlation radius, λ_{min} will rise to a value of 0.5 at a distance, h_{min} from the well. h_{min} is the ratio of the mean distance between fringe nodes times the number of fringe nodes to the number of wells time the number of simulations, I.

Figures 2 and 3 show an example of the information worth function, γ_f for an investigation with 10 monitoring wells. Each well's location is shown by the local minimums where γ_f equals 0. The fringe is shown in figure 3 by the locus of x's. The fringe's effect can clearly be seen by observing the rate of rise in γ_f as a function of distance from the fringe.

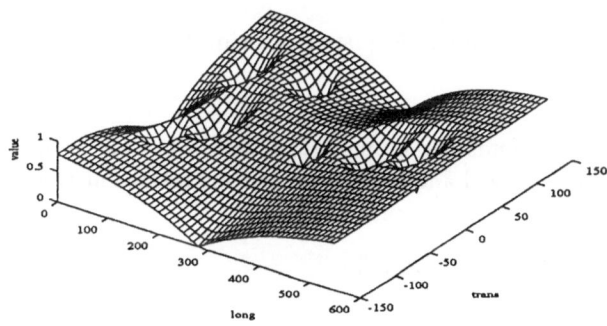

Figure 2: Information Worth function, γ_f

Figure 4 shows the adjusted decision variable field δ_A which was generated by multiplying the values of the decision variable field, δ_f of figure 1 by those of the information worth function, γ_f of figures 2 and 3. This is shown mathematically in equation 8. The next monitoring well is placed at the location of the maximum value of the adjusted decision field, $\mathbf{x}(\delta_{A_{max}})$.

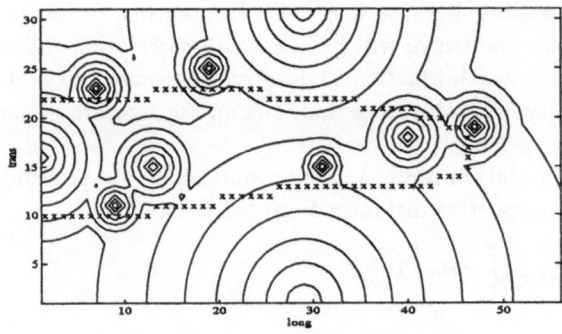

Figure 3: Information Worth Function, γ_f with Fringe

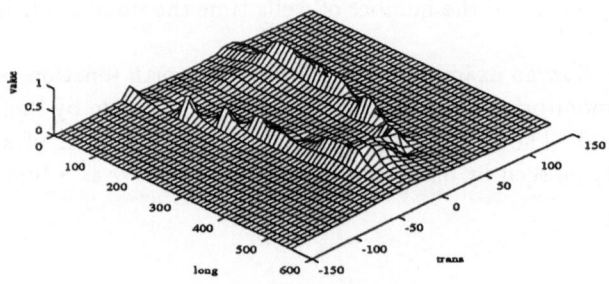

Figure 4: Adjusted Decision Field, δ_A

Closest Fit Method

The closest fit method finds the plume, P_{min}, which has the minimum one-norm of differences between observed and calculated concentrations at the monitoring wells.

$$L1 = \sum_{j=1}^{\text{num. wells}} \| c_j - c_{j \in P_{min}} \| . \tag{10}$$

Then, the action contour, or fringe of P_{min} is found. The next monitoring well is located at the point on the fringe which has the maximum information worth, γ_o. Note that the information worth function, γ, is not the same field as in previously shown in figures 2 and 3 from the reduced plume population method, because the fringe in this analysis is different.

RESULTS & CONCLUSIONS

Investigations were performed to find the boundary of a test plume which was calculated using a random, correlated hydraulic conductivity field with random boundary conditions of head. The conductivity, head, and concentration data were saved for use as sample data.

A small domain, 550 meters long and 300 meters wide was discretized into 10 meter by 10 meter elements, creating 1736 nodes. A constant contaminant source at the center of the left-hand boundary of the domain was assumed. All simulations were calculated for an 8 year time period.

For each independent investigation, 100 conductivity fields were generated and used to simulate 100 plumes. The closest-fit method was used for locating subsequent monitoring wells, while the reduced plume population method reduced each set of 100 plumes to 10, and used them to generate the statistical parameters for the random variogram and boundary conditions.

Table 1 shows the results of the investigation as wells were sequentially placed. The first 5 'initial wells' were assumed in place prior to investigation, and their data were used to initialize the procedure. Thus, the first well placed in the investigation was well number AW-1.

TABLE 1: INITIALIZING DATA & SEQUENTIAL RESULTS					
Initial Well	Added Well	K, ft/day	h, ft	c/c_o	L1 norm
IW1	-	105.5	2.00	1.000	-
IW2	-	102.2	1.50	0.000	-
IW3	-	102.3	2.50	0.000	-
IW4	-	102.1	6.00	0.149	-
IW5	-	103.7	2.00	0.383	0.0501
-	AW-1	104.4	1.43	0.043	0.0707
-	AW-2	103.0	3.46	0.030	0.0782
-	AW-3	101.7	9.74	0.162	0.0785
-	AW-4	101.4	6.47	0.079	0.1994
-	AW-5	104.4	1.92	1.157	0.1994

The investigation was stopped after well AW-5 was placed, because the plume boundary was sufficiently delineated. This can be seen in figure 5, which shows the locations of the original 5 wells (marked with o's) and the sequentially placed investigation wells (marked with x's) with respect to the unknown test plume.

ACKNOWLEDGEMENTS

Partial funding for this work was provided by the U.S. Department of Energy and
Lawrence Livermore National Laboratory under Subcontract B239648.

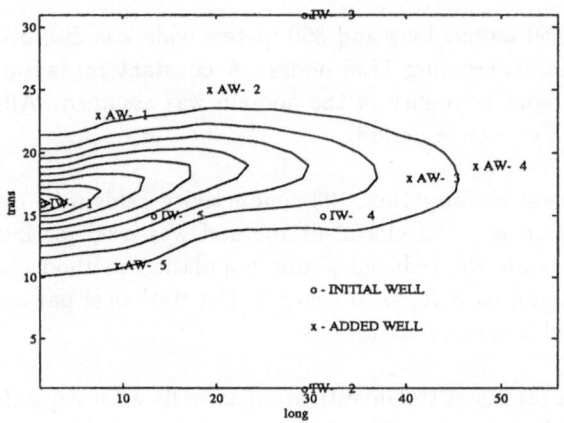

Figure 5: Test Plume With Well Locations

REFERENCES

Delhomme, J.P. (1979) "Spatial Variability and Uncertainty in Groundwater Flow
Parameters: A Geostatistical Approach," *Water Resour. Res.*, **15**(2), 269-280.

Imhoff, P.T., Pinder, G.F., Jaffe, P.R. Guarnaccia, G.F., (1988) "Well Location
Strategies for Groundwater Contaminant Plume Delineation," Dept. of Civil Engineering and Operations Research, Princeton University, Draft.

Smith, L. and Freeze, A. (1979) "Stochastic Analysis of Steady-State Groundwater
Flow in a Bounded Domain: 1. One-Dimensional Simulation" *Water, Resour. Res.*,
15(3), 521-528.

Smith, L., and Schwartz, F. (1981) "Mass Transport: 3. Role of Hydraulic Conductivity Data in Prediction," *Water Resour. Res.*, **17**(5), 1463-1479.

Tompson, A.F.B., Ababou, R., and Gelhar, L. (1989) "Implementation of the Three-
Dimensional Turning Bands Random Field Generator," *Water Resour. Res.*, **25**(10),
2227-2243.

A COMPARISON OF STRONGLY-CONVERGENT SOLUTION SCHEMES FOR SHARP-FRONT INFILTRATION PROBLEMS

C.T. MILLER
CB# 7400, 104 Rosenau Hall
Department of Environmental Sciences and Engineering
University of North Carolina
Chapel Hill, NC 27599-7400
USA

C.T. KELLEY
CB# 8205, 331 Harrelson Hall
Department of Mathematics
North Carolina State University
Raleigh, North Carolina 27695-8205
USA

Accurate and economical numerical solution of sharp-front problems remain fundamental concerns in water resources. Richards' equation (RE) was considered as a test problem for media properties and auxiliary conditions that resulted in the development and propagation of a sharp-front through a domain. Conventional modified Picard iteration (MPI) and Newton-Raphson iteration (NRI) methods were found to be unreliable, yielding non-convergent solutions for many discretization schemes. Strongly-convergent schemes were formulated based upon modifications to the NRI method, including method of lines formulations based upon ordinary differential equation or differential algebraic equation forms of the problem. Strongly-convergent methods were found to be much more robust than MPI or NRI methods. A model formulation based upon a transformed equation approach, which reduced the original nonlinearity present in the system, was found to offer further advantages over conventional approaches for some situations.

INTRODUCTION

The accurate and economical solution of sharp-front problems remains an important challenge for the simulation of flow and transport phenomena in subsurface systems. Transient wetting-phase displacement of a non-wetting fluid for a system initially at hydrostatic equilibrium is an example of a physical system that leads to sharp fronts in space and time, especially for uniform, coarse-grained media. The physics of this system leads to a system of highly nonlinear partial differential

A. Peters et al. (eds.), Computational Methods in Water Resources X, 325–332.
© 1994 Kluwer Academic Publishers. Printed in the Netherlands.

equations, which have been solved typically using Picard iteration or NRI methods. RE is a simplified formulation, which exhibits sharp-front behavior similar to the more general two-phase problem.

Objectives of this work were: (1) to document convergence problems associated with solving RE using conventional approaches; (2) to formulate strongly-convergent methods for approximating RE; (3) to investigate a transformed-equation method to reduce the nonlinearity in the system; (4) to investigate computational issues involved with developing economical solution algorithms; and (5) to compare performance among the various methods in terms of both economy and robustness.

MOTIVATION

Consider two common compressible forms of RE, the ψ-based form

$$[S_s S_a + c(\psi)] \frac{\partial \psi}{\partial t} = \nabla \cdot [\mathbf{K}(\psi) \cdot (\nabla \psi + \mathbf{z})] \tag{1}$$

and the mixed-form

$$S_s S_a \frac{\partial \psi}{\partial t} + \frac{\partial \theta}{\partial t} = \nabla \cdot [\mathbf{K}(\psi) \cdot (\nabla \psi + \mathbf{z})] \tag{2}$$

where S_s is a specific storage coefficient accounting for fluid compressibility (L^{-1}), S_a is the aqueous-phase saturation, $c(\psi) = \partial \theta / \partial \psi$ is the specific moisture capacity (L^{-1}), θ is the volumetric fraction of the aqueous phase, ψ is the pressure head (L), t is time (T), $\mathbf{K}(\psi)$ is the hydraulic conductivity tensor (LT^{-1}), and $\mathbf{z}$ is a unit vector, oriented positive upward (L). Equation (2) is the preferable formulation because of its mass conserving properties (e.g., Milly, 1985; Allen and Murphy, 1986), but requires reduction to a single dependent variable, which may be accomplished using a Newton-like method, or solving an equation to enforce mass conservation.

Constitutive theory is needed to close Equation (1) or (2), which is supplied in the form of pressure-saturation-conductivity (p-S-K) relations. The van Genuchten and Mualem p-S-K expressions are common choices for these relations (Kool and Parker, 1987)

$$S_e = \frac{\theta - \theta_r}{\theta_s - \theta_r} = [1 + |\alpha \psi|^n]^{-m} \tag{3}$$

and

$$K(\psi) = K[S_e(\psi)] = K_s Se^{1/2} \left[1 - \left(1 - S_e^{1/m} \right)^m \right]^2 \qquad (4)$$

where S_e is the effective saturation, θ_r is a residual wetting-phase saturation fitting parameter, θ_s is a maximum wetting-phase saturation, α is a model parameter related to the mean pore size (L^{-1}), n is a model parameter related to the variance of pore size distribution, $m = 1 - 1/n$, and K_s is the saturated hydraulic conductivity (LT^{-1}), assumed scalar. A strict imbibition path is further assumed, although hysteretic versions of the above constitutive relations are available (Kool and Parker, 1987).

Typical approaches for approximation of the derivatives in such problems include low-order finite difference discretization in time (e.g., implicit, Crank Nicolson), and low-order finite difference or finite element discretization in space (e.g., Abriola and Lang, 1990; Celia and Bouloutas, 1990), resulting in a system of nonlinear algebraic equations. Nonlinearities are typically resolved using MPI (e.g., Celia and Bouloutas, 1990) or NRI (e.g., Abriola and Lang, 1990). The resulting system of linear algebraic equations has traditionally been solved using direct methods, stationary iterative methods, or in a few cases nonstationary iterative methods.

These conventional methods do not provide a robust solution for common sharp-front infiltration problems, although problems with these approaches are not well documented in the literature. To illustrate this point, consider an implicit in time (first-order, backward difference), centered in space (second-order, central difference) finite difference approximation to Equation (2), which may be written in matrix form for a MPI scheme as

$$[A_p(\psi)]^{l+1,m} \{\Delta\psi\}^{l+1,m+1} = \{b_{pl}\}^l + \{b_{pm}\}^{l+1,m} = \{b_p\}^{l+1,m} \qquad (5)$$

where $[A_p(\psi)]$ is a MPI coefficient matrix; l is a known time level; $l+1$ is an unknown time level; m is a known iteration level; $m+1$ is an unknown iteration level; $\{b_{pl}\}$ is a vector depending upon information known at time level l; $\{b_{pm}\}^{l+1,m}$ is a vector depending upon information at time level $l+1$ and the iteration level m; and $\{b_p\}$ is the total right-hand-side vector, which has components that vary with time level and with iteration level.

Using an identical finite difference scheme, but a standard NRI approach yields a similar matrix form problem

$$[A_n(\psi)]^{l+1,m} \{\Delta\psi\}^{l+1,m+1} = -\{f\}^{l+1,m} \tag{6}$$

where $[A_n(\psi)]$ is the Jacobian matrix resulting from differentiation of the difference equation vector $\{f\} = \{0\}$ with respect to ψ_j in the usual manner

$$[A_n(\psi)]^{l+1,m} = \left[\frac{\partial f_i}{\partial \psi_j}\right], \qquad \text{for } i,j = 1,...,n_n \tag{7}$$

where i is a row index, j is a column index, and n_n represents the number of nodes in the system.

Test problems considered used the following auxiliary conditions

$$\psi(z, t = 0) = -z \tag{8}$$
$$\psi(z = 0, t) = 0 \tag{9}$$
$$\psi(z = 10, t > 0) = 0.1 \tag{10}$$

for a spatial domain $z \in [0, 10]$ meters, and a temporal domain $t \in [0, 0.20]$ days. These conditions physically represent drained to equilibrium conditions initially and constant head conditions starting at $t = 0$, with a water table boundary at $z = 0$ for all time.

Test problems were run using dune sand media parameters summarized in Table 1, which were taken from Kool and Parker (1987), using finite difference methods described above for a mixed-form formulation, subject to the above noted auxiliary conditions; MPI and NRI nonlinear solution schemes; and a direct tridiagonal solver. These media properties and auxiliary conditions yield a sharp infiltration front, which propagates about halfway through the domain at $t = 0.18$ days.

Results shown in Table 2 summarize convergence problems encountered when applying conventional numerical schemes to this problem to resolve nonlinearities. Application of a MPI scheme resulted in a spatial and temporal discretization region over which convergence was achieved, while NRI failed to converge for all discretization regions investigated.

Table 1. Media Properties

Property	Rideau Clay Loam	Dune Sand
K_s (m/day)	2.298	5.040
θ_s	0.416	0.301
θ_r	0.279	0.093
α (1/m)	6.12	5.47
n	1.914	4.264

Table 2. Time Step Convergence Summary

Spatial Step (m)	MPI (days)	NRI (days)	NRILS (days)
5.00×10^{-2}	$7 \times 10^{-4} \leq \Delta t \leq 4 \times 10^{-3}$	—	$\Delta t \leq 5 \times 10^{-2}$
1.25×10^{-2}	$1 \times 10^{-5} \leq \Delta t \leq 1 \times 10^{-3}$	—	$\Delta t \leq 2 \times 10^{-3}$

STRONGLY-CONVERGENT METHODS

Several strategies are possible to resolve the convergence problems demonstrated in the previous section: (1) alternate nonlinear solver methods, (2) time adaptive schemes, (3) spatial adaptive approaches, or (4) combinations of the above methods. Efforts have been made in each of these areas, with a summary of methods and results given for methods (1) and (4) in this work.

A well-known approach to enhance the robustness of NRI, is to accept the direction represented by solution of Equation (6), $\{\Delta\psi\}^{l+1,m+1} / \| \{\Delta\psi\}^{l+1,m+1} \|_2$, but not necessarily the step-size magnitude. This modification was implemented using a cubic line-search modification of NRI (NRILS) (Dennis and Schnabel, 1983).

While robustness of solution schemes can be improved using strongly-convergent schemes noted above, questions of solution accuracy remain. Approaches taken to date to solve RE usually do not address questions of solution accuracy as a routine and integral part of an approximation scheme. Common practice is to prove convergence, perhaps only heuristically, but solution methods for RE typically neither compute nor limit truncation error during a simulation. Method of lines (MOL) approaches provide a means for partially addressing this problem, through inclusion of control on temporal truncation error, typically applied as a relative or absolute error limit over a given solution step.

Two method of lines (MOL) approaches were investigated: a MOL/ordinary differential equation method (MOL/ODE), and a MOL/differential algebraic equation method (MOL/DAE) (Brenan et al., 1989). For the MOL/ODE approach, spatial derivatives were replaced with the low-order finite difference approximations noted above, yielding a system of ODE's of the form

$$\left\{ \frac{d\psi}{dt} \right\} = \{b_{mo}\} \tag{11}$$

which were solved using variable-order backward difference methods (BDM's) using VODE (Brown et al., 1989).

For the MOL/DAE approach, valid equations may be either ODE's or algebraic equations, conforming to the general vector form

$$\mathbf{f}(t, \psi_j, \psi_j') = 0 \qquad \text{for } j = 1, ..., n_n \tag{12}$$

where ψ' represents temporal differentiation. Substituting BDM's for the temporal derivatives yields

$$\mathbf{f}(t, \psi_j, \beta_1 \psi_j + \beta_2) = 0 \qquad \text{for } j = 1, ..., n_n \tag{13}$$

where β_1 and β_2 are constants.

Equation (13) may be solved using NRILS methods with variable-order BDM's for time integration to provide error estimation. This approach was implemented using DASSL (Brenan et al., 1989).

RESULTS AND DISCUSSION

NRILS methods yielded results shown in Table 2 for the sand test problem. Robust convergence was achieved over a wider Δt range with the NRILS method than with the MPI method, which was only bounded by a maximum step size for the problems investigated. Along these same lines, alternate nonlinear approaches would likely prove even more robust than NRILS (e.g., perturbed quadratic approximations, trust region methods, homotopic approaches).

Comparisons were performed between MOL/ODE and MOL/DAE methods for both test problems noted in Table 1. For both methods, time integration was restricted to second-order BDM's and the resultant linear system was solved using a direct method. Results for $\Delta z = 0.0125$ m shown in Table 3 are for CPU times on a Cray Y-MP for both ψ-based standard equation (SE) and transformed equation (TE) approaches, where the hyperbolic sine transformation of Ross (1990) was used in the TE approach.

Table 3. MOL CPU Time Result Summary

Media	SE ODE (secs)	TE ODE (secs)	SE DAE (secs)	TE DAE (secs)
Sand	—	11.9	6.2	5.9
Loam	4.3	7.3	3.7	3.9

MOL/DAE methods based upon DASSL yielded solutions of equivalent accuracy to the MOL/ODE approaches in less CPU time and converged in all cases. The MOL/ODE method failed to converge for the SE approach and the sand media. The TE approach made little difference for the MOL/DAE method but improved convergence and significantly decreased computational effort for the MOL/ODE approach. CPU times for the MOL/DAE method were competitive with the less robust MPI method.

Experiments were also performed using an iterative Jacobi preconditioned GMRES solver in place of the direct cyclic reduction solver used initially in the MOL/DAE approach (Brown et al., 1993). The iterative method showed good convergence properties, but as expected by the tridiagonal matrix structure of the problem, the direct method was in all cases significantly more efficient. This trend will be reversed for higher-dimension problems.

ACKNOWLEDGMENTS

The research of the first author was supported by Army Research Office grant DAAL03-91-G-0155. The research of the second author was supported by National Science Foundation grant DMS-9024622. Both authors acknowledge the support of the North Carolina Supercomputer Center (NCSC) and Dr. Bruce Loftis, a Computational Scientist at NCSC.

REFERENCES

Abriola, L.M. and Lang, J.R. (1990) Self-adaptive hierarchic finite element solution of the one-dimensional unsaturated flow equation, Int. J. Num. Meth. Fluids 10, 227–246.

Allen, M.B. and Murphy, C.L. (1986) A finite-element collocation method for variably saturated flow in two space dimensions, Wat. Resour. Res. 22, 1537–1542.

Brenan, K.E., Campbell, S.L. and Petzold, L.R. (1989) Numerical Solution of Initial-Value Problems in Differential Algebraic Equations, Elsevier, New York.

Brown, P.N., Byrne, G.D. and Hindmarsh, A.C. (1989) VODE: A variable coefficient ode solver, SIAM J. Sci. Stat. Comp. 10, 1038–1051.

Brown, P.N., Hindmarsh, A.C. and Petzold, L.R. (1993) Using Krylov methods in the solution of large-scale differential-algebraic systems, Tech. Report 93-37, Computer Science Department, University of Minnesota.

Celia, M.A. and Bouloutas, E.T. (1990) A general mass-conservative numerical solution for the unsaturated flow equation, Wat. Resour. Res. 26, 1483–1496.

Dennis, J.E. and Schnabel, R.B. (1983) Numerical Methods for Unconstrained Optimization and Nonlinear Equations, Prentice-Hall, Inc., Englewood Cliffs.

Kool, J.B. and Parker, J.C. (1987) Development and evaluation of closed-form expressions for hysteretic soil hydraulic properties, Wat. Resour. Res. 23, 105–114.

Milly, P.C.D. (1985) A mass-conservative procedure for time-stepping in models of unsaturated flow, Adv. Wat. Resour. 8, 32–36.

Ross, P.J. (1990) Efficient numerical methods for infiltration using Richards' equation, Wat. Resour. Res. 26, 279–290.

INTEGRATION ELMS OR INTERPOLATION ELMS?

ANABELA OLIVEIRA and ANTÓNIO M. BAPTISTA
Center for Coastal and Land-Margin Research and Department of Environmental Science and
Engineering, Oregon Graduate Institute of Science & Technology, P.O. Box 91 000, Portland OR
97 291-1000, USA .

Recent finite element Eulerian-Lagrangian Methods (ELMs) have traded the classical notion of
interpolation at the feet of the characteristic lines for that of integration. In this paper, the stability
and accuracy of selected integration and interpolation ELMs is formally analyzed and compared.
We show that integration generally leads to better accuracy, but some integration ELMs are only
conditionally stable. Stability criteria are established, on the basis of Courant number, diffusion
number, and type and number of quadrature points.

INTRODUCTION

Eulerian-Lagrangian Methods have long been recognized as an attractive technique for the solution of advection-dominated transport (Holly and Polatera 1984, Baptista et al. 1984, Neuman
1984). In particular, ELMs overcome the Courant number restrictions that are associated with the
traditional Eulerian methods.

The basic concept of ELMs is simple: the transport equation is solved in its Lagrangian form
"along characteristic lines", effectively decoupling advection and diffusion. Advection is solved
by the backward method of characteristics, while diffusion is solved either by finite elements or
finite differences. ELM solutions typically include three basic steps: (a) definition of the characteristic lines that follow the flow backwards, (b) determination of the concentration at the feet of the
characteristic lines, and (c) solution of the diffusion equation, with concentrations at the feet of
the characteristic lines as initial conditions.

The definition of initial conditions ("advective concentrations") for the diffusion equation has
been identified as a major source of numerical diffusion. In the past, the definition of the advective concentration was considered an *interpolation* problem. Many interpolators were proposed
and formally studied (Leith 1965, Holly and Preissmann 1977, Baptista 1987), none of which
proved optimal (Baptista 1986 and 1987). In the presence of sharp gradients, the necessary grid
refinement for acceptable accuracy can still lead to very high computational costs.

Coincidentally with the search for optimal interpolators, the use of ELMs in a finite element context (FE-ELMs) increased. In an attempt to develop a more accurate and efficient technique, several FE-ELMs proposed in the last decade (Hasbani et al. 1983, Russell 1985, Yeh et al.1992) share
a new concept: they deal with the definition of the initial conditions for the diffusion equation as
an *integration* rather than an interpolation problem. By doing so, these FE-ELMs become fundamentally different from the corresponding finite difference ELMs.

A. Peters et al. (eds.), Computational Methods in Water Resources X, 333–340.

If the Courant number is not an integer, the concentration between two consecutive feet of characteristic lines is a piecewise function. While interpolation methods approximate the concentration with polynomial shape functions, integration ELMs - piecewise ELMs and quadrature ELMs - propose potentially more accurate approaches.

Piecewise ELMs seek an exact evaluation of the integrals, taking into account the piecewise shape of the concentration at the feet of the characteristic lines. This is achieved by sub-dividing the domain in regions where the integrands are C^∞ functions. This sub-regions are defined both by the nodes and other notable points (e.g., discontinuities in the first derivative of concentration), leading to a final concentration that is a piecewise function inside each element (Yeh et al. 1992). Although numerical tests performed with this original formulation showed excellent accuracy, the computational costs associated with it may become unsustainable for practical applications, as the number of notable points needed to define the concentration in each element increases rapidly in time. In an attempt to overcome this problem, a simpler implementation of the piecewise integration concept was proposed, eliminating the concept of notable points. This formulation (pi-ELM) provided very good results in simple numerical tests (Oliveira 1994).

Quadrature ELMs (qu-ELMs) handle the definition of the initial conditions for the diffusion equation from a numerical integration perspective: quadrature points, defined at the time level in which the equation is being solved, are backtracked instead of the nodes. The quadrature points are then used in the numerical integration at the feet of the characteristic lines, providing a straightforward evaluation of the integrals. Quadrature ELMs can be very accurate (Hasbani et al. 1983), but are only conditionally stable (Morton et al, 1988). One of the most attractive features of these methods is their simple implementation in multiple dimensions.

This paper examines the stability and accuracy of integration methods performing both piecewise integration and quadrature integration, and compares them with a reference interpolation method. Both Fourier and truncation error analysis are employed. For simplicity, this study is conducted in one dimension and for constant coefficients, which allows us to characterize the methods in a systematic and general way.

INTEGRATION AND INTERPOLATION ELMS: BASIC CONCEPTS

A Galerkin FE-ELM formulation starts with the Lagrangian form of the transport equation, which is discretized in time as:

$$\frac{Dc}{Dt} = D\frac{\partial^2 c}{\partial x^2} \quad \Rightarrow \quad \frac{c^{n+1}-c^\xi}{\Delta t} - \alpha D\frac{\partial^2 c}{\partial x^2}\bigg|_{n+1} - (1-\alpha) D\frac{\partial^2 c}{\partial x^2}\bigg|_\xi = 0 \tag{1}$$

where c is the concentration, D is the diffusion coefficient, α is the time discretization weight and ξ denotes the feet of the characteristic lines that follow the flow backwards.

Standard application of a weak Galerkin weighted residual finite element formulation leads to:

$$\sum_{k=1}^{N_{elems}} \{ \frac{1}{\Delta t}\left(\Psi_{kj}^{n+1} - \Psi_{kj}^\xi \right) + D\left[\alpha\Phi_{kj}^{n+1} + (1-\alpha)\,\Phi_{kj}^\xi \right] \} + Boundary\ terms = 0 \tag{2}$$

with:

$$\Psi_{kj}^m = \int_{\Omega_k} c^m\phi_j dx \qquad \Phi_{kj}^m = \int_{\Omega_k} \frac{\partial c}{\partial x}\bigg|^m \frac{\partial\phi_j}{\partial x}dx \tag{3}$$

where ϕ_j are weighting functions that coincide, on an elemental basis, with the shape functions.

At this stage, choices need to be made regarding (a) the location of the heads of the characteristic lines, (b) the strategy for interpolation of concentrations and concentration derivatives at ξ, and (c) the strategy for evaluation of integrals at time n.

In interpolation ELMs, the nodes at time $n+1$ are tracked backwards, and the concentration and its first derivative at the feet of the characteristic lines are obtained by interpolation. The interpolated values are then assigned to shape functions that define the initial conditions for the diffusion equation.

In piecewise integration ELMs, nodes at time $n+1$ are tracked backwards and nodes that are found between two consecutive feet of characteristic lines at time n are tracked forward. This way, the initial conditions for the diffusion equation have discontinuous first derivatives inside each element. For a constant velocity problem, since only one node is found between two consecutive feet of characteristic lines, the advective concentration is defined as (in local coordinates):

$$c^{\xi}(r) = \begin{cases} \dfrac{c_{int}-c_1}{r_{int}+1}r + c_1 + \dfrac{c_{int}-c_1}{r_{int}+1}, & r \in [-1, r_{int}] \\[2ex] \dfrac{c_{int}-c_2}{r_{int}-1}r + c_2 - \dfrac{c_{int}-c_2}{r_{int}-1}, & r \in [r_{int}, 1] \end{cases} \tag{4}$$

where c_1 and c_2 are the concentrations at the feet of characteristic lines of the back-tracked nodes and c_{int} and r_{int} are, respectively, the concentration and location at time $n+1$ of the forward-tracked node.

In quadrature integration ELMs, quadrature points, rather than nodes, are backtracked from $n+1$ to n. Once the concentration and its first derivative at the feet of the characteristic lines of the quadrature points are interpolated, the evaluation of the integrals at time n becomes trivial:

$$\Phi_{kj}^{\xi} = \int_{\Omega_k} (1-\alpha)D\frac{\partial\phi_j}{\partial r}\frac{\partial c}{\partial r}\bigg|^{\xi} dr = (1-\alpha)D\sum_{i=1}^{nqp} w_i\frac{\partial c_i}{\partial r}\frac{\partial\phi_j}{\partial r}(r_i)$$

$$\Psi_{kj}^{\xi} = \int_{\Omega_k} \phi_j\frac{c^{\xi}}{\Delta t}dr = \frac{1}{\Delta t}\sum_{i=1}^{nqp} w_i c_i^{\xi}\phi_j(r_i) \tag{5}$$

where c_i^{ξ} and $\partial c_i/\partial r$ are the concentration and its first derivative at the feet of the characteristic line for quadrature point i, r_i and w_i are the location (at time $n+1$) and weight of point i, $\phi_j(r_i)$ is the value of the weight function at quadrature point i and nqp is the number of quadrature points. The initial conditions for the diffusion equation are thus defined as a polynomial function specified by the number and type of quadrature points.

A non-compact cubic interpolator (Baptista 1987) was selected to be our reference interpolation ELM. It was chosen for its accuracy, which bears some similarity to the accuracy of a widely used method - the quadratic compact interpolator (Baptista et al 1984). The use of the non-compact cubic interpolator, rather than the quadratic compact interpolator, guarantees time-independent propagation errors.

STABILITY

The stability of the selected ELMs is characterized through the analysis of both propagation and truncation errors. First, Fourier analysis is used to study the influence of both Courant number (β) and dimensionless wavelength ($L_m/\Delta x$) on stability, for the pure advection case. Then, truncation error analysis completes the previous study, by examining the influence of diffusion.

The Fourier analysis allows us to write the solution of the transport equation as a summation of Fourier components and to study the effect of a numerical approximation on each of those components. The error introduced by a numerical approximation on the concentration of component m (G_m) can be defined as (see Oliveira 1994 or Lapidus and Pinder 1982 for further details):

$$G_m = \frac{\tilde{c}_m}{c_m} = |G_m| exp(\Phi_m I) \tag{6}$$

where $I = \sqrt{-1}$, c_m is the exact concentration, $\tilde{c}_m$ is the concentration given by the numerical algorithm, $|G_m|$ is the amplification factor, and Φ_m is the phase error.

Since all selected methods have linear core elements, the propagation error is time-independent (Baptista 1987). Therefore, it is enough to examine the amplification factors after a single time step and the stability condition, for pure advection, is:

$$|G_m| \leq 1 \tag{7}$$

Amplification factors for the pi-ELM and the qu-ELM are presented in Figs. 1 and 2, for a range of β from 0 to 1 and a $L_m/\Delta x$ from 2 to 34. The values of β were limited to 1 because the errors per time step in ELMs with linear core elements are independent of the integer part of β (Baptista 1987, Oliveira and Baptista 1994).

Amplification factors show that the pi-ELM is unconditionally stable, while the qu-ELM is unstable for some ranges of β, for both 3 and 6 gauss points. Fig. 2 suggests that the number of instability regions is related to the number of quadrature points: for m gauss quadrature points, m unstable regions are detected. To further investigate this relationship and to examine the influence of diffusion on stability, we analyzed the truncation errors for the qu-ELM. Since we are interested in stability, we concentrate on the numerical diffusion associated with the second order derivative of concentration.

Figure 1- Amplification factors for the pi-ELM

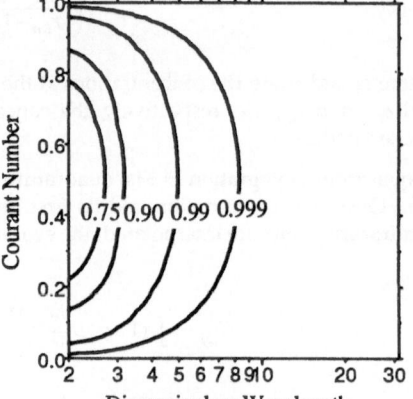

In order to analyze the influence of dimensionless numbers - Courant number and Diffusion number (D) - it is convenient to define the effective diffusion number (Υ):

$$\Upsilon(\beta, D) = \frac{\Delta t}{\Delta x^2}(D + \tau) \tag{8}$$

where Δt is the time step, Δx is the grid spacing and τ is the numerical diffusion coefficient. A method will be stable for positive values of Υ, and unstable otherwise. For the qu-ELM with a generic quadrature and a generic number of quadrature points:

$$\Upsilon = \sum_{i=1}^{nqp} \left[\frac{w_i}{2}\left(\frac{2K_i^2 + 1}{4} - \frac{K_i \upsilon_i}{2} \right) - \frac{w_i r_i}{8}(2K_i - \upsilon_i) \right] - \left(\frac{\beta^2}{2} + \frac{1}{6} \right) + D \qquad (9)$$

where υ_i is the distance between the left node and the foot of the characteristic line of point i, in local coordinates [-1,1]. K_i is defined as:

$$\begin{cases} int\left(\beta - \dfrac{r_i}{2} - \dfrac{1}{2} \right) + 1 & if \ \beta - \dfrac{r_i}{2} - \dfrac{1}{2} > 0 \\[2mm] 0 & otherwise \end{cases} \qquad (10)$$

The effective diffusion number for the qu-ELM with 3 Gauss points, is shown in Fig. 3, for the previously selected range of β and for a range of D from 0 to 0.1.

Figure 2 - Amplification factors for the qu-ELM with 3 and 6 Gauss points.

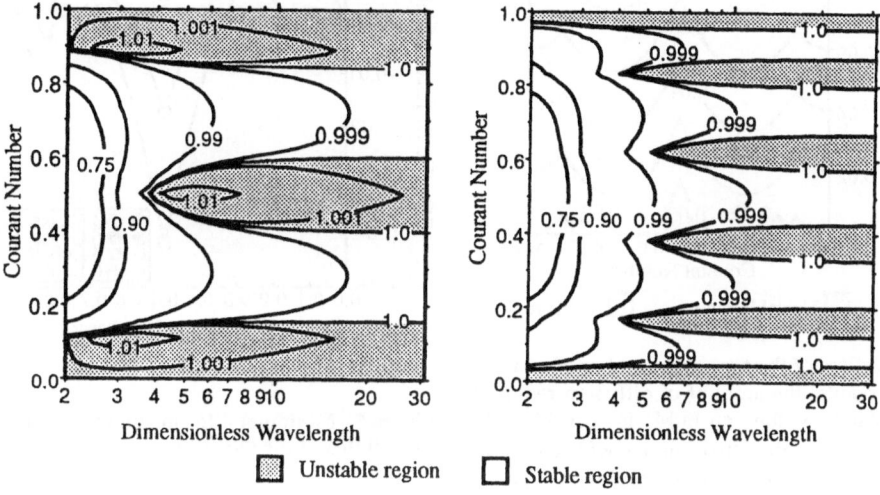

Fig. 3 confirms the relationship between the number of quadrature points and unstable regions suggested by the Fourier analysis. In order to examine this relationship in further detail, Υ was plotted against β, for $D = 0$ (Fig. 4). Fig. 4 shows that the number of local minima in Υ is equal to the number of quadrature points. It also shows that the first derivative of Υ is continuous everywhere, except at the local minima. Since an unstable region must include a local minimum, the definition of a first derivative discontinuity will also identify a potentially unstable region. By examining Equations (9) and (10), it can be recognized that the discontinuities in the first derivative are generated by K_i. A discontinuity occurs when K_i is increased by one, for a β given by:

$$fractional\left(\beta_i - \frac{r_i}{2} - 0.5 \right) = 0 \quad \Rightarrow \quad \beta_i = \frac{r_i}{2} + 0.5 + n \quad (\forall n \in N) \qquad (11)$$

Equation (11) shows that the Courant number that leads to a maximum of negative diffusion is also related to the location of the quadrature points: a minimum in Υ will occur when the foot of the characteristic line of any quadrature point coincides with a node. For gauss quadrature, each point identifies an unstable region, thus the number of quadrature points coincides with the number of instability regions. The feet of the characteristic lines of the extreme Lobatto points only coincide with the nodes when the Courant number is an integer, thus ELMs are exact (Baptista 1987). Therefore, m Lobatto points will lead to m-2 instability regions.

Fig. 3 also shows that the conditional stability of the qu-ELM can be eliminated by a very small amount of diffusion. The minimum diffusion number required for stability is presented in Fig. 5, for both Gauss and Lobatto points. Fig. 5 shows that the amount of diffusion for stability decreases as the number of quadrature points is increased, and is smaller for gauss quadrature than for Lobatto quadrature, for an equal number of quadrature points.

Figure 3 - Effective diffusion number for the qu-ELM with 3 Gauss points

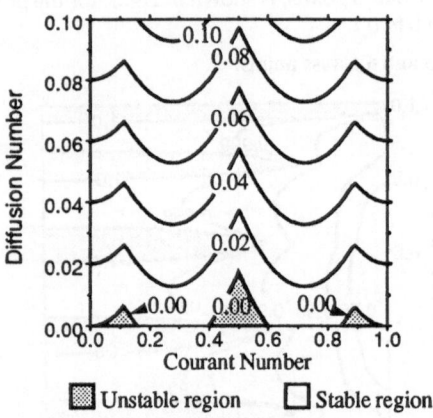

Unstable region ▨   Stable region ☐

Figure 4 - Comparison of effective diffusion numbers for $D = 0$

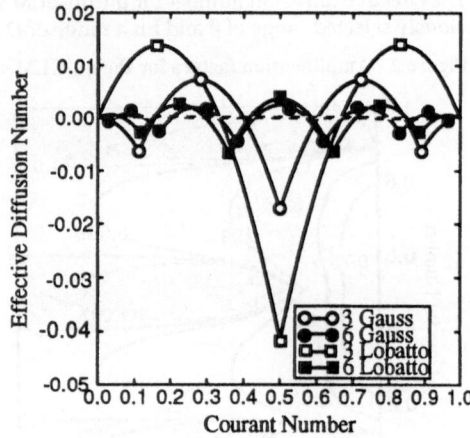

Regardless of the type and number of quadrature points, the amount of diffusion required to stabilize the qu-ELM is considerably smaller than the dispersion coefficients currently used in numerical simulations. Therefore, the conditional stability does not necessarily preclude the practical interest of the qu-ELM.

ACCURACY

In this section, we compare the accuracy of the selected methods. This comparison is based on the analysis of the previously presented amplification factors. Although an accuracy analysis usually requires the study of both amplification factors and phase errors, earlier work showed that amplification errors are dominant over phase errors for ELMs with linear core elements and centered interpolators (Baptista 1986, Oliveira 1994).

Figure 5 - Minimum diffusion number required for stability of the qu-ELM

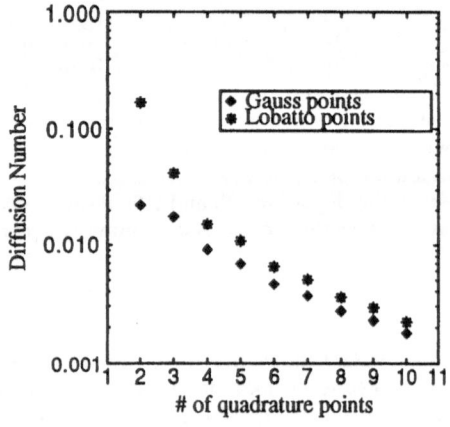

Figs. 1 and 6 show that the pi-ELM provides a considerable improvement in accuracy over the reference method, its amplification error being very small for the entire range of β. While both the pi-ELM and the reference method present a similar dependence of the amplification factor with β (a single maximum at β = 0.5), the amplification factor of the qu-ELM depends strongly both on the fractional part of β and on the number and type of quadrature points. The qu-ELM with three gauss points (Fig. 2a) is only slightly more accurate than the reference, while the amplification factor for 6 Gauss points (Fig. 2b) is as small as the pi-ELM's. An analysis performed with Lobatto points showed that Lobatto points are slightly less accurate than Gauss points, for an equal number of points (Oliveira 1994).

Figure 6 - Amplification factors for the reference interpolation ELM.

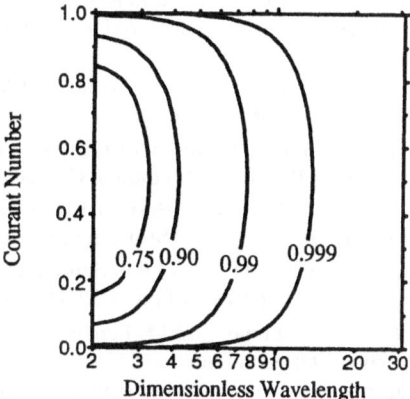

Statistical analysis of the amplification factor over the range of β showed that qu-ELMs have large standard deviation (Oliveira and Baptista 1994). Under a non-constant flow field, this strong dependence on β may lead to spurious oscillations, as consecutive nodes can have very distinct errors.

FINAL CONSIDERATIONS

Although no method emerged as optimal, our comparative analysis of integration and interpolation ELMs revealed important differences between methods that are worth taking into account when developing FE-ELMs transport models. Integration ELMs emerge as very attractive techniques, that can provide a considerable gain in accuracy over the ELMs currently used in application-oriented models, for an equal number of nodes. However, our analysis was done in a very simple context, and other key factors relevant in real world simulations (feasibility and computational cost) must also be considered.

Finally, the inherently non-conservative nature of ELMs is still one of their major drawbacks (Baptista 1987, Dimou 1992). Mass conservation problems have been detected in many applications, which may compromise the use of ELMs for long term simulations and when complex processes are present. Non-conservative flow fields, errors in the evaluation of the characteristic lines and the treatment of boundary conditions have been identified as major sources of mass imbalances (Oliveira 1994, Russell 1989). ELLAMs (Celia et al 1990, Russell 1989) provide an efficient solution to handle the mass problems associated with the implementation of boundary conditions, but an efficient solution for the mass problems due to the other two factors is still to be found.

ACKNOWLEDGMENTS

We thank André Fortunato, of OGI, for very useful discussions. The first author was sponsored by Junta Nacional de Investigação Científica e Tecnológica (Portugal), under grants BM-1354/91-RN and BD-2775/93.

REFERENCES

Baptista, A.M., E.E. Adams, and K. D. Stolzenbach (1984) Eulerian- Lagrangian analysis of pollutant transport in shallow water, *Technical Report no. 296*, MIT R.M. Parsons Laboratory, Cambridge, Mass.

Baptista, A. M. (1986) Accurate Numerical Modeling of Advection-Dominated Transport of Passive Scalars: A contribution, Laboratório Nacional de Engenharia Civil, Lisbon, Portugal.

Baptista, A. M. (1987) Solution of Advection-Dominated Transport by Eulerian-Lagrangian Methods using the Backwards Method of Characteristics, Ph.D. Dissertation, MIT, Cambridge, Mass.

Celia, M. A., T. F. Russell, I. Herrera, R. E. Ewing (1990) An Eulerian-Lagrangian Localized Adjoint Method for the Advection-Diffusion Equation, *Advances in Water Resources*, 13(4), 187-206.

Dimou, K. (1992) 3-D Hybrid Eulerian-Lagrangian / Particle Tracking Model for Simulating Mass Transport in Coastal Water Bodies, Ph.D. Dissertation, MIT, Cambridge, Mass.

Hasbani, Y., E. Livne, and M. Bercovier (1983) Finite Elements and Characteristics Applied to Advection-Diffusion Equations, *Computers and Fluids*, 11, 71-83.

Holly, F. M. and A. Preissmann (1977) Accurate Calculation of Transport in Two Dimensions, *Journal of the Hydraulics Division, ASCE*, 103, 1259-1278.

Holly, F. M. and J. U. Polatera (1984) Dispersion Simulation in Two-Dimensional Tidal Flow, *Journal of Hydraulic Engineering, ASCE*, 110(7), 905-926.

Leigh, C. E. (1965) Numerical Simulation of the Earth's Atmosphere, *Meth. in Comp. Phys*, 4, 1-28.

Lapidus, L. and G.F. Pinder (1982) Numerical solution of partial differential equations in Science and Engineering, Wiley-Interscience Publ.

Morton, K. W., A. Priestley, and E. Suli (1988) Stability of the Lagrange-Galerkin Method with Non-exact Integration, *Mathematical Modelling and Numerical Analysis*, 22(4), 625-653.

Neuman, S. P. (1984) Adaptive Eulerian-Lagrangian Finite Element Method for Advection-Dispersion, *International Journal for Numerical Methods in Engineering*, 20, 321-337.

Oliveira, A. (1994) A comparison of Eulerian-Lagrangian methods for the solution of the transport equation, M.S. Thesis, Oregon Graduate Institute of Science and Technology, Portland, Or.

Oliveira, A. and A.M. Baptista (1994) Comparison of integration and interpolation Eulerian-Lagrangian methods (in preparation).

Russell, T. F. (1985) Time Stepping along characteristic with incomplete iteration for a Galerkin approximation of Miscible Displacement in Porous Media, *SIAM J. of Numerical Analysis*, 22(5), 970-1013.

Russell, T. F. (1989) Eulerian-Lagrangian Localized Adjoint Methods for Advection-Dominated Problems, *Proceedings of the 13th Biennial Conference on Numerical Analysis*, Pitman, Dundee, Scotland.

Yeh, G. T., J. R. Chang, and T. E. Short (1992) An Exact Peak Capturing and Oscillation-Free Scheme to Solve Advection-Dispersion Transport Equations, *Water Resources Research*, 28(11), 2937-2951.

FINITE DOMAIN ANALYTIC SOLUTION
OF THE DIFFUSION - CONVECTION EQUATION

O.O. ONYEJEKWE
National University Of Science And Tech.
Bulawayo, Zimbabwe

A new numerical formulation based on the finite analytic (FA) method is used to solve the diffusion-convection (DC) equation which is a linear parabolic nonsymmetrical equation. This method utilizes local analytic solutions obtained from small domains which comprise the total region of the governing partial differential equation. The assembly of all the composite solutions then constitutes the solution of the problem. Results from the present scheme are compared with the analytic solution of the classic one-dimensional diffusion-convection equation for various values of Peclet number. The method is found to perform satisfactorily even for convection dominated flows.

INTRODUCTION

For more than two decades the numerical solution of the DC equation has received considerable attention from numerical analysts. A concise treatment of these solution methods is given by (Anderson 1979).

A major numerical difficulty is experienced in simulating convection dominated flows with the DC equation. When the convection term is dicretized with central difference type expressions, the resulting coefficient matrix is no longer diagonally dominant ,the solutions turn out to be physically unrealistic and dislpay spurious oscillations of the dependent variable.

Most of the current literature on the DC equation is replete with several methods specifically designed to deal with oscillations arising from the convection term. These techniques follow two major routes, namely: (i) Introduction of artificial diffusion in such a way as to smoothen or balance both the undershoot and overshoot inherent in the discretization of the convection term (Hughes and Brooks 1979). (ii) Concentration of grid points optimally in those areas where the gradient of the scalar profile changes rapidly (Onyejekwe 1987). While each of these techniques has always resulted in satisfactory results; the gain has often been at the expense of increased computational effort; in the form of large number of elements, time steps, and in some cases a major alteration of the governing equation.

A. Peters et al. (eds.), Computational Methods in Water Resources X, 341–347.

The present scheme aims at deriving an analytic solution from the differential equation in a small region of the problem domain. It does not tamper with the differential, hence it minimizes the truncation errors inherent in the finite difference representation whereby the diffrential operators are approximated by difference forms via the Newton divided differece or the Taylor series expansion. It is also free from the error due to the finite element method in which an approximate solution function is first chosen to satisfy the governing differential equation.

This method of solution was first introduced by Chen and Li (1979) and was further improved and applied to practical engineering problems.(Chen et al 1981)

NUMERICAL FORMULATION

We consider the one-dimensional DC equation expressed in dimensionless variables:

$$\phi_t + Pe\,\phi_x = \phi_{xx} \tag{1}$$

where

ϕ is the concentration of a scalar variable
Pe is the Peclet number, and shows the relative
importance of difffusion and convection
t, x are time and distance parameters respectively

We seek the analytic solution of equation (1) in a small rectangular region as shown in fig.(1), which is dimensioned as 2h x τ, where h = Δx , τ = Δt and Δt = t_n - t_{n-1}. The interior nodal point 'P' is sorrounded by five nodal points identified as SW (south west), SC (south center), SE (south east), WC (west center) and EC (east center). In order for the DC equation to be well posed in the local element, plausible initial and boundary conditions must have to be specified. Since the boundary conditions are not known a-priori, they must be given through a proper assumed profile of the dependent variable.(Chen et al 1990).

Equation (1) is put in the form:

$$B\,\phi_t + 2A\,\phi_x = \phi_{xx} \tag{2}$$

where   A = 0.5Pe and B = 1

Equation (2) is expressed with the following initial and boundary conditions:

$$\phi(x,0) = a_s(e^{2Ax} - 1) + b_s x + c_s \tag{3}$$

$$\phi(-h,t) = a_w + b_w t \tag{4}$$

$$\phi(h,t) = a_E + b_E t \tag{5}$$

Where

$$a_s = \phi_{SE} + \phi_{SW} - 2\phi_{SC}/(4\sinh^2 Ah), \quad c_S = \phi_{SC} \tag{6}$$

$$b = \phi_{SE} - \phi_{SW} - \coth Ah(\phi_{SE} + \phi_{SW} - 2\phi_{SC})/2h \tag{7}$$

$$a_W = \phi_{SW}, \qquad b_W = (\phi_{wc} - \phi_{sw})/\tau \tag{8}$$

$$a_E = \phi_{SE}, \qquad b_E = (\phi_{EC} - \phi_{SE})/\tau \tag{9}$$

The analytic solution of eqaution(2) is found for the nth time step, and should be a function of $\phi^n(h,t^n)$ which can be expressed without the superscript n as:

$$\phi(h) = \phi(\phi_{EC} \ \phi_{wc} \ \phi_{SC} \ \phi_{SW} \ \phi_{S\dot{E}} \ h, A, B) \tag{10}$$

Since equation (10) is valid at any point (h,t^n) within the problem domain, it can be evaluated by an algebraic equation for the point 'P' and is given by:

$$\phi_P = c_{wc}\phi_{wc} + c_{EC}\phi_{EC} + c_{sw}\phi_{sw} + c_{SE}\phi_{SE} + c_{SC}\phi_{SC} \tag{11}$$

equation (11) relates the value of the scalar at the interior node 'P' to its five neighboring nodal values. The coefficients c_{wc}, c_{se}, etc are finite analytic coefficients whose values are obtained by the solution of equation (10) and are given by:

$$c_{WC} = e^{Ah}S_1, \quad c_{EC} = e^{-Ah}S_1, \quad c_{SW} = e^{Ah} S_2 \tag{12}$$

$$c_{SE} = e^{-Ah} S_2, \quad c_{SC} = 4Ah \cosh Ah \coth ah P_2 \tag{13}$$

$$S_1 = Bh^2/\tau(P_2 - Q_2) + Q_1, \tag{14a}$$

$$S_2 = Bh^2/\tau(Q_2 - P_2) - 2Ah \coth AhP_2 \tag{14b}$$

$$P_2 = \sum_{m=1}^{\infty} \frac{-(-1)^m \lambda_m}{[(Ah)^2 + (\lambda_m h)^2]^2} \frac{he^{-2F_m\tau}}{} \tag{15}$$

$$Q_i = \sum_{m=1}^{\infty} \frac{-(-1)^m \lambda_m h}{[(Ah)^2 + (\lambda_m h)^2]^i} \qquad i = 1,2 \tag{16}$$

$$2F_m\tau = (Ah)^2 + (\lambda_m h)^2/(Bh^2/2\tau) \tag{17a}$$

$$\lambda_m h = (m - 0.5)\pi \tag{17}$$

Using the coefficients given above, equation(11) is solved for each node oι the problem domain. An assemblage of all the elements results in a system of algebraic equations which are solved iteratively for a given time level, knowing the specified values at a previous time level.

DISCUSSION OF RESULTS

Equation (1) is specified with the following initial and boundary conditions:

I.C.: $\phi(x,0) = 0$ (18)

B.C1 : $\phi(0,t) = 1$ (19)

B.C2 : $\phi(1,t) = 0$ (20)

The analytical solution of the above problem is given by, Farrukh et al (1983) as:

$$\phi \exp(-Pex/2) = \sinh[Pe/2(1-x)]/\sinh(Pe/2) - 2\pi(4)/Pe^2$$

$$x \sum \frac{n}{1 + n^2\pi^2(4)/Pe^2} \ x \ \sin(n\pi x)$$

$$x \ \exp[-(1+4n^2\pi^2/ \ Pe^2)t] \tag{22}$$

Fig.2 illustrates the behavior of the maximum errors of the FA numerical solution with increasing values of Peclet number for a mesh width of 0.1. Maximum error is defined as the maximum of the absolute value of deviations of the numerical solution from the exact solution at all the points where computation is carried out. Results are obtained with double precision arithmetic.

The scheme is tested for Pe = 0.1, 1, 10, and 50 to illustrate the ability of the scheme to handle various modes of transport. Numerical results obtained for Pe = 0.1 and 1 are almost identical with the analytic and display negligible percentage errors. As the Peclet number increases, the numerical solution is found to deviate from the analytical especially in those areas of the problem domain where there is a sudden change in the spatial distribution of the dependent variable. This is found most noticeable in regions between the Dirichlet boundaries. It is likely that more of these deviations will occur as the Peclet number increases. It is gratifying to note that a maximum error of only 0.4 percent was obtained for this test; and for all the cases considered, the numerical results are almost identical with the analytical in regions close to the two boundaries. These observations suggest that the scheme is competitive and can be used with confidence to model a wider range of flows indicative of higher Peclet numbers.

CONCLUSION

The FA solution for the unsteady one-dimensional DC equation has been derived and applied to examples illustrating different modes of scalar transport involving both diffusion and convection. For every element of the problem domain, a 5 - point FA algebraic equation is obtained from the analytic solution of the dicretized

DC equation with the element boundary approximated by a combination of linear and exponential functions. The method is found to perform satisfactorily for the different cases studied, and can therefore be extended to flows in porous media, and to other flows where convection is the dominant mode of transport. However the test can not be considered complete without the addition of sources and sinks. The only drawback to this scheme is the rigor involved in the analytic solution of the governing equation in a finite element of the problem domain.

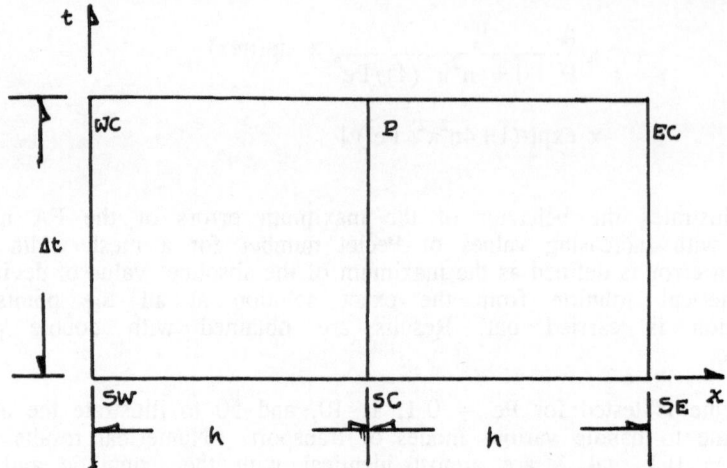

FIG: 1

A FINITE ANALYTIC COMPUTATIONAL SUBREGION

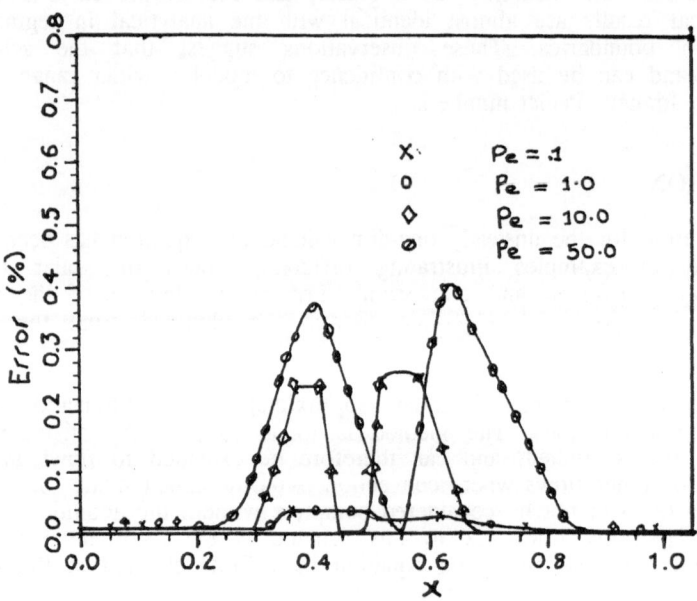

FIG: 2

PERCENTAGE ERROR DISTRIBUTION FOR VARIOUS VALUES
OF PECLET NUMBER

REFERENCES

1. Anderson, M.P. (1979) 'Using models to simulate the movement of contaminant through groundwater flow systems' CRC Crit. Rev. Environ. Contr, 9(2) 97-156

2. Farrukh,M. and Baruch, M.A.(1983) 'An analytical solution of diffusion-convection equation over a finite domain' Appld. Math. Modelling. 7, 285-287

3. Chen,C.J. and Li P. (1979) 'Finite differential method in heat conduction - Application of analytic solution technique' ASME paper 79-WA/HT-50

4. Chen, C.J. Naseri-Neshat, H. and HO K.S. (1981) 'Finite analytic numerical solution of heat transfer in two - dimensional cavity flow' Jnl.of Num.Heat Transf. 4, 179-197

5. Chen,C.J. Bravo,R. and Sheiskholeslami, M.Z. (1990) ' The finite analytic method and its application' IIHR report No. 334-II vol.(II) , Iowa inst. of hydraulics research, university of Iowa, Iowa city, 52242 U.S.A.

6. Hughes T.J.R. and Brooks.A.N. (1979) ' A multi dimensional upwind scheme with no cross wind diffusion ' in Hughes T.J.R.(ed) Finite element methods in convection dominated flows, AMD vol 34 ASME U.S.A.

7. Onyejeke O.O. (1987) 'A node moving method of solution with application to the convection -diffusion equation' Proceedings of heat transfer and fluid mechanics instituteMay 28-29 Sacramento Calif. U.S.A. 81-87

3D SOLUTE TRANSPORT SIMULATIONS OF STRUCTURED SOILS

E. Priesack, M. Thoma and R. Dinauer
GSF-Institut für Bodenökologie
Postfach 1129, Oberschleißheim, 85758
Germany

Soil structures as given by non-capillary sized macropores or large fractions of stones can strongly influence transport processes in soils resulting in preferential movement of water and solutes. 3D-simulation of water and solute transport provides a tool to characterize the impact of different geometric arrangements of macropores or stones on these transport processes. Using different modelling approaches to describe water flow in macropores and around stones, pore water velocities and resident concentration of tracer chemicals are calculated. For saturated soils with water filled macropores a 3D two domain fluid dynamic model is proposed. It allows to investigate the interaction of Darcy flow in porous regions and Newtonian flow in macropores. The momentum, continuity and convection dispersion equations are solved by the finite element method. The interaction of macropores is quantified for different model geometries, showing the importance of the contribution of a small proportion of the total soil porosity to solute transport. To determine the effects of stones, movement of water and chemicals is simulated solving the 3D Richards and the convection dispersion equations for unsaturated soils by the finite element method. As input information on the location, form and size of stones is used and knowledge of the soil hydraulic properties of the soil without stones is required. Corresponding to the lower water filled pore volume, simulated breakthrough was earlier for stony soils. Simulations show zones of retarded solute transport above and underneath stones. They indicate that zones of almost immobile water and solute retardation underneath stones are larger under more saturated conditions. In conclusion simulations are very useful in giving new insight into three-dimensional behaviour of transport in structured soils

INTRODUCTION

Soil structures as given by noncapillary macropores or high volume fractions of stones may cause preferential flow and early breakthrough of solutes (Beven and Germann 1982, Edwards et al. 1993, Schulin et al. 1987). Thus considerable amounts of applied water and solutes may bypass the upper soil zone of highest biotic activity where contaminants may be sorbed and degraded by microbes and plants. To better understand the possible effects of geometric arrangements of macropores or stones on solute transport 3D simulations were performed for soil columns with embedded macropores or stones.

A. Peters et al. (eds.), Computational Methods in Water Resources X, 349–356.

MACROPOROUS SOIL

Using an example of simple geometry the importance of connectivity between macropores is studied. Assuming steady state saturated water flow driven by force of gravity solute transport in a soil column with embedded macropores is simulated. After a sudden increase of the input concentration at the upper boundary resident solute concentrations are calculated for the soil matrix and the macropores. The rectangular soil column contains two cylindrical macropores, one macropore at the left hand side of the column with open top and bottom dead end, the other macropore at the right hand side of the column with top dead end and open bottom end (fig. 1).

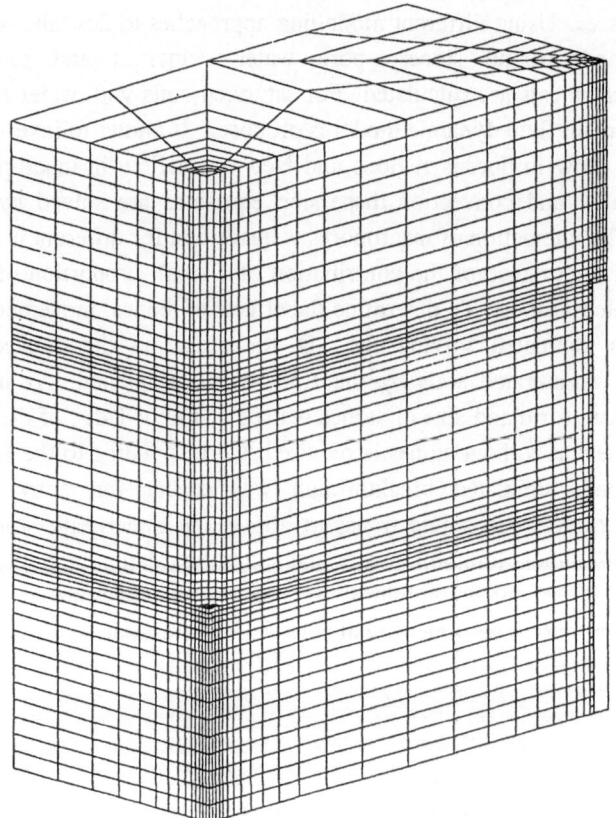

Fig. 1: Finite element grid of the soil column
a) for the porous domain and

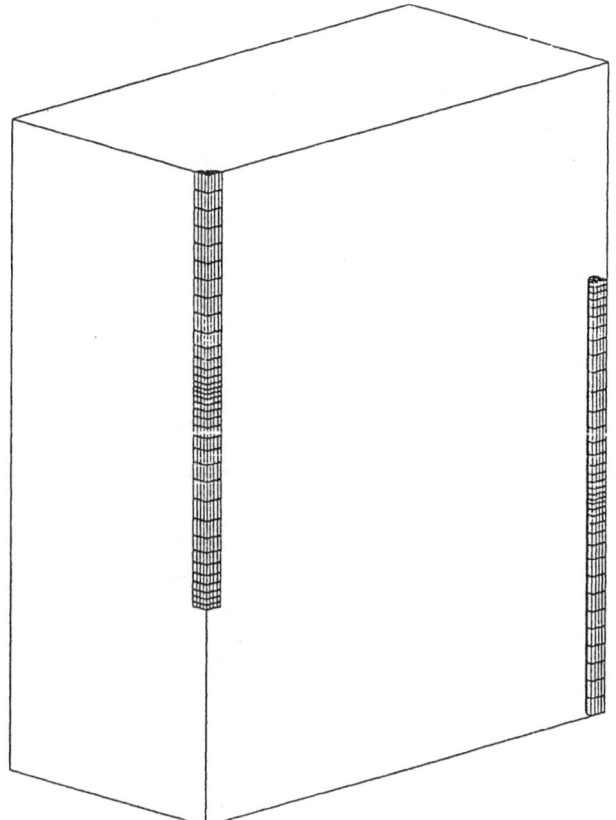

b) for the macropore domain

Inside the macropores, where there is no solid matter, water flow is considered as Newtonian incompressible laminar isothermal flow governed by the continuity equation and Navier Stokes equations (Thoma and Priesack 1993). For the matrix region, which microscopically consists of a granular rigid phase and a fluid phase in the micropores between, water flow obeys averaged macroscopic conservation laws for mass and momentum (Du Plessis and Masliyah 1991). At the interface between macropores and soil matrix continuity of pressure and discharge is assumed (Thoma and Priesack 1993). Steady state water flow is calculated considering no normal flux conditions at the sides and stress free conditions at top and bottom. Given the flow velocities solute transport is described by the convection dispersion equation using a constant dispersion coefficient D_0 for transport within the macropores and a diagonal dispersion tensor $D_1 = (D + a\ v)\ E$, where D is an effective diffusion coefficient, a is the dispersivity and v the absolute value of the filter velocity in the porous matrix.

Fig. 2: Contour plot of resident solute concentration between macropores
a) overlapping and b) non overlapping case
for different times t=500 s, 2000 s, 4000 s, 6000 s and 8000 s after increase of input
concentration. (macropore radius r=0.073 cm, distance between centers of
macropores d=2.1 cm, length of macropores for a) l=40 cm and for b) l=30 cm).

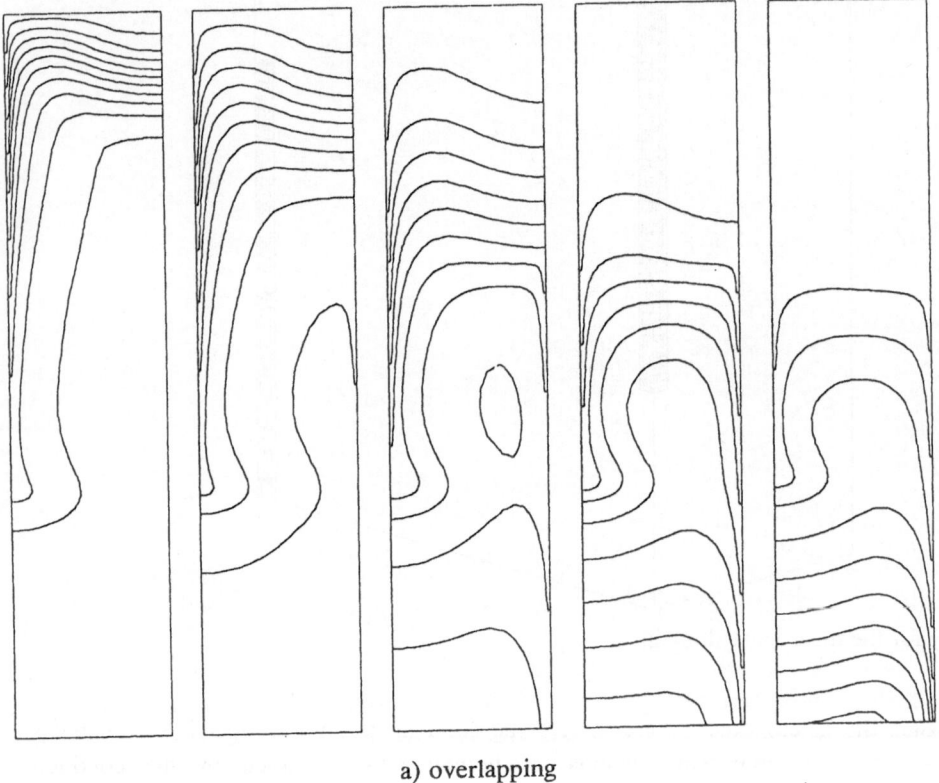

a) overlapping

Figure 2 a) shows the resulting contour plots of resident concentrations in the soil
column for overlapping macropores, i.e. if the dead end of the macropore at the top is
lower than the dead end of the macropore open to the bottom. Figure 2 b) represents
a case of no overlap. Calculation of solute breakthrough curves indicates faster
breakthrough if the dead ends are at the same height rather than if the macropores are
overlapping.

and b) non overlapping case

STONY SOILS

To study the effect of stones on solute transport three different cases were considered. A cube, 3 plates or 27 cubes are embedded in a cylindrical soil column representing stones which for each case fill the same volume fraction of the soil column (Hantschel et al. 1992). For steady state water flow conditions the transport and breakthrough of a short pulse of solute at the upper boundary of the soil column is simulated (Dinauer et al. 1993). Water flow is calculated by the 3D equation for stationary unsaturated flow (Istok 1989, appendix I) and solute transport by the 3D dispersion advection equation (Istok 1989, appendix II) assuming identical values for transversal and longitudinal dispersivity.

Fig: 3: Contour plots of resident solute concentrations in cylindrical soil columns with stones at times t=4 d, 8 d and 12 d after application of a solute pulse at the upper boundary for an average pore water velocity of 0.4 cm/d.

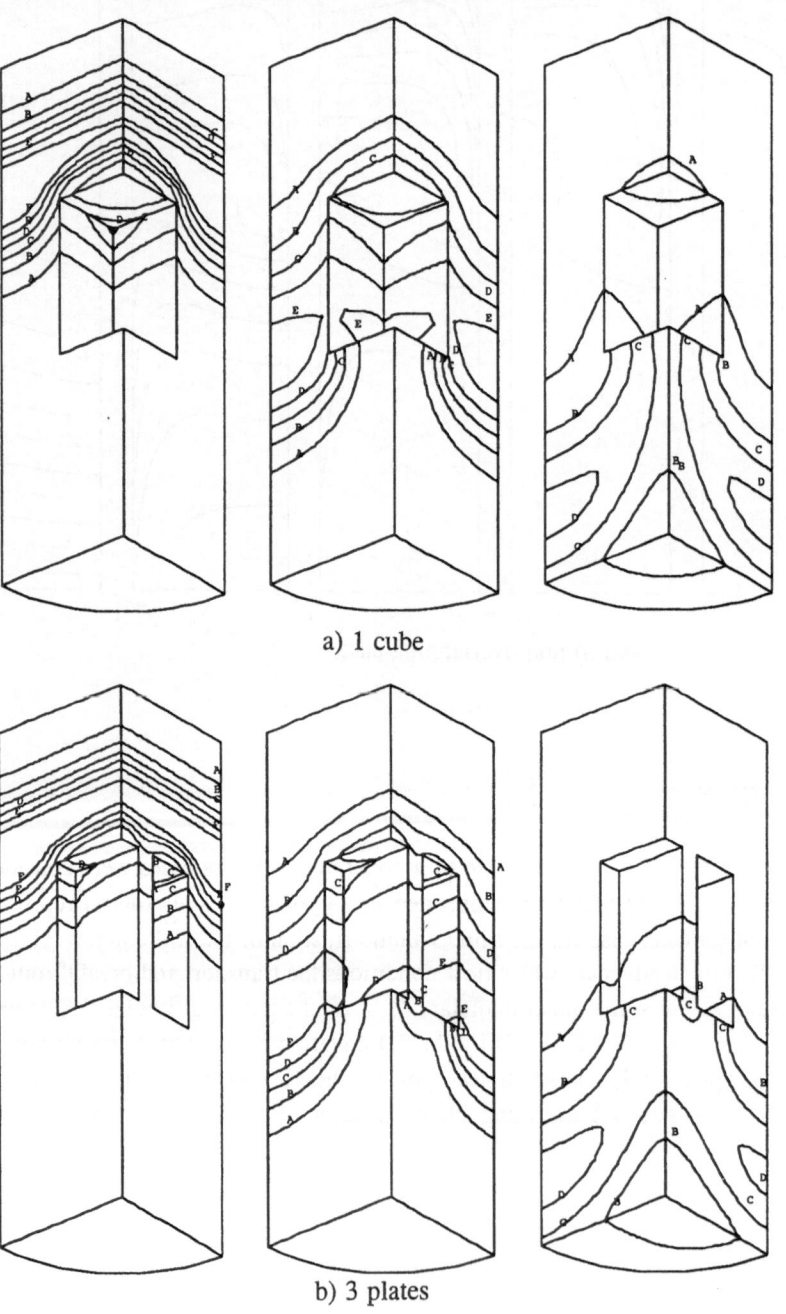

a) 1 cube

b) 3 plates

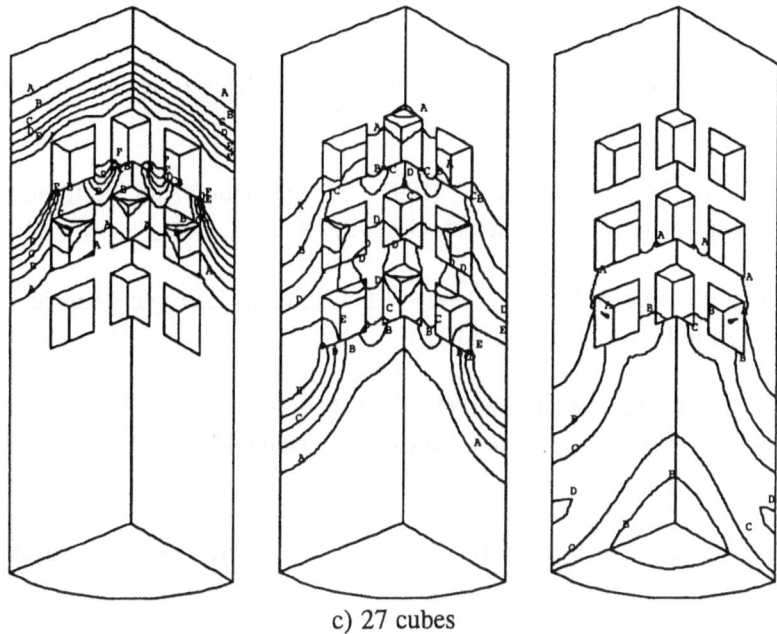

c) 27 cubes

Simulation results (fig. 3) show faster solute transport between stones and column walls and slower transport above and underneath stones. According to the smaller pore volume accessible to solute transport, breakthrough is earlier for stony soils. Since the shapes of the calculated curves for the three stony soils are very similar to the shape of the breakthrough curve obtained for a soil column without stones, simulations of breakthrough for stony soils might be well achieved using the classical one-dimensional convection dispersion equation by accounting for the smaller pore volume.

NUMERICAL SOLUTION

The numerical solution of the transport equation is based on the finite element method using a discretization of the flow domains into 3D bricks with trilinear basis functions. Calculations were performed using the Fluid Dynamics Analysis Package FIDAP (Fluid Dynamics International Inc. 1991)

REFERENCES

Beven, K. and Germann P.F. (1982) "Macropores and water flow in soils", Water Resour. Res. 18, 1311-1325.

Dinauer, R., Thoma, M. and Priesack, E. (1993) "3D Simulation zum Stofftransport in Böden mit Steinen", Mitt. Dtsch. Bodenkd. Ges. 72, 89-92.

Du Plessis, J.P. and Masliyah, J.H. (1991) "Flow through isotropic granular porous media", Transport in Porous Media 6, 207-221.

Edwards, W.M., Shipitalo, M.J. and Owens L.B. (1993) "Gas, water and solut transport in soils containing macropores: a review of methodology", Geoderma 57, 31-49.

Fluid Dynamics International Inc. (1991) FIDAP User's Manual 6.01, Evanston, Illinois, USA.

Hantschel, R., Beese, F. and Hoeve, R. (1992) "Bedeutung von Steinen für den Wassertransport. Ein idealisierter Laborversuch" Mitt. Dtsch. Bodenkd. Ges. 67, 79-82

Istok, J., Groundwater Modeling by the Finite Element Method, Water Resources Monograph 13, American Geophysical Union, Washington DC.

Schulin, R., Wierenga P.J., Flühler, H. and Leuenberger J. (1987) "Solute transport through a stony soil" Soil Sci. Soc. Am. 51, 36-42.

Thoma, M. and Priesack, E. (1993) "Coupled porous and free flow in structured media" Z. angew. Math. Mech. 73, T566-T569.

SOLUTION OF GROUNDWATER TRANSPORT EQUATION USING A COMBINATION OF DISCONTINUOUS FINITE ELEMENTS AND MIXED FINITE ELEMENTS

SIEGEL P., MOSE R.
Institut de Mécanique des Fluides
Université Louis Pasteur - URA CNRS 854
2 rue Boussingault, Strasbourg 67000
France

A new numerical method is developed for the solution of the contaminant transport equation in groundwater. A discontinuous finite element method is employed for the discretization of the advective terms. The dispersive term is discretized using a mixed hybrid finite element method. With this approach, numerical oscillations are completely avoided for a full range of cell Peclet numbers. Numerical tests show good agreement with a 2D analytical solution. This new approach is compared to different numerical methods, mixed finite element method and finite volume approach with high-resolution upwind terms. Regular and irregular meshes are used for the numerical tests to study the mesh effects on the numerical results. In all cases this approach performs well.

Introduction

In groundwater, classical methods like finite elements and finite differences are used to solve the contaminant transport equation. When the transport is advection dominated, this equation becomes hyperbolic. The resolution of such equation by the classical methods introduces spurious oscillations and artificial diffusion. Special schemes have been developed to solve hyperbolic equations with finite differences or finite elements. Upwind schemes have been introduced in finite differences. This technique gives first order accurate stable scheme (without oscillations), but with too much numerical diffusion smearing the front. To overcome this problem, high-order accurate and nonoscillatory finite difference upwind schemes have been developed (Van Lerr, 1977). These schemes are generally constructed through a discontinuous piecewise polynomial representation of the solution, linear or quadratic. These methods are stabilized with slope limiters. The numerical flux at an interface is obtained by solving a local Riemann problem. The extension of these schemes to two dimensions is obtained by writing the one-dimensional schemes in directions parallel to the axes using quadrilateral control volumes. With these elements it becomes difficult to describe complex real area. Chakravarthy and Osher (1985) extended such schemes to triangular

A. Peters et al. (eds.), Computational Methods in Water Resources X, 357–364.

elements using a finite volume approximation. Putti *et al.* (1990) applied this method to the resolution of the general transport equation in groundwater. This scheme is second order accurate for equilateral triangles. High-order schemes with multidimensional slope limiters have been developed using discontinuous finite elements by G. Chavent and J. Jaffré (1986). In the paper, this method is adapted to triangular elements for the discretization of the advective term. The dispersive term is approximated by a mixed finite element method.

The discontinuous finite element scheme in one-dimension

The advection-dispersion equation in groundwater can be written as (Bear, 1979):

$$\frac{\partial c}{\partial t} = -\nabla.(\bar{v}c) + \nabla.(\mathbf{D}\nabla c) + G \tag{1}$$

where c is the concentration of the contaminant, $\bar{v}$ is the velocity vector, $\mathbf{D}$ is the dispersion tensor and G is the source or sink term.

A discontinuous finite element method is used for the discretization of the advective term. Without the dispersive term, the equation (1) becomes a hyperbolic equation :

$$\frac{\partial c}{\partial t} = -\nabla.(\bar{v}c) \tag{2}$$

The one-dimensional space interval [a,b] is discretized with a set of elements $K = [x_i, x_{i+1}]$ and the nodes $x_1 = a <...< x_{i+1} <...< x_{I+1} = b$. We denote Δx_K the size of the element K. The concentration c is approximated in a space of discontinuous linear functions $M^1 = \{f|_K \in (m_{K1}, m_{K2})\}$. In other words, the restriction of f over K is a linear combination of m_{K1} and m_{K2}. The functions of M^1 are discontinuous at nodes of discretization so that we denote by f_i^{in} and f_i^{out} respectively the inside and outside limits of a function in M^1 at the node x_i with respect to the element K. The linear variation of c over K is defined as follow : $c_K(x) = m_{K1}(x)c_i^{in} + m_{K2}(x)c_{i+1}^{in}$.

By writting a variational form of the hyperbolic equation and using the Green formula, we obtain over each element K :

$$\int_K \frac{\partial c_K}{\partial t} m\, dx = \int_K \bar{v}c_K \cdot \nabla m\, dx - Q_{K,i+1} c_{i+1}^{in\ or\ out} m(x_{i+1}) + Q_{K,i} c_i^{in\ or\ out} m(x_i) \tag{3}$$

where m $(\in M^1)$ is a test function over K and $Q_{K,i}$ is the flux at the node i for the element K. The out-flux is positive defined and the in-flux is negative defined.

To preserve the mass balance, the advective flux have to be uniquely defined at interface of two elements. The advective flux is obtained by solving a standard Riemann problem in the linear case (Godunov, 1959). In other word, the numerical advective flux is calculated with the upstream value of c. The choice between the discontinuous values

c^{in} and c^{out} depend of the sign of Q ($c_i = c_i^{in}$ if $Q_{K,i} > 0$ and $c_i = c_i^{out}$ if $Q_{K,i} < 0$). In order to solve the Riemann problem, an explicit time discretization is used. Denote by $0 = t_0 \leq t_1 \leq \ldots \leq t_n \ldots$ with $\Delta t_n = t_{n+1} - t_n$ the time discretization, the equation (3) becomes:

$$\int_K \frac{c_K^{n+1} - c_K^n}{\Delta t_n} m\, dx = \int_K \bar{v} c_K^n \cdot \nabla m\, dx - Q_{K,i+1} c_{i+1}^{in\ or\ out,n} m(x_{i+1}) + Q_{K,i} c_i^{in\ or\ out,n} m(x_i) \quad (4)$$

By using successively m_{K1} and m_{K2} as test functions and solving the Riemann problem at the interfaces, we obtained a system of two unknowns per element $c_{i+1}^{in,n+1}$ and $c_i^{in,n+1}$ (compounds of c_K^{n+1} in the basis m_{K1} and m_{K2}). This scheme does not have good stability properties and the calculated solution oscillates. To increase the time discretization order, an intermediate time step at time $t_{n+1/2}$ is presented. The advective fluxes are calculated using the concentration values defined in the element K.

Step 1 : calculation of $c^{n+1/2}$ with local values:

$$\int_K \frac{c_K^{n+1/2} - c_K^n}{1/2 \Delta t_n} m\, dx = \int_K \bar{v} c_K^n \cdot \nabla m\, dx - Q_{K,i+1} c_{i+1}^{in,n} m(x_{i+1}) + Q_{K,i} c_i^{in,n} m(x_i) \quad (5)$$

Step 2 : calculation of c^{n+1^*} by solving a Riemann problem :

$$\int_K \frac{c_K^{n+1^*} - c_K^n}{\Delta t_n} m\, dx = \int_K \bar{v} c_K^{n+1/2} \cdot \nabla m\, dx - Q_{K,i+1} c_{i+1}^{in\ or\ out,n+1/2} m(x_{i+1}) + Q_{K,i} c_i^{in\ or\ out,n+1/2} m(x_i) \quad (6)$$

To stabilize this scheme, we introduced a slope limiting step. The slope limiting operator L associates to each function c^{n+1^*} ($\in M^1$) a function $L(c^{n+1^*}) = c^{n+1}$ ($\in M^1$), which satisfies the following conditions :
-we preserve the mass balance by

$$\overline{c_K^{n+1}} = \int_K \frac{c_K^{n+1}}{\Delta x_K} dx = (c_i^{in,n+1} + c_{i+1}^{in,n+1})/2 = \overline{c_K^{n+1^*}} = (c_i^{in,n+1^*} + c_{i+1}^{in,n+1^*})/2 \quad (7)$$

- we limit the slope by setting an upper and a lower limit values to $c_i^{in,n+1}$ and $c_{i+1}^{in,n+1}$
  (which have not to be greater or smaller than the mean concentration of the element formed by these nodes)

$$\min(\overline{c_{K-1}^{n+1^*}}, \overline{c_K^{n+1^*}}) \leq c_i^{in,n+1} \leq \max(\overline{c_{K-1}^{n+1^*}}, \overline{c_K^{n+1^*}}) \quad (8)$$

$$\min(\overline{c_{K+1}^{n+1^*}}, \overline{c_K^{n+1^*}}) \leq c_{i+1}^{in,n+1} \leq \max(\overline{c_{K+1}^{n+1^*}}, \overline{c_K^{n+1^*}}) \quad (9)$$

- if $c_K^{n+1^*}$ is a local minimun or maximun of c, than $c_i^{in,n+1}$ and $c_{i+1}^{in,n+1}$ are defined by :

$$c_{i+1}^{in,n+1} = c_i^{in,n+1} = \overline{c_K^{n+1^*}} \quad if\ \overline{c_K^{n+1^*}} \geq \max(\overline{c_{K-1}^{n+1^*}}, \overline{c_{K+1}^{n+1^*}})\ or\ \overline{c_K^{n+1^*}} \leq \min(\overline{c_{K-1}^{n+1^*}}, \overline{c_{K+1}^{n+1^*}}) \quad (10)$$

If $\overline{c_K^{n+1^*}}$ is not in the case (10), note that c_K^{n+1} is not uniquely defined by (7) (8) (9). To define it uniquely, we impose that c_K^{n+1} be as close as possible to $c_K^{n+1^*}$. A minimization problem of dimension 2 has to be solved for the element K. To determine c_K^{n+1} an objective function J is defined :

$$J(c_i^{in,n+1}, c_{i+1}^{in,n+1}) = \left\| c_i^{in,n+1} - c_i^{in,n+1^*} \right\|^2 + \left\| c_{i+1}^{in,n+1} - c_{i+1}^{in,n+1^*} \right\|^2 \tag{11}$$

$c_i^{in,n+1}$ and $c_{i+1}^{in,n+1}$ are estimated by minimizing the objective function J and satisfying the constraints (7),(8) and (9). In one dimension, this method is very similar to high-order finite difference method (Van Leer, 1977). The discontinuous finite element method is more expensive compared to high-order finite difference scheme in one dimension, but it can be easily extended in truely multidimensional schema.

Solution of the advection-dispersion equation in 2D with triangular elements

The two-dimension space is discretized with triangular elements K. We consider the triangular element K formed by the nodes A,B,C and the edges E_i (i = 1,..,3). The concentration c is approximated in a space of discontinuous linear functions $M^1 = \left\{ f|_K \in (m_{K1}, m_{K2}, m_{K3}) \right\}$. The function of M^1 are discontinuous at nodes of discretization.

$c_K(x,y) = m_{K1} c_A^K + m_{K2} c_B^K + m_{K3} c_C^K$ (variation of c over K where c_A^K is the value of c at the node A in the element K).

We denote by c_{AB}^{in} the linear variation of c over the edge AB in the element K and by c_{AB}^{out} the linear variation of c over the edge AB in the element adjacent to K. The dispersive term $T_{D,K}^n$ is determined by a mixed approximation (Raviart and Thomas, 1977; Chavent and Roberts, 1991). The evaluation of this term needs an approximation of c per edge. This concentration is determined from the $\overline{c_K}$ ($\overline{c_k} = (c_A^K + c_B^K + c_C^K) / 3$ is the average of c over K) by solving a linear equation system whose associated matrix is positive definite.

<u>Step 1 : the calculation of $c^{n+1/2}$ with local values:</u>

The dispersive terms obtained with the mixed approximation $T_{D,K}^n$ are introduced :

$$\int_K \frac{c_K^{n+1/2} - c_K^n}{1/2\Delta t_n} m = \int_K \vec{v} c_K^n . \nabla m - Q_{K,AB} \int_{AB} \frac{c_{AB}^{in,n} m}{AB} ds - Q_{K,BC} \int_{BC} \frac{c_{BC}^{in,n} m}{BC} ds - Q_{K,CA} \int_{CA} \frac{c_{CA}^{in,n} m}{CA} ds$$
$$+ T_{D,K}^n \int_K m + G_K \int_K m \tag{12}$$

where m is successively the three basis functions of M^1 over K and $Q_{K,AB}$ is the flux through the edge AB with respect to $\bar{n}_{K,AB}$. $\bar{n}_{K,AB}$ is the unit exterior normal vector of this edge for the element K $(\text{sign}(Q_{K,AB}) = \text{sign}(\bar{n}_{K,AB} \cdot \bar{v}))$. The value at time $t^{n+1/2}$ at each node in K is easily calculated by solving a local system. The concentrations per edge $\overline{TC^{n+1/2}}$ are calculated from $\overline{c_K^{n+1/2}}$ by solving the mixed system.

Step 2 : calculation of c^{n+1^*} by solving the Riemann problem :

$$\int_K \frac{c_K^{n+1^*} - c_K^n}{\Delta t_n} m = \int_K \bar{v} c_K^{n+1/2} . \nabla m - Q_{K,AB} \int_{AB} \frac{c_{AB}^{\text{in or out},n+1/2}}{\overline{AB}} m \, ds - Q_{K,BC} \int_{BC} \frac{c_{BC}^{\text{in or out},n+1/2}}{\overline{BC}} m \, ds$$
$$- Q_{K,CA} \int_{CA} \frac{c_{CA}^{\text{in or out},n+1/2}}{\overline{CA}} m \, ds + T_{D,K}^{n+1/2} \int_K m + G_K \int_K m \tag{13}$$

where m is successively the three basis functions of M^1. The value at each node in K are easily calculated by solving a local system.

Step 3 : slope limitation : The scheme is stabilized with multidimensional slope limiter $L(c^{n+1^*}) = c^{n+1}$ satisfying following conditions :

- $\overline{c_K^{n+1}} = \overline{c_K^{n+1^*}}$ in order to preserve mass balance. $\tag{13}$
- in order to limit the slope,

$$\min(A) \le c_A^{K,n+1} \le \max(A), \quad \min(B) \le c_B^{K,n+1} \le \max(B)$$

and $\min(C) \le c_C^{K,n+1} \le \max(C)$ $\tag{14}$

where min(A) is the minimum and max(A) is the maximum of the average concentration $\overline{c_K^{n+1^*}}$ of the elements having A for node.

$$c_A^{K,n+1} = c_B^{K,n+1} = c_C^{K,n+1} = \overline{c_K^{n+1^*}} \quad \text{if} \quad \overline{c_K^{n+1^*}} \ge \max_{K'}(\overline{c_{K'}^{n+1^*}}) \text{ or } \overline{c_K^{n+1^*}} \le \min_{K'}(\overline{c_{K'}^{n+1^*}}) \tag{15}$$

where K' represents all the elements having a common node with the element K.

If $\overline{c^{n+1^*}}$ is not in the case (15), c_K^{n+1} is not uniquely defined. We impose that c_K^{n+1} be as close as possible to $c_K^{n+1^*}$. A minimization problem of dimension 3 has to be solved for each element to determine c_K^{n+1}. With this method, the mass balance is exact over each element K. The advective fluxes are uniquely defined by solving a Riemann problem at the interface of two elements. Moreover, the dispersive flux is continuous from one element to the adjacent one.

Comparison studies

This technique is compared to a classical mixed finite element method and to the high-order finite volume method developed by Putti *et al.* (1990) using the following two dimensional analytical solution given by Leij and Dane (1990) (figure 1).

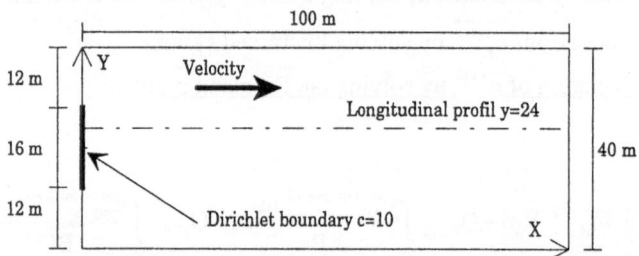

Fig. 1 : The two-dimensional convection-diffusion problem.

This problem is solved for different values of the cell-Peclet number at time t=50 s. The parameters used in the different cases are reported in the following table.

Case	v(x), m/s	v(y), m/s	D_L, m^2/s	D_T, m^2/s	Δx, m	Δy, m	Pe
1	1.0	0.0	2	0.2	2.0	2.0	1
2	1.0	0.0	0.02	0.002	2.0	2.0	100

Regular and irregular meshes (figure 2) are used to study the mesh effects on the numerical results.

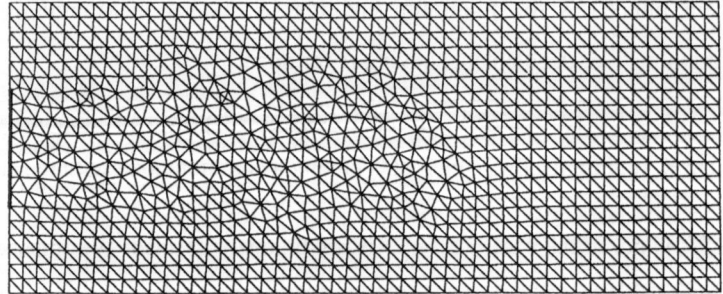

Fig. 2 : The irregular mesh.

In the case one, the transport is diffusion dominated (Pe ≤ 2). There is no difficulties to solve this case with the classical methods (F.E. or F.D.). Moreover, the solution obtained with the mixed approximation (figure 4) agrees well with the analytical solution (figure 3). The discontinuous finite element method gives accurate results for the regular (figure 5) and the irregular (figure 6) meshes compared to the analytical solution. The second case is another advection dominated case with a Peclet number equal to

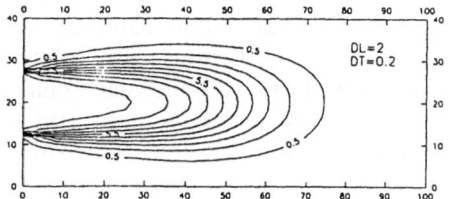

Fig. 3 : The analytical solution for the case 1.

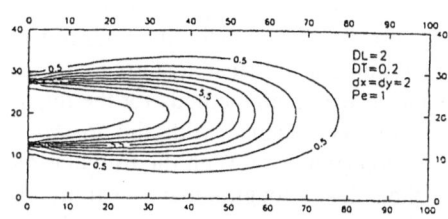

Fig. 4 : Computation results from case 1: Mixed approximation with the regular mesh.

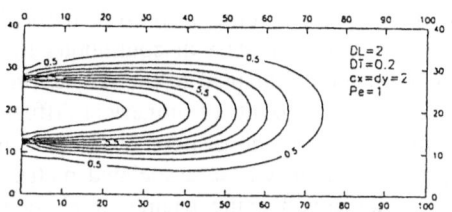

Fig. 5 : Computation results from case 1: Discontinuous F.E. with the regular mesh.

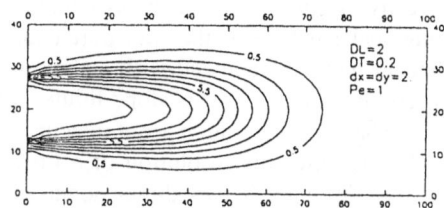

Fig. 6 : Computation results from case 1: Discontinuous F.E. with the irregular mesh.

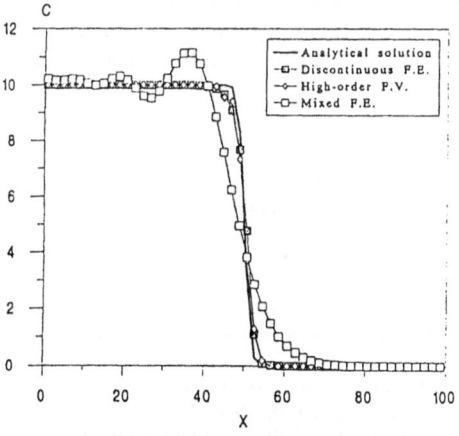

Fig. 7 : Comparison of computation results from the different numerical methods with the regular mesh using the profile y=24 : case 2.

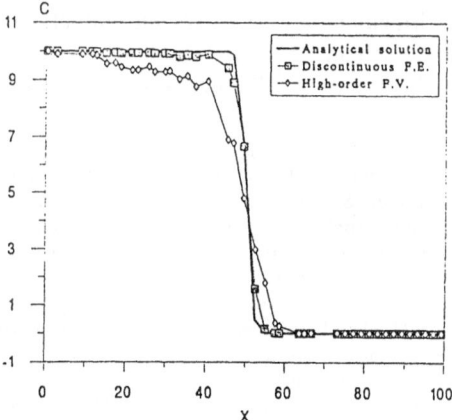

Fig 8: Comparison of computation results from the different numerical methods with the irregular mesh using the profile y=24 : case 2.

hundred. Longitudinal profiles at y=24 (figure 7 and 8) are used to see the effect of the transverse diffusion. With the high-order methods (finite volume and discontinuous F.E.) small numerical diffusion starts to appear for the regular mesh (figure 7). The solutions obtained with both methods for the regular mesh are very close to the analytical solution. The results obtained with the mixed approximation show oscillations and artificial diffusion (figure 7). With the irregular mesh, the solution of the high-order finite volume method shows longitudinal and transverse diffusion (figure 8). The discontinuous F.E. method gives the same accurate results with the irregular mesh (figure 8).

Conclusion

A method, using discontinuous finite elements for the discretization of the advective term and mixed approximation for the dispersive term, is presented for the numerical solution of the transport equation in groundwater. The discontinuous finite element method stabilized with a slope limiter is especially adapted to the resolution of hyperbolic equations. This slope limiter introduces small amounts of numerical diffusion when sharp concentration fronts occur. The introduction of the dispersive term in this method requires the resolution of a linear equation system, whose associated matrix is positive definite. A two dimensional scheme is developed using triangular elements. With this approach, numerical oscillations are completely avoided for a full range of cell Peclet numbers. Numerical tests show good agreement with 2D analytical solutions. This new approach is compared to different numerical methods (mixed finite element method and finite volume approach with high-resolution upwind terms). Regular and irregular meshes are used for the numerical tests to study the mesh effects on the numerical results. The irregular mesh destroys the global second order accuracy of the high-order finite volume scheme. The discontinuous finite element method performs well with triangles of any shape and for full range of Peclet numbers.

References :

Bear J., Hydraulics of Groundwater, McGraw-Hill, New York, 1979, 764 p.
Chakravarthy S. and Osher S., Computing with high resolution: Upwind schemes for hyperbolic equations, Lect. Notes Appl. Math., Vol. 22, pp. 57-86, 1985.
Chavent G. and Chaffré J., Mathematical models and finite elements for reservoir simulation, North Holland, Amsterdam, 1986.
Chavent G., Roberts J.E., A unified physical presentation of mixed, mixed hybrid finite elements and standard finite difference approximations for the determination of velocities in waterflow problems, Adv. in Water Resources, Vol. 14, n°6, pp. 329-348, 1991.
Leij Feike J. and Dane J. H., Analytical solution of the one-dimensional advection equation and two- or three-dimensional dispersion equation, Water Resour. Res., Vol. 26, No. 7, pp. 1475-1482, 1990.
Godunov S., Finite difference methods for numerical computation of discontinuous solutions of the equations of fluid dynamics, Math. SborniK, Vol. 47, pp. 271-306, 1959.
Putti M. W.-G. Yeh W., Mulder W. A., A triangular finite volume approach with high-resolution upwind terms for the solution of groundwater transport equations, Water Resour. Res., Vol. 26, No. 12, pp. 2865-2880, 1990.
Raviart P.A., Thomas J.M., A mixed finite method for the second order elliptic problems, Mathematical Aspects of the Finite Element Method, Lecture Notes in Mathematics, Springer-Verlag, New York, 1977
Van Leer B., Towards the ultimate conservative scheme : IV. A new approach to numerical convection, J. Comp. Phys., Vol. 23, pp. 276-299, 1977.

A COMPARISON OF FINITE ELEMENT AND FINITE VOLUME SOLUTIONS FOR UNSATURATED FLOW AND CONTAMINANT TRANSPORT IN GROUNDWATER

F.T. TRACY
Information Technology Laboratory
US Army Engineer Waterways Experiment Station
Vicksburg, MS 39180
USA

Finite volume (FV) computer programs to model unsaturated flow and contaminated transport in porous media have been written, and results from these codes have been compared with finite element (FE) results for selected test problems. The problems include laboratory tests as well as analytical solutions. The unusual characteristic of the FV codes is that results are computed at the grid points or nodes instead of the centers of the elements. The disadvantage is that partial cells exist at the boundaries, but the advantage is that the identical grid, initial conditions, and boundary conditions can be used with both the FE and FV programs. Also, Picard and Newton iteration techniques are compared for an unsaturated flow problem. Finally, an implementation of a Lagrangian-Eulerian (LE) method in the contaminant transport FV code is described.

INTRODUCTION

There is much discussion regarding the appropriate computational engine to use in groundwater modeling. There are those with backgrounds in aerospace engineering that prefer the FV method, and there are still others from the geotechnical perspective who prefer the FE method. This study will not answer this question, but it will give results obtained for some particular problems that should help in understanding how to improve either the FV or FE method that one prefers to use.

FV METHOD

The standard FV strategy is to consider each computational cell a finite volume with the unknown variables to be computed evaluated at the center of the cell. However, the FE method usually has the value of the unknowns computed at the grid points (node points). So, as shown in Figure 1, a modified way of defining

A. Peters et al. (eds.), Computational Methods in Water Resources X, 365–372.
© 1994 *Kluwer Academic Publishers. Printed in the Netherlands.*

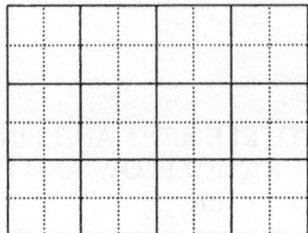

Figure 1. 2-D Computational Cells

the finite volumes (McCormick 1992) is to use the dotted lines to form cells around the node points. This has the advantage of allowing the identical grid to be used for FE or FV, but a disadvantage is the cells along the boundary have reduced size by one-half or one-fourth. This problem has, however, been minimized by a judicious software design of subroutines.

UNSATURATED FLOW

Unsaturated flow is modeled using the governing equation

$$\nabla \cdot (k_r \boldsymbol{k_s} \cdot \nabla \phi) + q = F \frac{\partial \phi}{\partial t} \tag{1}$$

where ϕ is the total head or potential, $\boldsymbol{k_s}$ is the saturated hydraulic conductivity tensor, k_r is the relative hydraulic conductivity, and q is a source or sink. ϕ is given by

$$\phi = h + z - \phi_d \tag{2}$$

where h is the pressure head, and ϕ_d is a datum. F is the water capacity given by

$$F = \frac{d\theta}{dh} \tag{3}$$

where θ is the moisture content. F and k_r are functions of h in an unsaturated zone, making Equation 1 highly nonlinear. Transforming Equation 1 into a (ξ, η, ζ) curvilinear coordinate system yields

$$\sum_{i=1}^{3} \sum_{j=1}^{3} \frac{\partial}{\partial \xi_i} \left(\nabla \xi_i \cdot \left[k_r k_s \cdot \frac{\partial}{\partial \xi_j} (\Phi \nabla \xi_j) \right] \right) + Jq = F\frac{\partial \Phi}{\partial t} \tag{4}$$

where ξ_i, i = 1, 2, 3 represent ξ, η, and ζ, respectively, and

$$\Phi = J\phi \tag{5}$$

where J is the determinant of the transformation matrix between the two coordinate systems.

2-D FV implementation

A 2-D FV implementation of Equation 4 becomes

$$-\left[(A^+ + A^-) \cdot \nabla \xi + (B^+ + B^-) \cdot \nabla \eta + \frac{1}{\Delta t} F^{n+1} \right] \Phi_{i,j}^{n+1}$$

$$+ A^- \cdot \nabla \xi \, \Phi_{i-1,j}^{n+1} + A^+ \cdot \nabla \xi \, \Phi_{i+1,j}^{n+1} + B^- \cdot \nabla \eta \, \Phi_{i,j-1}^{n+1} + B^+ \cdot \nabla \eta \, \Phi_{i,j+1}^{n+1}$$

$$+ \left(\frac{A^- \cdot \nabla \eta + B^- \cdot \nabla \xi}{2} \right) \Phi_{i-1,j-1}^{n+1} - \left(\frac{A^+ \cdot \nabla \eta + B^- \cdot \nabla \xi}{2} \right) \Phi_{i+1,j-1}^{n+1} \tag{6}$$

$$- \left(\frac{A^- \cdot \nabla \eta + B^+ \cdot \nabla \xi}{2} \right) \Phi_{i-1,j+1}^{n+1} + \left(\frac{A^+ \cdot \nabla \eta + B^+ \cdot \nabla \xi}{2} \right) \Phi_{i+1,j+1}^{n+1}$$

$$= - Jq - \frac{1}{\Delta t} F_{i,j}^{n+1} \Phi_{i,j}^{n}$$

with the flux related coefficients on the sides of the cells given by

$$A^{\pm} = (k_r k_s \cdot \nabla \xi)_{i \pm \frac{1}{2}, j}^{n+1} \qquad B^{\pm} = (k_r k_s \cdot \nabla \eta)_{i, j \pm \frac{1}{2}}^{n+1} \tag{7}$$

The A's and B's are determined by first computing information at the nodes and them taking an average using the proper nodes to get the flux terms at the faces of the cells (Finlayson 1992). Now, summarize Equation 6 using C coefficients by

$$\sum_{m=-1}^{1} \sum_{n=-1}^{1} C_{i+m, j+n}^{n+1} \Phi_{i+m, j+n}^{n+1} = - Jq - \frac{1}{\Delta t} F_{i,j}^{n+1} \Phi_{i,j}^{n} \tag{8}$$

Let the function f be defined using Equation 8 as follows:

$$f(\Phi^{k+1}) = \sum_{m=-1}^{1} \sum_{n=-1}^{1} C_{i+m,j+n}^{k+1} \Phi_{i+m,j+n}^{k+1} + Jq + \frac{1}{\Delta t} F_{i,j}^{k+1} \Phi_{i,j}^{k} \qquad (9)$$

Then a modified Newton iteration for solving Equation 8 is

$$\sum_{m=-1}^{1} \sum_{n=-1}^{1} C_{i+m,j+n}^{k} \Delta \Phi_{i+m,j+n}^{k+1} = - f(\Phi^{k}) \qquad (10)$$

Equation 10 can be solved by any number of direct or iterative techniques, such as forms of Gauss Elimination or the preconditioned conjugate gradient method, and then the dependent variables can be updated with the process continuing until convergence. However, what worked well for the problems tested was an under-relaxation method. Equation 10 can be iterated several times before updating the C's and F, but it was more efficient to update all data after each iteration.

Laboratory test problem

Results from an experimental study of 2-D transient unsaturated/saturated flow with water table recharge (Vauclin, Khanji, and Vachaud 1979) will now be compared with results from both the FV formulation and a FE code. The problem, as shown in Figure 2, consists of flow in a homogeneous soil of saturated hydraulic conductivity 35 cm/hr in a tank 600 cm long, 200 cm tall, and 5 cm thick with an impervious bottom. Because of symmetry, only 300 cm of the tank are modeled with the center line (AF in Figure 2) being treated as an impervious boundary. A constant pool elevation of 65 cm is maintained along BC in Figure 2 with the boundary CDE covered to avoid evaporation. EF is initially covered and the tank is allowed to completely settle. Then EF is uncovered and a flow rate of 14.8 cm/hr is applied to the system for 8 hr while holding BC at the constant total head of 65 cm. The relative hydraulic conductivity in the unsaturated region was determined experimentally to be

$$k_r = \frac{A}{A + (-h)^B} \qquad (11)$$

where A = 2.99 X 10^6, and B = 5.0. The moisture content equation was also determined experimentally and by the use of a least-squares fit to be

$$\theta = \theta_s \frac{\alpha}{\alpha + (-h)^\beta} \qquad (12)$$

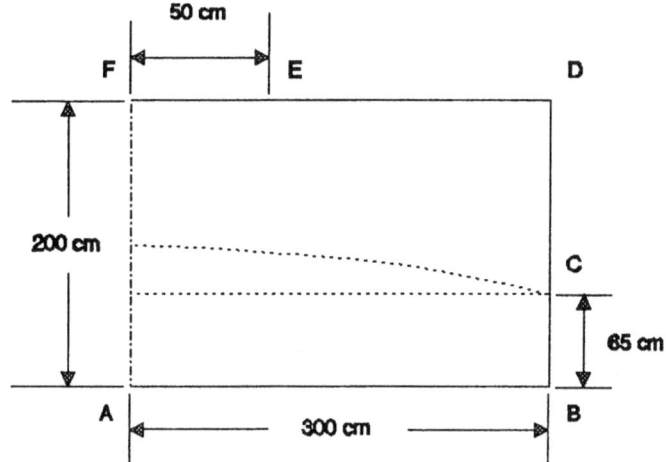

Figure 2. Laboratory Problem

where θ_s = 0.30, α = 40.00, and β = 2.90. The grid consists of a 16 X 16
structured grid with the intervals slightly different to align with key points. Δt was
set to 0.05 hr and allowed to grow 20% a time step until a maximum of Δt = 1 hr
was reached. 20 time steps were done for a total of eight hours. The proper
subroutine was modified to incorporate Equations 11, 12, and the derivative of
Equation 12 (F), and the problem was run on both the FE and FV codes.
Because of the nature of the water capacity curve, the FE program scheme would
not converge, and the FV solution converged only by using a small relaxation
coefficient of 0.1 for the first few time steps. The FE algorithm uses a Picard type
iteration strategy, while the FV scheme uses the Newton type iteration as
previously described. However, tabular forms of Equations 11, 12, and F (27 data
points) with F not allowed smaller than 0.001 for large -h were used with the FE
code, and convergence was then achieved. A Picard type algorithm was
implemented in the FV code, and the same lack of convergence was observed.
The results for 40 time steps with Δt = 0.1 hr for both the FV and FE solutions
were then obtained, and the respective free surfaces were compared at t = 4 hr
with the laboratory results as shown in Figure 3. The dissipative error in the FE
solution is more than that of the FV implementation. F being modeled more
accurately in the FV code could have had an impact as well.

CONTAMINANT TRANSPORT

Contaminant transport in unsaturated porous media is modeled by

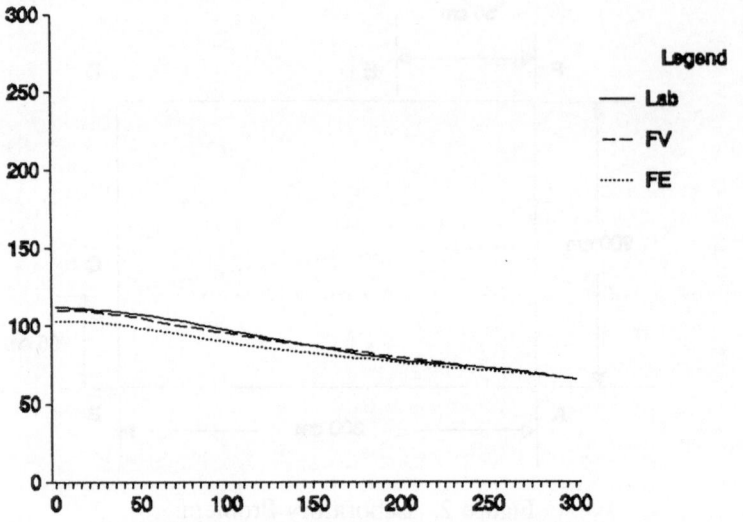

Figure 3. Comparison of Laboratory, FE, and FV Results

$$\frac{\partial (\theta C)}{\partial t} + \rho_b \frac{\partial S}{\partial t} + \nabla \cdot (vC) = \nabla \cdot (\theta D \cdot \nabla C)$$
$$- \lambda (\theta C + \rho_b S) + Q C_{in} \qquad (13)$$

where C is the material concentration in aqueous phase, ρ_b is the bulk density of the medium, S is the material concentration in adsorbed phase, v is the discharge velocity vector, D is the dispersion coefficient tensor, λ is the decay constant, Q is the source rate of water, and C_{in} is the material concentration in the source. For this study, the linear isotherm model for adsorption is used, which is

$$S = K_d C \qquad (14)$$

where K_d is the distribution coefficient. Also, unsaturated flow is modeled by

$$\frac{\partial \theta}{\partial t} + \nabla \cdot v = Q \qquad (15)$$

Using the Lagrangian approach let the velocity of the grid points be

$$v = \frac{v}{\theta + \rho_b K_d} \qquad (16)$$

Also define

$$\frac{DC}{Dt} = \frac{\partial C}{\partial t} + \boldsymbol{v} \cdot \nabla C \qquad (17)$$

Putting Equations 14, 15, 16, and 17 into Equation 13 gives the final equation

$$(\theta + \rho_b K_d) \frac{DC}{Dt} = \nabla \cdot (\theta \boldsymbol{D} \cdot \nabla C) - [\lambda (\theta + \rho_b K_d)$$
$$+ Q] C + Q C_{in} \qquad (18)$$

Rather than use a moving grid, however, back-track a node point P to P' using

$$\frac{1}{2} (\boldsymbol{v}'^n + \boldsymbol{v}'^{n+1}) \Delta t = \boldsymbol{r}_p - \boldsymbol{r}' \qquad (19)$$

where $\boldsymbol{r}'$ is the position of P', $\boldsymbol{v}'$ is the grid velocity at P', and $\boldsymbol{r}_P$ is the position of point P. $\boldsymbol{r}'$ can be computed by using the FE interpolation functions for the ' variables in adjacent elements giving in 3-D a set of three equations and three unknowns for curvilinear coordinates (ξ', η', ζ'). If ξ', η', and ζ' lie within the appropriate range (typically between -1 and 1), then the correct element has been found and $\boldsymbol{r}'$ and a corresponding concentration C^* can be computed. The C^*'s become an intermediate solution at the nodes with Equation 18 being used to disperse them in a similar way as done in Equations 4-10.

The analytical, LE FV, and LE FE solutions to the problem of saturated flow in a rectangular vertical cross-section of sand of size 100.0 X 20.0 that is initially clean until a spill occurs on the top of the sand (y = 20.0) are computed and compared. A concentration 1.0 in a strip of length 20.0 in the middle of the top of the sand is maintained for a time 10.0, and then it decays expedientially with a decay constant 1.0. Water is flowing in the +x direction with a discharge velocity 2.0. However, no contaminant passes at x = 100. Adsorption into the medium of bulk density 1.2 occurs linearly with a distribution coefficient of 0.2. Maximum moisture content is 0.4, and dispersion coefficients D_1 and D_2 are 50.0 and 5.0, respectively. Figure 4 shows results for position (100.0, 18.0) for 10 time steps with Δt = 1.0. The FV solution is slightly better for the particular algorithms tested.

REFERENCES

Finlayson, Bruce A. (1992) Numerical Methods for Problems with Moving Fronts,

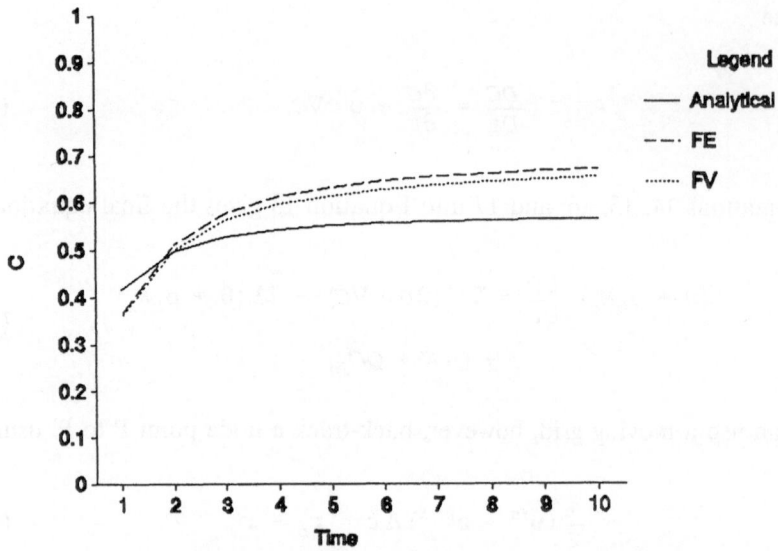

Figure 4. Contaminant Transport Results

Ravenna Park Publishing, Seattle, pp. 75-76, 496-541.

McCormick, S. (1992) "Finite volume element and multilevel adaptive methods", In T.F. Russell, R.E. Ewing, C.A. Brebbia, W.G. Gray, and G.F. Pinder (eds.), Computational Methods in Water Resources IX, Vol. 1: Numerical Methods in Water Resources, Computational Mechanics Publications, New York, pp. 539-553.

Putti, M. and Paniconi, C. (1992) "Evaluation of the Picard and Newton iteration schemes for three-dimensional unsaturated flow", in T.F. Russell, R.E. Ewing, C.A. Brebbia, W.G. Gray, and G.F. Pinder (eds.), Computational Methods in Water Resources IX, Vol. 1: Numerical Methods in Water Resources, Computational Mechanics Publications, New York, pp. 529-536.

Vauclin, M., Khanji, D., and Vachaud, G. (1979) "Experimental and numerical study of a transient two-dimensional unsaturated-saturated water table recharge problem", Water Resources Research, 15, 1089-1101.

Yeh, G. T, (1990) "A Lagrangian-Eulerian method with zoomable fine-mesh approach to solving advection-dispersion equations", Water Resources Research, 26, 1133-1144.

COMPUTER EVALUATION OF HIGH-ORDER NUMERICAL SCHEMES TO SOLVE ADVECTIVE TRANSPORT PROBLEMS

DOROTHEA YEH and GOUR-TSYH YEH
Department of Civil and Environmental Engineering
The Pennsylvania State University
University Park, PA 16802
USA

This study evaluates the performance of three high-order schemes in terms of their ability to resolve spurious oscillation, numerical dispersion, and peak clipping for two simple one-dimensional benchmark problems. Three models, namely QUICKEST, ULTIMATE, and ENO were constructed to represent the classical high-order schemes without a flux limiter, TVD with a flux limiter, and TVB schemes, respectively. Three sets of results generated by QUICKEST, ULTIMATE, and ENO were compared with the analytical solutions. The first set indicated that none of these high-order schemes could yield satisfactory simulations when the gird size and time-step size specified by the benchmark problems were used. The second set showed that all three numerical schemes generated excellent computations when the grid size and time-step size were reduced by a factor of 10 and 5, respectively. The third set demonstrated that the results obtained by these schemes deteriorated when a 100 folds of simulation times was conducted. The ENO and ULTIMATE schemes successfully eliminated spurious oscillations for all cases as expected. The QUICKEST scheme alleviated the problem of spurious oscillations only when the reduced grid and time-step sizes were used. In terms of numerical dispersion and peak clipping, none of the three schemes produced satisfactory results unless the reduced grid and time-step sizes were used. Peak clipping poses a more severe problem for these high-order schemes than numerical dispersion.

INTRODUCTION

Numerical problems associated with advection dominated transport include spurious oscillation, numerical dispersion, peak clipping, and grid orientation. Total-variation-diminishing (TVD) and total variation-bounded (TVB) schemes are considered among the best methods to address the difficulties of numerical dispersion and spurious oscillation. There are many TVD schemes that were constructed by a variety of researchers (Roe, 1981; Harten, 1984; Sweby, 1984; Osher and Chakravarthy, 1986; Leonard, 1988). From the paper by Shu and Osher (1988), we observed that third order schemes work quite well while a fifth order scheme losses accuracy in a fairly large region near discontinuities.

Although there is no rigorous theory about TVB methods, many models have been constructed according to this general class of numerical methods. A notable subset of TVB methods is the essentially no oscillation (ENO) schemes. A number of

373

A. Peters et al. (eds.), Computational Methods in Water Resources X, 373–380.

researchers have been working on numerical algorithm according to the ENO schemes. Analysis and numerical experiments are found in the literature (Harten et al., 1987; Shu and Osher, 1988). These ENO schemes have been demonstrated to be very useful and the numerical simulation seems to be accurate. The oscillating characteristic was reduced enormously.

The above high-order TVD and TVB schemes have been successfully applied to many initial and boundary-value problems to show their accuracy using the grid and time-step sizes at the pleasure of the authors. Unfortunately, no attempt has been made to evaluate these high-order schemes with a common set of benchmark problems in which a reasonable large grid size was specified. In practical situations, the region of interest is often large and it is impractical to use a small grid size. In this study, the capability of simulating problems with these high-order TVD and TVB schemes using a reasonable large grid size is tested. The performance of these models for a long real-time simulation is also tested.

REPRESENTATIVE TVD AND TVB SCHEMES

Three models which include QUICKEST (Quadratic Upstream Interpolation for Convective Kinematics with Estimated Streaming Terms), ULTIMATE (Universal Limiter for Transient Interpolation Modeling of the Advective Transport Equations), and ENO were constructed to represent the classical high-order schemes without a flux limiter, TVD with a flux limiter, and TVB schemes, respectively. The detail derivation, numerical algorithms, and numerical analyses for QUICKEST, ULTIMATE, and ENO are described in Leonard (1979), Leonard (1988), and Shu and Osher (1988), respectively. Computer implementation of these models are straightforward.

THE BENCHMARK PROBLEMS

The three high-order numerical models described in the previous section were applied to two simple one-dimensional problems to test their accuracy. These two problems are a subset of the five benchmark problems specified in the convection-diffusion forum in conjunction with the 7th International Conference on Computational Methods in Water Resources held at Massachusetts Institute of Technology on June 13-17, 1988.

Problem No. 1 is concerned with the solution of the transport of concentration hills in a one-dimensional uniform flow. This represents an initial-value problem. Case 1A serves as a reference for pure advection transport. Cases 1B and 1C involve the diffusion transport as well, and are not included here. Case 1D and 1E are applied to test the performances of the schemes when the base width of the initial peak concentration is widened. Case 1F tests the performances of the models under a transient flow condition. Case 1G and 1H are used to review the behavior of the models when triangular hills with different base width are encountered. Case 1I is used to test the capability of the models when using variable grid size. Case 1J is used to investigate the performances of the schemes when a double-peak solution is present in the problem: one peak from the initial condition and the other peak from a transient boundary condition imposed at the upstream. Case 1K and 1M are applied to see the effect of Courant number on the models. Case 1L has a mesh Courant number greater than 1, hence it is not included. This problem could be a vehicle to evaluate the ability of a numerical scheme to alleviate the problem of numerical dispersion, spurious oscillation, and peak clipping, in particular the peak clipping behavior.

Problem No. 2 is concerned with the solution of an advancing front in a one-dimensional flow. This represents a boundary-value problem. Like case 1A, case 2A serves as the reference of pure advection transport for Problem No. 2. Cases 2B and 2C involve both advection and diffusion transport, hence they are no included here. Case 2D tests the models ability to deal with nonuniform grid size. Case 2E is not included here because it has a mesh Courant number greater than 1. Case 2F tests the performances of the models when reducing time step size. This problem serves to evaluate a numerical scheme's ability to overcome spurious oscillation and numerical dispersion. A description of these two sets of benchmark problems is summarized in Table 1 and Table 2.

Table 1 - List of Cases for Example 1: Initial Value Problems

Case	Velocity	Initial Condition	Δt	Δx
1A	0.5	$x_0=2000$, $\sigma_0=264$	96	200
1D	0.5	$x_0=2000$, $\sigma_0=320$	96	200
1E	0.5	$x_0=2000$, $\sigma_0=400$	96	200
1F	Transient	$x_0=2000$, $\sigma_0=264$	96	200
1G	0.5	$x_0=2000$, $l_0=800$	96	200
1H	0.5	$x_0=2000$, $l_0=1000$	96	200
1I	0.5	x_0 at node 16, $\sigma_0=264$	96	Variable
1J	0.5	$x_0=600$, $\sigma_0=264$, $C_u(t)$	96	200
1K	0.5	$x_0=2000$, $\sigma_0=264$	192	200
1M	0.5	$x_0=2000$, $\sigma_0=264$	48	200

Table 2 - List of Cases of Example 2: Boundary Value Problems

Case	Velocity	Δt	Δx
2A	0.5	96	200
2D	0.5	96	Variable
2F	0.5	48	200

PERFORMANCE EVALUATION

The results generated by QUICKEST, ULTIMATE, and ENO are compared with the exact solutions for all cases of Problems No. 1 and No. 2, but only Cases 1A and 2A will be illustrated here. Figures 1 and 2 compare the results for cases 1A and 2A, respectively. They show that the ability of ULTIMATE and ENO to eliminate spurious oscillations is excellent as expected. On the other hand, the oscillation produced by the classical high-order schemes without a flux limiter, QUICKEST, is obviously not satisfactory.

Figure 1 clearly shows that the peak values generated by any of the three schemes are not acceptable. The information of peak clipping for all cases in Problem No. 1 are summarized in table 3. It is clear that all three high-order schemes produce very inaccurate solutions with respect to peak clipping. In general, the results generated by QUICKEST and ENO have less peak clipping than those by the ULTIMATE model. Yet, as far as the oscillatory characteristic is concerned, the ULTIMATE model has the least oscillation among these three models. As an overall quality, considering both peak clipping and oscillation, the ENO model gives the most satisfactory results among these three models.

Figure 2 indicates that none of these models could resolve numerical dispersion to a reasonable degree of satisfaction, even though all three models are third-order accurate in spatial discretization. One may argue that the inability to overcome the problem of numerical dispersion is due to the first-order time discretization in QUICKEST and ULTIMATE; but ENO has also the third-order time discretization. A better way to explain the inability of high-order spatial and temporal discretization to resolve numerical dispersion can be found in terms of a Fourier analysis. In a Fourier analysis, every numerical frequency can have a phase error, an amplitude error, or both. Numerical dispersion could result even if the scheme does not contain any error in amplitude for all frequencies. Similarly, peak clipping could result if the scheme has phase error for some frequencies even though it may not have any amplitude error for all frequencies.

REDUCTION OF GRID SIZE

Although the ULTIMATE and ENO schemes can eliminate oscillation with the grid size specified by the benchmark problems, they could not satisfactorily resolve numerical dispersion and peak clipping. In particular, the peak clipping errors are too severe even for a very forgiven author. Therefore, a sensitivity study on the reduction of grid sizes was performed on Cases 1A and 2A for all three schemes to investigate the improvement of accuracy of the results. In the mean time, in order to keep the Courant number less than 1.0, the time step size (Δt) had to be reduced corresponding to the reduction of grid sizes. All these schemes generate excellent results for Case 1A after the grid size was reduced by a factor of ten and the time step size was reduced by a factor of five (Fig. 3). For Case 2A, considerable improvement on numerical dispersion was obtained for all three schemes although QUICKEST still had problem of oscillation (Fig. 4). The dispersion-range reduces from approximately 3500 before grid-seize reduction to within 700 after the grid-size reduction. These results are much are closer to exact solutions.

LONGER SIMULATION TIME

Two examples with longer simulation time were also performed on Cases 1A and 2A with the reduced grid size stated in last section. Instead of printing out the results at time equals 9,600, the results are printed out at time equals 960,000. Since the velocity remains the same, 0.5, the total running distance is therefore 100 times longer. The results for a longer simulating time are plotted in Figures 5 and 6, respectively, for Cases 1A and 2A. As can be seen from the figures, the accuracy of the generated results deteriorated for all three models. All three schemes produced a fair amount of peak clipping again. The peak clipping by the ENO model is compensated by an obvious numerical dispersion (Figs. 5 and 6). Furthermore, the peak generated by the ENO model shifts from the grid point located at x = 482,000 (which is where it is supposed to be) to the next grid point at x = 48,2020. This implies that ENO may have a grid orientation problem when two- and three-dimensional problems are encountered.

Table 3 - Peak values for all cases in Problem No. 1

	Peak Values			
Scheme	Exact	QUICKEST	ULTIMATE	ENO
Case 1A	1.0	0.693	0.640	0.703
Case 1D	1.0	0.776	0.731	0.808
Case 1E	1.0	0.858	0.819	0.902
Case 1F	1.0	0.683	0.633	0.564
Case 1G	1.0	0.765	0.720	0.793
Case 1H	1.0	0.830	0.796	0.862
Case 1I	0.983	0.703	0.655	0.703
Case 1J	1.0 1.0	0.639 0.661	0.609 0.611	0.692 0.703
Case 1K	1.0	0.733	0.687	0.698
Case 1M	1.0	0.680	0.626	0.704

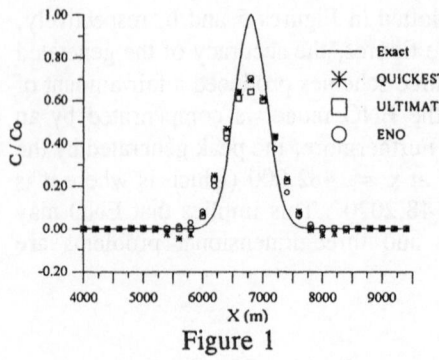

Figure 1

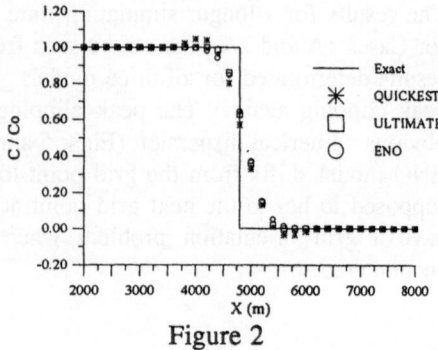

Figure 2

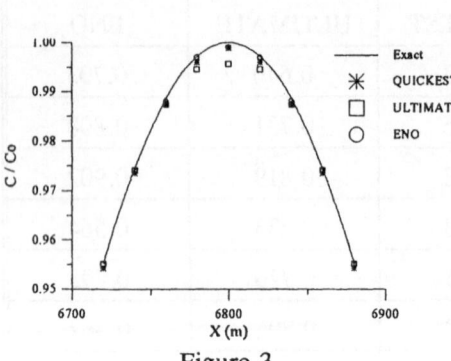

Figure 3

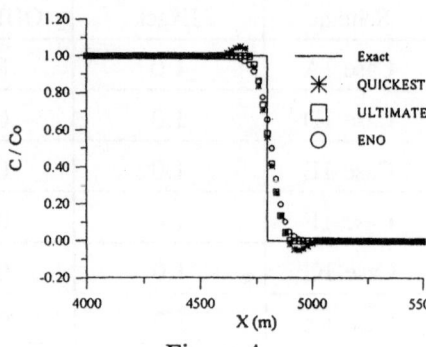

Figure 4

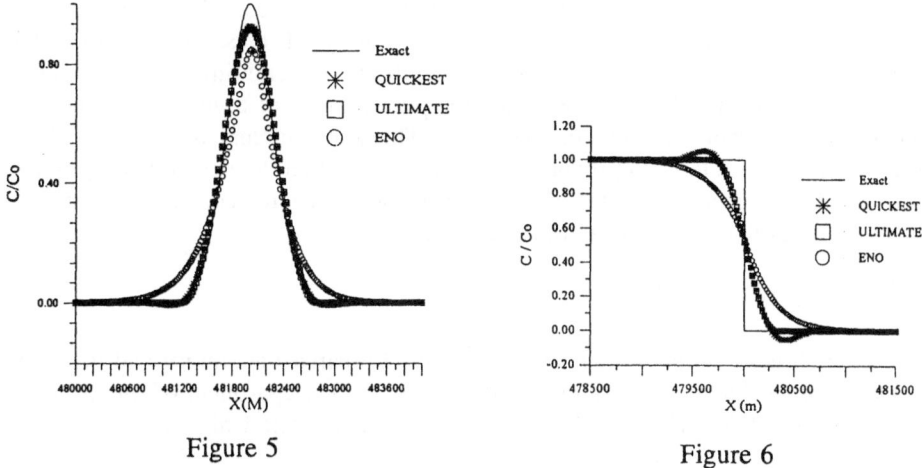

Figure 5

Figure 6

CONCLUSION AND DISCUSSIONS

This study has evaluated three high-order models in terms of their ability to resolve the problems of spurious oscillation, numerical dispersion, and peak clipping using two benchmark problems with a variety of cases. The ULTIMATE model completely eliminated spurious oscillation and the ENO scheme produced essentially no oscillation solutions for all cases as expected. The QUICKEST model generated oscillation results for all cases. All three schemes created comparable, unsatisfactory numerical dispersion for all cases in benchmark Problem No. 2. None of these schemes could resolve the problem of peak clipping. They all generated severe peak clipping for all cases in benchmark Problem No. 1.

When the grid size was reduced by a factor of ten and the time-step size was reduced by a factor of five, all three schemes gave very accurate solutions with respect to spurious oscillation, numerical dispersion, and peak clipping for Problem No. 1. For Problem No. 2, QUICKEST still produced numerical oscillation. All three schemes greatly reduced numerical dispersion but not to a degree one would like to.

A longer real-time simulation was also applied to the models using the reduced space-grid and time-step sizes. All three schemes again rendered unacceptable peak clipping for Problem No. 1 and yielded unsatisfactory numerical dispersion again.

Numerous high-order TVD and TVB models have been developed to successfully alleviate the problems of spurious oscillation around sharp-fronts. Few has attempted to address the problem of peak clipping which is perhaps much more difficult to deal with than spurious oscillation and numerical dispersion. Numerical dispersion has been normally considered along with spurious oscillation by almost all of the high-order TVD and TVB schemes. Unfortunately, the ability of these models to reduce numerical dispersion have been demonstrated with the grid and time-step sizes chosen at the pleasure of modelers. To evaluate the performance of a numerical scheme, a common set of benchmark problems is needed. This paper attempted this approach and the authors hope that this can provoke a further investigation on many other high-order TVD and ENO schemes by others engaging in the search for accurate and efficient numerical algorithms to solve transport equations.

ACKNOWLEDGEMENT

This research is supported in part by the Applied Research Laboratory and in part by the Office of Health and Environmental Research, U. S. Department of Energy under Grant No. DE-FG02-91ER61197 with the Pennsylvania State University.

REFERENCES

Harten, A. (1984) "On a Class of High Resolution Total-Variation-State Finite Difference Schemes", SIAM Journal of Numerical Analysis, Vol. 21, No.1, 1-23.
Harten, A., B. Engquist, S. Osher and S. Chakravarthy (1987) "Uniformly High Order Accurate Essentially Non-Oscillatory Schemes, III", Journal of Computational Physics, Vol. 71, No.2, 231-303.
Leonard, B. P. (1979) "A stable and accurate convective modeling procedures based on quadratic upstream interpolation", Computer Methods Applied Mechanics and Engineering, Vol. 19, No. 1, 59-98.
Leonard, B. P. (1988) "Universal Limiter for Transient Interpolation Modeling of the Advective Transport Equations: The ULTIMATE Conservative Difference Scheme", NASA Technical Memorandum 100916, ICOMP-88-11.
Osher, S. and S. Chakravarthy (1986) "Very High Order Accurate TVD Schemes", IMA Volumes in Mathematics and its Applications, Vol. 2, Springer-Verlag, 229-274.
Roe, P. L. (1981) "Numerical Algorithms for the Linear Wave Equation", Royal Aircraft Establishment Technical Report 81047.
Shu, C. W. and S. Osher (1988) "Efficient Implementation of Essentially Non-Oscillatory Shock-Capturing Schemes", Journal of Computational Physics, Vol. 77, No. 2, 439-471.
Sweby, P. K. (1984) "High Resolution Schemes Using Flux Limiters for Hyperbolic Conservation Laws", SIAM Journal of Numerical Analysis, Vol. 21,No. 5, 995-1011.